T0203847

International Environmental Standards Handbook

Scott S. Olson

CRC Press
Taylor & Francis Group
Boca Raton London New York

CRC Press is an imprint of the
Taylor & Francis Group, an **informa** business

CRC Press
Taylor & Francis Group
6000 Broken Sound Parkway NW, Suite 300
Boca Raton, FL 33487-2742

First issued in paperback 2020

© 1999 by Taylor & Francis Group, LLC
CRC Press is an imprint of Taylor & Francis Group, an Informa business

No claim to original U.S. Government works

ISBN-13: 978-0-367-57916-6 (pbk)
ISBN-13: 978-1-56670-270-6 (hbk)

This book contains information obtained from authentic and highly regarded sources. Reasonable efforts have been made to publish reliable data and information, but the author and publisher cannot assume responsibility for the validity of all materials or the consequences of their use. The authors and publishers have attempted to trace the copyright holders of all material reproduced in this publication and apologize to copyright holders if permission to publish in this form has not been obtained. If any copyright material has not been acknowledged please write and let us know so we may rectify in any future reprint.

Except as permitted under U.S. Copyright Law, no part of this book may be reprinted, reproduced, transmitted, or utilized in any form by any electronic, mechanical, or other means, now known or hereafter invented, including photocopying, microfilming, and recording, or in any information storage or retrieval system, without written permission from the publishers.

For permission to photocopy or use material electronically from this work, please access www.copyright.com (http:// www.copyright.com/) or contact the Copyright Clearance Center, Inc. (CCC), 222 Rosewood Drive, Danvers, MA 01923, 978-750-8400. CCC is a not-for-profit organization that provides licenses and registration for a variety of users. For organizations that have been granted a photocopy license by the CCC, a separate system of payment has been arranged.

Trademark Notice: Product or corporate names may be trademarks or registered trademarks, and are used only for identification and explanation without intent to infringe.

Visit the Taylor & Francis Web site at
http://www.taylorandfrancis.com

and the CRC Press Web site at
http://www.crcpress.com

Library of Congress Card Number 99-13383

Library of Congress Cataloging-in-Publication Data

Olson, Scott S.
    International environmental standards handbook / Scott S. Olson
        p. cm.
    Includes bibliographical references and index.
    ISBN 1-56670-270-4 (alk. paper)
    1. Environmental management. I. Title.
GE300.038   1999
363.7'05—dc21   99-13383

"If Delicate Arch has any significance it lies, I will venture, in the power of the odd and unexpected to startle the senses and surprise the mind out of their ruts of habit, to compel us into a reawakened awareness of the wonderful—that which is full of wonder."

Edward Abbey

Author

Scott Olson is Director of the Colorado Environmental Training Center and former Chair of the Environmental, Health and Safety Department at Red Rocks Community College in Lakewood, Colorado. He is a faculty instructor and coordinator of the Water Quality Management and Environmental and Safety Technology Degree programs. Mr. Olson is also a program chair and instructor for the Rocky Mountain Education Center, OSHA Training Institute at Red Rocks Community College. He is a member of the Board of Directors for the Rocky Mountain Hazardous Materials Association representing training and education. His company, Altitude Training Associates, provides environmental, health and safety training to a variety of organizations.

Contents

Part III–Environmental Standards

Part IV–Environmental Protection in the United States

Part V–Sources of Additional Information

List of Abbreviations and Acronyms

ACM	Asbestos Containing Material
ACP	Area Contingency Plan
AEPS	Arctic Monitoring and Assessment Program
AIR	Automobile Repair and Inspection
AIRS	Aerometric Information Reporting System
ALARA	As Low As Reasonably Achievable
ANSI	American National Standards Institute
APEN	Air Pollution Emission Notice
ASQC	American Society of Quality Control
AST	Above Ground Storage Tank
ASTM	American Society for Testing and Materials
ATP	Anaerobic Thermal Processor
BAT	Best Available Technology
BATNEEC	Best Available Technology Not Entailing Excessive Cost
BS	British Standard
BSI	British Standards Institute
CAA	Clean Air Act
CAAA	Clean Air Act Amendments (of 1990)
CAS	Chemical Abstract System
CBT	Chemical and Biological Treatment
CDPHE	Colorado Department of Public Health and Environment
CDPS	Colorado Discharge Permit System
CE	Categorical Exclusion
CEI	Continuous Environmental Improvement
CEN	Comite European de Normalisation
CEQ	Council on Environmental Quality
CERCLA	Comprehensive Environmental Response, Compensation and Liability Act
CERCLIS	Comprehensive Environmental Response Compensation and Liability Information System
CERES	Coalition on Environmentally Responsible Economies
CFC	Chloroflurocarbon
CFR	Code of Federal Regulations
CHMR	Center for Hazardous Materials Research
CIA	Chemical Industries Association
CITES	Convention on International Trade in Endangered Species
CMA	Chemical Manufacturers Association
CORACT	RCRA Corrective Action Records
CSO	Combined Sewer Overflows
CWA	Clean Water Act
CZM	Coastal Zone Management
DIN	Deutches Institut for Normung e.V.
DIS	Draft International Standards
DOT	Department of Transportation
DRT	Disaster Recovery Team

EA	Environmental Assessment
EC	European Community
EEA	European Environment Agency
EEZ	Exclusive Economic Zone
EHS	Environmental Health and Safety
EIA	Environmental Impact Assessment
EINECS	European Register of Chemical Substances
EIS	Environmental Impact Statement
EMAS	Eco-Management and Audit Scheme
EMINWA	Environmentally Sound Management of Inland Water
EMS	Environmental Management Systems
EPA	Environmental Protection Agency
EPADKT	EPA Civil Enforcement Docket
EPCRA	Emergency Planning and Community Right-to-Know Act
ER	Emergency Response
ERNS	Emergency Response Notification System
ERT	Emergency Response Team
EU	European Union
FDSITE	Federal Site Information
FINDS	EPA Facility Index System
FONSI	Finding Of No Significant Impact
FR	Federal Register
GAC	Granular Activated Carbon
GATT	General Agreement on Tariffs and Trade
GCPA	Global Climate Protection Act (U.S.)
GEMS	Global Environment Monitoring System
GRID	Global Resource Information Database
HAP	Hazardous Air Pollutant
HAZWOPER	Hazardous Waste Operations and Emergency Response
HFC	Hydrofluorcarbons
HMTUSA	Hazardous Materials Transportation and Uniform Safety Act
HMW	Hazardous Material and Hazardous Waste
HR	Human Resources
IARC	International Agency for the Research on Cancer
IC	Incident Commander
ICOLP	Industry Cooperative for Ozone Layer Protection
ICS	Incident Command System
IEC	International Electrotechnical Commission
IECA	International Erosion Control Association
IGO	Intergovernmental Organization
ILEV	Inherently Low-Emitting Vehicles
I/M	Inspection/Maintenance
IPC	Integrated Pollution Control (Ireland, England)
IPM	Integrated Pest Management
IPPC	Intergrated Pollution Prevention and Control (EC)
IRPTC	International Register of Potentially Toxic Chemicals
ISO	International Organization for Standardization
ITFM	Intergovernmental Task Force on Monitoring (Water Quality)
JTPC	Joint Technical Programming Committee
LCA	Life Cycle Assessment
LD50	Lethal Dose 50% (median lethal dose)

LDR	Land Disposal Restrictions
LEPC	Local Emergency Planning Committee
LSA	List of CFR Sections Affected
LUST	Leaking Underground Storage Tank
MAC	Maximum Admissible Concentration (European Union)
MARPOL	Convention on the Prevention of Marine Pollution by Dumping of Wastes and Other Matter
MCL	Maximum Contaminant Level (U.S.)
MMPA	Marine Mammal Protection Act
MRL	Maximum Residue Limits
MSDS	Material Safety Data Sheet
NAAQS	National Ambient Air Quality Standards
NAFTA	North American Free Trade Agreement
NCP	National Contingency Plan
NCPDI	National Coastal Pollutant Discharge Inventory
NEFCO	Nordic Environmental Finance Corporation
NEP	National Estuaries Program
NEPA	National Environmental Policy Act
NESHAP	National Emission Standard for Hazardous Air Pollutants
NFPA	National Fire Protection Association
NIOSH	National Institute for Occupational Safety and Health
NGO	Nongovernmental Organization
NOEL	No Observable Effect Level
NOV	Notice of Violation
NPDES	National Pollutant Discharge Elimination System
NPLDSC	National Priorities List Descriptions
NTP	National Toxicology Program
NWI	National Wetlands Inventory
OCS	On-Scene Coordinator
ODC	Ozone Depleting Compound
ODS	Ozone Depleting Substances
OHSMS	Occupational Health and Safety Management System
OPA	Oil Pollution Act of 1990
ORD	EPA's Office of Research and Development
OSHA	Occupational Safety and Health Administration
OSHAIR	Occupational Safety and Health Administration Inspection Reports
PA	Preliminary Assessment
PACM	Presumed Asbestos Containing Material
PAH	Polycyclic Aromatic Hydrocarbon
PBB	Polybrominated Biphenyls
PEL	Permissible Exposure Limit (OSHA)
PERT	Project Evaluation and Review Technique
PPE	Personal Protective Equipment
ppm	Parts Per Million
PRP	Potentially Responsible Party
PSM	Process Safety Management
QC	Quality Control
RAB	Registration Accreditation Board
RACT	Reasonably Achievable Control Technology
RCRA	Resource Conservation and Recovery Act
RCRIS	Resource Conservation and Recovery Information System

REA	Registered Environmental Assessor
RIFS	Remedial Investigation Feasibility Study
RMHMA	Rocky Mountain Hazardous Materials Association
RMPP	Risk Management Prevention Plan
ROD	Record of Decision
RODS	EPA Superfund Record of Decision
RRT	Regional Response Team
RQ	Reportable Quantity
SACM	Superfund Accelerated Cleanup Model
SAGE	Strategic Advisory Group on the Environment
SARA	Superfund Authorization and Reauthorization Act
SBUV	Solar Backscatter Ultraviolet
SC	Subcommittee For ISO 14000
SCBA	Self-Contained Breathing Apparatus
SDWA	Safe Drinking Water Act
SEA	Single European Act
SEM	Strategic Environmental Management
SERC	State Emergency Response Commission
SHEMS	Safety, Health and Environmental Management Systems
SI	Site Investigation
SIP	State Implementation Plan
SITE	Superfund Innovative Technology Evaluation
SME	Solar Mesospheric Explorer
SONS	Spills of National Significance
SOP	Standard Operating Procedure
SUB TAG	Sub Technical Advisory Group for ISO 14000
SWCP	State Wetland Conservation Plans
TAG	Technical Advisory Group
TC	Technical Committee
TC	Toxicity Characteristic
TCDD	Tertrachlorodibenzo-para-dioxin
TCLP	Toxicity Characteristic Leaching Procedure
TLV	Threshold Limit Value (ACGIH)
TMB	Technical Management Board
TOMS	Total Ozone Mapping Spectrometer
TPQ	Threshold Planning Quantity
TSCA	Toxic Substance Control Act
TQ	Threshold Quantity
TQEM	Total Quality Environmental Management
TRI	Toxic Release Inventory
TRIS	Toxic Release Information System
TSDF	Treatment, Storage and Disposal Facility
UN	United Nations
UNCED	United Nations Conference on the Environment and Development
UNDP	United Nations Development Programme
UNEP	United Nations Environment Programme
USC	United States Code
USEPA	United States Environmental Protection Agency
UST	Underground Storage Tank
UV	Ultraviolet (radiation)
VOC	Volatile Organic Compound

WG	Working Group for ISO 14000
WHO	World Health Organization
WRI	World Resources Institute
WTO	World Trade Organization

Part I

Introduction to International Environmental Standards

1 Introduction

"Man has the fundamental right to freedom, equality and adequate conditions of life, in an environment of a quality that permits a life of dignity and well-being, and he bears a solemn responsibility to protect and improve the environment for present and future generations."

Principle 1, The Stockholm Declaration

Significant changes in the environmental field have required a re-evaluation of the environmental protection programs in place around the world. Although reductions in pollution levels have been realized in certain areas, necessary expectations for environmental protection have not been met in spite of extensive regulatory programs and cost expenditures. This book examines new approaches to environmental protection, including the emerging field of international environmental law and environmental management systems (EMSs). This book is intended for environmental professionals, regulators, students, educators, lawyers, business leaders, organizers, and others. It is intended to provide information on environmental protections programs, international treaties, national law, environmental management systems, and codes of business practice. As a student, use this book to assist in the research of international environmental law, regulatory compliance, or environmental management. Learn about existing principles recommended for businesses to be good stewards of the environment while improving efficiency. As a government official, use this book to stay informed about emerging strategies to address environmental problems from both business and other nations. As a business person or environmental professional, this book provides a host of references and materials to assist in program development. Educators may use this book for program delivery and reference. Individuals may learn the national and international efforts to improve the environment and determine the impetus for local programs. For a company to meet its environmental policy, they must urgently comply with environmental laws while developing and implementing an EMS.

As the EMS movement grows, students of the subject need a variety of resources to understand this new paradigm. Some people have found it difficult if not impossible to think beyond compliance and embrace the EMS or other system to improve environmental performance. This work identifies the important drivers for such an approach. Corporate environmental policies go beyond a system to comply with legal requirements into strategies to reduce operating costs and improve sales (Figure 1). Jennifer Kraus of Global Environmental Consulting Company, Inc. has provided insight into ISO 14000 conformance based on her many years of experience working with companies on EMS development and implementation. John Grosskopf, PE, DEE, president of Environmental Resources Engineering, has provided guidance to the approach of U.S. federal agency actions regarding the EMS and regulatory programs. Nancy Montgomery of Environmental Outsource effectively introduces the subject in Chapter 2. I wish to thank all the contributing authors for providing their valuable expertise.

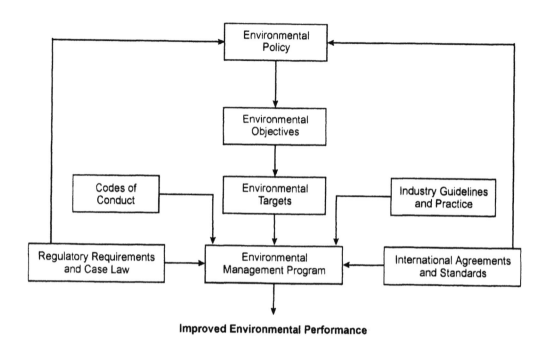

Figure 1 Environmental Performance Chart

Transboundary movement of pollution and the desire to balance economic growth and environmental protection have stimulated a variety of actions from businesses, countries, states, and organizations around the world. Some involve national laws and regulations and others are in the form of treaties for environmental protection or promotion of balanced trade. Gatherings of country representatives in Stockholm and Rio de Janeiro have produced principles and proclamations that recognize the need for environmental protection and the right of individual states to exploit their own resources without damaging the environment of other states. The text of many of these treaties has been provided here as well as summaries of actions taken by key organizations of the world, including the United Nations and the European Union. The framework for the development and implementation of EU legislation is especially important even for local U.S. businesses. The doctrine of direct effect should stimulate regulators and government representatives to implement directives in their respective countries.

The internet has become such an important tool for business and was instrumental in developing this book. A list of internet and e-mail addresses for contacts around the world is provided in Chapter 14.

Environmental problems are situational and typically are addressed on an as-needed basis. These problems require analysis, planning, and action. This book has organized many of the international agreements and requirements by topic, including air pollution, water quality, and hazardous waste. These requirements and recommendations include guidelines, standards, laws, regulations, codes of conduct, treaties, protocols, conventions, guidelines, case law, recommendations, declarations, policies, agreements, measures, principles, reports, directives, or orders. Use this information to improve environmental quality while meeting strategic objectives.

Scott S. Olson
Littleton, Colorado
November 1998

Part II

International Conventions, Treaties, and Agreements

2 An Introduction to International Environmental Controls

CONTENTS

The World Resources Institute (WRI) publishes an annual environmental almanac in which a brief description of a country and a brief description of that country's environmental problems are listed. For instance, the almanac noted that in Belgium, even though water protection laws have been in effect since 1971, these laws have not stopped steel factories from pouring waste effluent into the Meuse River. The river is the source of drinking water for 5 million people—yet it is highly polluted. The reason appears to be that the factories can afford to pay the penalties. "Fines are so low that companies would rather pay them than treat their effluent adequately (WRI 1994)."

Indeed, commitment to environmentally sound policy is as much a function of dollars as anything else. This is true in the United States. It is true in the international community. An examination of international environmental controls finds that the concerns and problems associated with finding solutions to environmental issues in the United States are essentially the same on an international level, although more complex. Why?

The elements that often influence our ability to find solutions to environmental problems include geography, politics, economics, and business. At the international level, these elements take on greater significance. For example, with regard to *geography*—Will a country located in a temperate climactic region understand the needs of a tropical region? For *politics*—Which countries are in a better position to influence the behavior of other countries? For *economics*—Should a poor, undeveloped country be expected to meet the same environmental standards as that of a vital, developed country? And for *business*—Can an industry in a country with environmental regulations (affecting the industry's bottom line) compete with the same industry in a country with no environmental regulations?

It would seem at first that the differences are too extreme at the international scale to produce effective international environmental controls. However, environmental management has always involved an assortment of diverse opinions, needs, and requirements. Through a combination of scientific study, regulations, lawsuits, trial-and-error, and creative solution, humans continue to deal constructively with environmental problems. In that sense, it is no different at the international level. The system of current international environmental controls has evolved from a series of agreements, disagreements, improved communication, improved technological advances, and a growing attempt to meet the needs of the global community.

In this chapter, we will get an overview of conventions held to address environmental problems and the treaties and agreements made between various countries as a result of these conventions. We will learn that there is a strong relationship between international trade and international environmental controls. We will see how one environmental law in the United States has been adapted worldwide to aid in making environmentally sound decisions. And we will preview a new tool for looking at the environmental impacts of producing international products and services. In subsequent chapters, how these treaties, agreements, controls, and tools may affect your organization's bottom line or mission will be explained.

AN OVERVIEW OF INTERNATIONAL CONVENTIONS AND AGREEMENTS

The past 30 years have brought a number of environmental issues to the forefront worldwide. These issues have been addressed, both generally and specifically, through a series of treaties and agreements between nations. In its 1992 report on world resources, the WRI observed that two main indications of any country's commitment to environmental protection are (1) whether it collects and disseminates environmental information effectively, and (2) its participation in relevant international agreements (WRI 1992).

Before any agreement can occur, there must be an understanding of the environment. For instance, the collection and distribution of environmental information in various countries and regions has been a priority of the United Nations. In conjunction with the United Nations Environment Programme (UNEP) and the United Nations Development Programme (UNDP), the WRI recently collected data worldwide to establish both a baseline of environmental conditions and notable trends for the world's natural resources (1992). This report focused on sustainable development in preparation for the then-upcoming 1992 *United Nations Conference on Environment and Development* in Rio de Janeiro, Brazil. The agenda items for this conference included degradation of the soil resource base, the greenhouse effect, and biodiversity.

In addition, other efforts to provide global baseline environmental information have been undertaken. The results of these efforts are found in publications such as the *Environmental Data Report* (which is published biannually by UNEP, WRI, and the United Kingdom) and the World Bank's *World Development Report* (which, in 1992, focused exclusively on the environment). WRI is also involved in maintaining and publishing the *World Resources Data Base* (which contains a 20-year data set, when possible, and is available on computer disk). Several countries have also collected and distributed their own baseline data. These sources can be used to provide the information essential to finding solutions to environmental problems—solutions that may take the form of treaties and agreements between nations.

Numerous international treaties and agreements exist. As of 1991, the U.S. International Trade Commission had identified 170 multilateral and bilateral agreements for the protection of the environment and wildlife (Kennedy 1994). Most of these agreements dated from the 1970s. The WRI researched current participation by countries in critical international conventions and regional agreements protecting the environment. To illustrate the sheer number of conventions and agreements, their geographical reach, their development, and evolution, the most notable treaties and agreements identified by the WRI are summarized below (WRI 1992).

The following conventions and agreements were put forth to protect *wildlife*, *wildlife habitat*, and the *oceans*:

- The *Antarctic Treaty and Convention* was held in 1959. It provided assurance that Antarctica would be used only for peaceful purposes, primarily cooperative and international scientific research. In 1980, the *Convention on the Conservation of Antarctic Marine Living Resources* amended the original agreement. The amendment was to safeguard the environment, retain the integrity of the ecosystem in the adjacent seas, and conserve the marine living resources.

- The 1971 *Convention on Wetlands of International Importance (Especially as Waterfowl Habitat)* sought to stop the encroachment of development on wetlands. A list of wetlands considered to be of international importance resulted. The parties agreed to establish wetland nature preserves and to consider their responsibilities in providing habitat for migratory waterfowl.
- The *Convention Concerning the Protection of the World Cultural and Natural Heritage* was held in Paris in 1972. It established a system for collective protection of noted sites considered to represent outstanding universal value.
- The *Convention on the Prevention of Marine Pollution by Dumping of Wastes and Other Matter* (MARPOL), in 1972, provided for controls on ocean dumping of certain materials (either through prohibition or regulation). It also established a mechanism for assessing liability and settling disputes.
- To protect endangered species from overexploitation, the 1973 *Convention on International Trade in Endangered Species* (CITES) provided controls for trade in live or dead animals and/or animal parts via a permit system.
- A 1973 convention to eliminate international pollution resulting from oil and other harmful substances was modified with the *Protocol of 1978 Relating to the International Convention for the Prevention of Pollution from Ships* to minimize accidental discharge.
- In 1979, the *Convention on the Conservation of Migratory Species of Wild Animals* used international agreements to protect wild animal species that migrated across international borders.
- Although only 43 of 60 nations have ratified it, the 1982 *United Nations Convention on the Law of the Sea* established a comprehensive legal regime for the seas and oceans. It established rules for environmental standards and enforcement provisions and developed international rules and legislation to prevent and control marine pollution.

Regarding the *atmosphere* and *hazardous substances*, the prominent conventions and agreements include the following:

- The *Treaty Banning Nuclear Weapon Tests in the Atmosphere, in Outer Space, and Under Water* in 1963 prohibited this testing and other nuclear explosions that would spread radioactive debris outside of the country conducting the test.
- In 1972, the *Convention on the Prohibition of the Development, Production, and Stockpiling of Bacteriological (Biological) and Toxin Weapons, and on Their Destruction* prohibited the acquisition and retention of biological agents and toxins that are not justified for peaceful purposes. It also prohibited the means for delivering these agents and toxins for hostile purposes or armed conflict.
- To minimize the consequences from transboundary radiological events, the *Convention on Early Notification of a Nuclear Accident* in 1986 called on countries to provide relevant information about their nuclear accidents as early as possible to potentially affected countries.
- Also in 1986, the *Convention on Assistance in the Case of a Nuclear Accident or Radiological Emergency* facilitated the prompt provision of assistance in the event of a nuclear accident or radiological emergency.
- The *Vienna Convention for the Protection of the Ozone Layer* was held in 1985. It launched an effort to conduct research on the earth's ozone layer to determine the adverse effects from chemical substances, primarily chlorofluorocarbons. As part of this research, the ozone layer would be monitored, alternative substances and technologies would be studied, and measures would be taken to control activities that produce adverse effects.
- The Vienna Convention was followed in 1987 by the *Protocol on Substances that Deplete the Ozone Layer* (known as "the Montreal Protocol"). The Montreal Protocol required

nations to (1) cut consumption of five chlorofluorocarbons and three halons by 20 percent of their 1986 level by 1994, and (2) by 50 percent of their 1986 level by 1999. Allowances were made for increases in consumption by developing countries.

- In 1989, the *Basel Convention on the Control of Transboundary Movements of Hazardous Wastes and Their Disposal* set forth controls to reduce transboundary movement of wastes. Participating countries were obligated to minimize the amount and toxicity of hazardous wastes generated, ensure the sound management of these wastes, and assist developing countries in the sound management of their hazardous wastes.

Conventions and agreements that have addressed *region-specific* environmental issues include the following:

- the *African Convention on the Conservation of Nature and Natural Resources* (1968),
- the *Convention on Conservation of Nature in the South Pacific* (1976),
- the *Treaty for Amazonian Cooperation* (1978),
- the *Convention on the Conservation of European Wildlife and Natural Habitats* (1979),
- the *Convention on Long-Range Transboundary Air Pollution* (1979), which was followed by specific protocols to the 1979 convention in 1984, 1985, and 1988,
- the *ASEAN Agreement on the Conservation of Nature and Natural Resources* (1985),
- the *Bamako Convention on the Ban of the Import into Africa and the Control of Transboundary Movements of Hazardous Wastes Within Africa* (1991), and
- the *Convention on Environmental Impact Assessment in a Transboundary Context* (1991).

In addition to the aforementioned conventions and agreements, the United Nations Environmental Programme (UNEP) also initiated the *Regional Seas Programme* in 1974. Since that time, the program has developed regional action plans for controlling marine pollution and managing marine resources in several geographical areas. The action plans have taken the form of regional environmental assessments, management plans, legislation, institutional arrangements, and financial agreements through the following regional conventions:

- the *Mediterranean Convention Against Pollution,*
- the *West and Central Africa Convention on Environmental Cooperation,*
- the *East African Convention on Environmental Protection,*
- the *Red Sea and Gulf of Aden Convention on Conservation,*
- the *Caribbean Convention on Environmental Protection,*
- the *South-East Pacific Convention on Environmental Protection,*
- the *South Pacific Convention on Environmental Protection,* and
- the *Kuwait Convention on Environmental Cooperation.*

Each of these conventions has, in turn, yielded various general and specific agreements and protocols throughout the past 30 years.

The two most significant documents produced through international conventions are the 1972 *Stockholm Declaration on the Human Environment* and the 1992 *Rio Declaration on Environment and Development.* According to the *Harvard Environmental Law Review,* these "... two closely related documents set forth the basic principles of international environmental law and policy and are recognized as the centerpieces of the field (Kennedy 1994)."

The focus of the Stockholm and Rio declarations was primarily environmental protection. It is notable, however, that in the 20 years that separated the conventions, the titles reflect the change in priority from just environment to an integration of environment and development. The Rio Declaration, in fact, "... gives trade issues primacy over environmental concerns (Kennedy 1994)." This primacy was as much in response to a concern over avoiding potential "eco-imperialism" by developed and industrialized countries as anything else (that is, the big guys should not tell the little guys what to do). Above all, however, it is noteworthy that the declarations do seek to balance the concepts of trade, economic growth, and environmental concerns.

These two declarations took a balanced approach to recognizing what can often be conflicting goals—maintaining both national sovereignty and international cooperation. In 1972, Principle 21 of the Stockholm agreement asserted the following:

> States have, in accordance with the Charter of the United Nations and principles of international law, the sovereign right to exploit their own resources pursuant to their own environmental policies, and the responsibility to ensure that activities within their jurisdiction or control do not cause damage to the environment of other States or of areas beyond the limits of national jurisdiction.

The *Harvard Environmental Law Review* noted that the principles of the Rio Declaration provide for states to enact their own legislation with regard to environmental damage, liability, and compensation. The declarations also call upon states to cooperate in the development of international conventions addressing transboundary pollution (Kennedy 1994). Significantly, this declaration also employs the "polluter pays principle," which is one of the cornerstones of much of the United States' environmental legislation.

THE RELATIONSHIP BETWEEN INTERNATIONAL TRADE AND ENVIRONMENTAL CONTROLS

Trade is a prominent issue in many of the aforementioned conventions and agreements. It is even featured in the title of some conventions. As we will see in later chapters, many of the environmental standards being promulgated today (such as ISO 14000) are intended to protect the environment while removing trade barriers.

Trade has also been used as a hammer. Unilateral trade sanctions have been used to promote the United States' environmental interests abroad. Kevin C. Kennedy, in a recent issue of the *Harvard Environmental Law Review*, pointed out that there is some fairness in using trade sanctions to promote environmental protection, especially in consideration that "... the markets created by open international trade are in large part responsible for the environmental damage [e.g., trade in toxic chemicals or endangered species] (1994)."

However, international trade and international environmental controls have not always been mentioned in the same breath, much less aligned. The laws that govern both have, in fact, proceeded on separate but parallel tracts over the years—even though the issues are related. Kennedy noted that, "as levels of world trade and investment have rapidly risen in the post-war era, so too has the level of environmental degradation (1994)."

Thus, there is a greater need to "integrate" both trade law and environmental law, since the two are not mutually exclusive. Kennedy submitted that it is not "liberal trade" that would reduce environmental standards, rather, "...the problem lies either in the market's failure to reflect environmental costs in prices or in government interference with the market through subsidization of polluting industries." He added that the market approach alone has not proven entirely effective because industries cannot count on their competitors in other countries to voluntarily act environmentally responsible, nor have unilateral trade sanctions proven to be reliable at effecting change

Thus, there seems to be a need for some level of environmental regulatory standards to which most of the international community can commit and meet. This, obviously, is the next goal for international environmental controls. Yet, due to the numerous multilateral and bilateral agreements for the protection of the environment and wildlife, Kennedy contended that international environmental law resembles a patchwork quilt. Kennedy believed the aforementioned goal is currently most attainable by using existing international trade law—such as the General Agreement on Tariffs and Trade (GATT). He contended that "… no single document emerges as the international environmental law constitution analogous to GATT in the trade field (1994)."

The GATT is both an agreement and an organization. In 1947, GATT was originally conceived as an interim treaty focused on trade in goods and was based on the economic principle of "liberal" trade. However, over the years, it has become one of the central world trade organizations, a hub for international trade dispute resolution, and has expanded to address trade in services, rules on investment, and the protection of intellectual property (Kennedy 1994). The organization is head-quartered in Geneva, Switzerland, with 111 member countries.

The protection of the environment was not an original intent of GATT. However, as environmental issues have increasingly been woven into the fabric of economics, politics, and business, this protection is also being addressed through established GATT principles. According to Kennedy, Article III of GATT provides "…a national treatment obligation that requires contracting parties to treat imports the same as like domestic products insofar as taxes and other domestic regulations are concerned (1994)." GATT also includes exemptions that have been construed to benefit environmental issues. Under Article XX, contracting parties are permitted to restrict imports under a number of specific exceptions. Of the exceptions enumerated under this Article, Kennedy stated that "… the public health and safety exception [oil pollution prevention], the customs enforcement exception [endangered species], and the exception for conservation of natural resources [fishing waters] touch most directly on the enforcement of environmental measures (1994)."

NEPA INFLUENCE ON INTERNATIONAL ENVIRONMENTAL CONTROLS

In addition to the various conventions and agreements, one environmental law passed in the United States has had an influence on the environmental controls of other nations—the National Environmental Policy Act (NEPA). What makes this law adaptable in other nations is that it is, above all, a *process*. To understand how the NEPA process may be used abroad, a brief explanation of the statute is provided.

The United States' National Environmental Policy Act of 1969 was one of the first environmental statutes. It is published in the United States Code under 42 USC Sections 4321 through 4370a. NEPA set a precedent in that it considered the environmental impacts of the proposed actions and projects of the federal government. The intent of the law was to promote better decision making. In the late 1960s, lawmakers and their constituents felt that environmental concerns were not being considered during the planning of these projects—to the point that the environment was being degraded and the public was bearing the cost of mitigation or reclamation after a project or action was completed. NEPA set up a *process* wherein significant impacts to natural resources, wildlife, and social entities would be considered *while* a project or action was being planned. As part of this process, the public would be allowed to express its concerns about the proposal.

Unlike most other U.S. environmental regulations, NEPA does not set quantity levels or treatment standards that must be met. Rather, because it is a *process*, it outlines *procedures* to be followed so that potential environmental problems are avoided or addressed. However, the law does not state *what* must be done about the problems. That is ultimately determined from the study, the public input, and by the project's proponent.

The procedures for conducting the NEPA process were established by the Council on Environmental Quality (CEQ). These procedures are called "implementing regulations" and are published in the Code of Federal Regulations (40 CFR 1500 through 1508). The regulations outline the procedure for conducting studies to determine environmental impacts.

Two main levels of study are specified by NEPA—the *Environmental Assessment* and the *Environmental Impact Assessment*. The Environmental Assessment (EA) is primarily an *initial* study used to determine whether an Environmental Impact Assessment (EIS) is required. The EA is similar in format to an EIS, although less detailed. The information gathered and analyzed for an EA study may result in either (1) a decision to prepare a more detailed EIS, or (2) a decision to issue a *Finding of No Significant Impact*. A Finding of No Significant Impact (FONSI) means that a project or action may begin without further study. When an EIS is determined to be necessary, a more comprehensive study is made.

NEPA requires that the EIS include a discussion of the following: (1) the purpose and need for the action; (2) the proposed action; (3) alternatives to the action (including "no action"); (4) the affected environment; (5) the environmental consequences; and (6) information and opinions garnered from public comments. Depending upon the size of the project or action, its potential environmental effects, or public concern, an EIS may range from about 30 pages to several volumes. The procedures give NEPA this flexibility.

In the beginning, NEPA regulated only projects and actions undertaken directly by the federal government. Over the past 20 years, however, the courts have determined that the statute is also applicable to indirect projects. Therefore, NEPA now covers most actions or projects conducted by other organizations that receive federal funding (such as, transmission lines constructed by electric cooperatives). In addition, several states have adopted what have become known as "little NEPAs" for actions taken at the state level. Other nongovernmental organizations also employ the procedures as a matter of course in the study of environmental impacts. It is not surprising that this methodology has also been adopted at the international level.

In his presentation before a conference on NEPA in 1993, N. A. Robinson asserted that the concept of environmental impact assessment (EIA) conceived in NEPA "... is today increasingly being established as a routine decision making technique worldwide," noting that some form of EIA "... has been required by law in more than 75 separate jurisdictions (1993)." Robinson uses the term "EIA" as a descriptive reference to the process, not specifically the statute's NEPA studies (that is, the EA and EIS).

The attraction to the EIA concept, according to Robinson, is that it is a "proven technique" that provides a way to institutionalize *foresight*, (that is, the ability to consider impacts prior to the final plan for a project or action). He also explained that, "while its essential structure is substantially the same wherever used throughout the world, the EIA is flexible and has been adapted successfully to operate within the cultural, political, and socioeconomic development conditions prevalent in each jurisdiction which has enacted EIAs (1993)."

Robinson observed that users of the NEPA process abroad have not just copied the statute and regulations but have adapted it to their specific needs (1994). The flexibility of the EIA procedures has allowed the users in various countries to integrate innovations (which improve upon the original techniques) or to adjust the methodology to the user's situation. Robinson pointed to the following characteristics that make the EIA procedure an appealing and functional import:

- The EIA works in all political systems. "It can be and has been established alike in common law, civil law, and socialistic law traditions. It is equally useful in developed and in developing countries (Robinson 1993)."
- The EIA is a relatively new "analytic tool for decision makers." And as with other fledglings, it can be molded. For example, Robinson observed that the EIA process works best when an independent authority provides oversight of the process. In the United States, the courts provide this type of NEPA oversight through judicial review. But even countries or jurisdictions without a similar judicial review are able to provide the necessary oversight with an administrative organization.
- The EIA process also provides local citizens with the opportunity to voice their opinion, introduce information, and participate in the decision making that may affect their environment. Robinson mentioned that this participation feature has been used even in

the former USSR, where residents were allowed to review and comment on plans for a power plant.

- The EIA process is also, according to Robinson, effective in locating, collecting, and analyzing environmental data for decision makers. The EIA process encourages communication and consultation between experts, the public, and governmental agencies. As a result, project proponents have the opportunity to incorporate data or information into their final plan that they may not have been aware of or previously considered before unintended environmental damage occurs.

There are, of course, drawbacks in the effective use of the EIA process at the international level. These drawbacks are similar to those associated with the application of the NEPA process in the United States and include the following:

- Robinson reported that the EIA procedure "... is not easy to establish at the outset (1993)." As with NEPA, the EIA process is "... often, almost always, resisted until decision makers and administrators become educated about its utility." The process sets forth only a framework for study and does not specifically spell out what is good or bad. It demands that participants in the process determine these effects within the context of the project or action. This is not always easy, and so may be rejected.
- As is also true in the United States, the tendency is to apply the EIA process only to large projects. Robinson observed that, "since environmental *significance* is not merely a function of *bigness,* the trends toward using lists [of applicable projects] and restricting EIAs to large projects do not assure an effective application of the EIA (1993)."
- Again, as with its application in the United States, Robinson acknowledged that "...it is rare to require post-project monitoring to find out whether all adverse impacts were accurately anticipated or whether mitigation plans in fact were successful. [...] There is a constant need to evaluate the effectiveness of each jurisdiction's EIA process, to improve, streamline, and weed out dysfunctional aspects."

Nevertheless, the EIA process has seen a prolific acceptance worldwide—EIA statutes can be cited in 41 countries. Robinson related that, "after enactment of NEPA in 1969, the EIA was quickly adopted by the mid-1970s in Australia, Canada, and New Zealand. Although the process is not yet widely used in many parts of Latin America, the Middle East, or Africa, it has since been instituted in the following countries and jurisdictions:

Argentina	Greece	Kuwait
Belgium	Hong Kong	Luxembourg
Brazil	India	Malaysia
China	Indonesia	Mexico
Columbia	Ireland	The Netherlands
Costa Rica	Israel	New Zealand
Denmark	Italy	Norway
France	the Ivory Coast	Pakistan
Gambia	Japan	Papua New Guinea
Germany	Korea	Peru
the Philippines	Spain	the United Kingdom
Portugal	Taiwan	the U.S.S.R
Sri Lanka	Thailand	Uruguay
South Africa	Turkey	Venezuela

Robinson suggested that "it is becoming a norm of customary international law that nations should engage in effective EIAs before taking action which could adversely affect shared natural resources, another country's environment, or the earth's commons...". This is a guideline promulgated in Principle 21 of the United Nations' *Stockholm Declaration on The Human Environment*. Thus, it is notable that many of the previously mentioned international agreements employ the EIA process, and many international organizations advocate its use as well.

For example, Robinson noted that during the 1970s, the North Atlantic Treaty Organization undertook to explain how to use the EIA procedures to its member nations. In addition, the Organization for Economic Cooperation and Development was involved early in the potential application of the process to its related projects and actions. According to Robinson, "these educational efforts led to early acceptance of EIAs in Western Europe (1993)."

The UNEP and the United Nations General Assembly have also endorsed the use of EIAs. Robinson stated that "the World Charter of Nature, adopted by the General Assembly, expressly calls for the use of EIAs. Article 206 of the Law of the Sea Convention provides for the use of EIAs, and agency practice among international organizations increasingly requires it (1993)." And he noted that "the World Bank has adopted a six-step process, not dissimilar to the EIA process of Canada or the United States...". The following nine organizations currently having international EIA provisions:

- the European Economic Community Directive,
- the Association of South East Asian Nations,
- the United Nations Economic Commission for Europe,
- the United Nations Environment Programme,
- the United Nations Environment Programme Regional Seas Conventions,
- the Organization for Economic Cooperations and Development,
- the World Bank,
- the United Nations General Assembly, and
- the United Nations Law of The Sea.

The EIA process is incorporated into organizational efforts in various ways. As an example, Robinson pointed out the following:

> Through the U.N. Economic Commission for Europe, a negotiation has proceeded to prepare a 'Convention on Environmental Impact Assessment Transboundary Context.' This treaty's preparation began with a Seminar on Environmental Impact Assessment in Warsaw, Poland in 1987 and culminated in signing a final agreement at Espoo, Finland on February 25, 1991. The parties agreed to 'prevent, reduce, and control significant adverse transboundary environmental impact from proposed activity.' To do so, an EIA is required for projects on a List of Activities included as Appendix I ... (1993).

Another vehicle for international NEPA influence is the provision in Section 102(2)(f) of the NEPA statute that United States federal agencies should "...recognize the worldwide and long-range character of environmental problems and where consistent with the foreign policy of the United States, lend appropriate support to initiatives, resolutions, and programs designed to maximize international cooperation in anticipating and preventing a decline in the quality of mankind's world environment."

Although United States foreign policy does not, in practice, view environmental impact as a priority, Robinson noted exceptions. "The U.S. Army has been distinguished by developing methodologies to comply with NEPA and Executive Order 12114; the Army's EIS for returning weapons, including chemical munitions, from Europe to the United States for dismantling is a useful example (1993)." The U.S. Army specifically addresses this issue in its regulations under *32 CFR Part 651 Subpart H: Environmental Effects Abroad of Major Army Actions*. Under this regulation, the U.S.

Army makes a concerted effort to act overseas in a manner comparable to the way it manages environmental issues within the United States.

The NEPA framework provides a way for other environmental regulations to be integrated into its process. This feature has been beneficial in accommodating new and more specific regulations that were put into effect *after* NEPA was passed. In following the spirit and intent of Section 102(2)(f) of the NEPA statute, the United States has likewise been able to incorporate subsequent treaties and agreements when using NEPA practices abroad. An example would be the *Convention on the International Trade in Endangered Species* (CITES). "In response to CITES and the U.S. Endangered Species Act, the U.S. Agency for International Development issued rules, for instance, which expressly require that foreign assistance programs consider how to protect endangered species (Robinson 1993)."

Robinson found that "the most explicit U.S. foreign policy directives for EIAs abroad are provided for in Executive Order 12114 (1993)." Issued under President Carter's Administration, Executive Order 12114 required the use of the EIA process in the following circumstances, according the Robinson:

- under NEPA for actions on the commons (e.g., the oceans and Antarctica);
- when an action will affect uninvolved nations (innocent bystander situations);
- when an action is strictly regulated in the United States (e.g., actions involving radioactive materials or toxic substances); and/or
- when the president or secretary of state designates a natural or ecological resource to be of global importance (1993).

NEPA could become an important tool in coping with global trends, according to Robinson, because the scientific analysis in the EIA process "...increasingly can and does identify impacts that are transnational and even global (1993)."

NEW TOOLS FOR PROVIDING PROACTIVE ENVIRONMENTAL CONTROL

Although NEPA is relatively new, the need for additional or different tools to study and evaluate environmental concerns and problems is ongoing and evolving. One of the most progressive and comprehensive tools is called "life cycle assessment." Life cycle assessment (LCA) may also be one of the most potentially applicable environmental tools at the international level.

A recent article in *Environmental Science and Technology* observed that "products have environmental impacts over the course of their existence, from extraction and procurement of raw materials to manufacturing, distribution, use, and disposal. Determining what is green or even what is environmentally preferable is a complex business, and the boundaries of such an assessment cannot be drawn at the factory door (Nash 1994)."

LCA is the tool that has been developed to address this complex issue of assessing product impact. Nash commented that LCA is receiving increased attention because of its ability "to identify and measure both 'direct' and 'indirect' environmental, energy, and resource impacts associated with a product or process." An example of a *direct impact* would be air emissions, effluent, and/or energy consumption generated from the manufacturing process. An example of an *indirect impact* would encompass the extraction of raw materials to make the product, run the manufacturing plant, and distribute the product, as well as consumer use and final disposition.

At one of the first conferences on LCA held at the Massachusetts Institute of Technology in 1993, the participants "examined practical lessons learned from LCA and addressed LCA theory, methodology, and regulatory implications (Nash 1994)."

One of the lessons learned through conducting LCAs is that indirect impacts (particularly postmanufacturing) may surpass direct impacts. Nash explained that "the American Fiber Manufacturers Association, for example, conducted an LCA of an 'average' polyester blouse and found that far more resources were used by consumers in washing and drying the blouse than were used in its manufacture." The example given did not indicate whether hazardous substances produced in the manufacturing process were of lesser or greater impact than those produced in the washing and drying process, although it is expected that LCA could be used to determine this related aspect.

Still, the example provides insight into the usefulness of LCA. Nash suggested that the lesson points "to a new direction for corporate environmental practice: Firms wishing to improve the environmental performance of their products may need to design them to reduce postmanufacturing impacts."

Another lesson revealed at the conference was that, "when indirect effects are taken into account, conventional wisdom about the environment—for example, the reduce, reuse, recycle hierarchy—may no longer apply." A number of presenters reported that "recycling may actually consume more resources than it saves, if, for example, recyclables must be transported long distances for processing or sale." Here again, LCA could integrate into the equation the cost and opportunity for landfilling over the consumption of resources.

Interestingly, Nash asserted that the use of LCA is growing in the private sector, especially in Europe—130 known LCAs were conducted in 1992 alone. Nash went on to point out the following:

> In Europe and the United States, interest in LCA as a regulatory tool is growing. The European Commission is planning to use life cycle studies as the basis for packaging recovery and recycling targets. The recently developed Belgian 'eco-tax' will be levied on products based on results of LCA studies. The European Commission will use LCA as the basis for awarding eco-labels to all products sold in Europe (including products manufactured elsewhere). This EC-wide approach will replace the national green labels of France, Germany, and other countries that were not based on LCA and often were awarded on the basis of a single environmental criterion (1993).

And, as we will see in a later chapter, the use of LCA is being incorporated into the ISO 14000 methodology.

CONCLUSION

An initial examination of international environmental law finds a diverse system of agreements, adapted statutes, and methodologies—the "patchwork quilt," as observed by Kennedy. The numerous international conventions and agreements range from general guidance documents to region-specific protocols to issue-specific prohibitions. These conventions and agreements may overlap or require consistent updating. In addition, the existing system of international environmental law often conflicts with trade and economic issues, which usually take priority.

Ideally, the international community would be better served by a more comprehensive and conventional system. However, as with the United States, when the legal framework for addressing environmental issues takes place over a number of years, it is influenced heavily by politics, economics, technology, and other environmental factors. For example, the *Emergency Planning and Community Right-to-Know Act* (EPCRA) is more of a "RCRA" issue than a "CERCLA" issue. But, in part because CERCLA was up for reauthorization at the time that EPCRA issues were in the forefront, legislation was attached to that statute.

International environmental control is being increasingly molded by the changing goals and complexities of the parties involved and the state of the world. As nations interact with each other more often, it is likely that the specifics of their agreements may reflect less suspect-induced clauses, and thus, more effective environmental regulation. And, as Kennedy's predictions on the utility of GATT and similar agreements to effect an integration of both trade and environmental policies prove viable, international environmental law may be considered as a matter of course, rather than as a nuisance. This type of attitude will likely provide an environment in which tools such as the EIA and the LCA are used more frequently, more effectively, and more innovatively.

But how does this patchwork quilt affect you now? Are there portions of these controls and agreements that directly relate to your bottom line or your mission? What does an agreement such as GATT look like? What is the status of the various treaties and agreements, especially in lieu of the creation of new, more effective and comprehensive tools? The following chapters will provide a road map as to how you can achieve compliance and locate additional information.

REFERENCES

Kennedy, Kevin C. "Reforming U.S. Trade Policy to Protect Global Environment: A Multilateral Approach." *The Harvard Environmental Law Review.* Volume 18, Number 1, 1994.

Nash, Jennifer and Mark D. Stoughton. "Learning to Live with Life Cycle Assessment." *Environmental Science and Technology.* Volume 28, Number 5, 1994.

Robinson, N.A. *EIA Abroad: The Comparative and Transnational Experience.* Proc. of a Conference published as *Environmental Analysis: The NEPA Experience.* Hildebrand, Stephen G. and Johnnie B. Cannon, Eds. Boca Raton: Lewis Publishers, 1993.

World Resources Institute. *World Resources: 1992-1993.* New York: Oxford University Press, 1992.

World Resources Institute. *The 1994 Information Please Environmental Almanac.* Boston: Houghton Mifflin Company, 1994.

3 The Influence of Command and Control Regulations

CONTENTS

INTRODUCTION

The United States has employed a system of environmental protection through the passage of laws enforced by state and federal government agencies (see listing). The U.S. Environmental Protection Agency (EPA) is responsible for the federal program and provides guidance and financial assistance to the states. Although other approaches have been employed, this system is primarily built on the command and control approach. That is, specific conditions or quantitative restrictions must be met or action can be taken against the violator such as the assessment of penalties. Another term for command and control might be specification standard as opposed to a performance-oriented standard designed to provide flexibility in its implementation. Upon passage of proposed legislation developed by congressional committees and subcommittees, the implementing regulations must be developed by the agency (Figure 2).

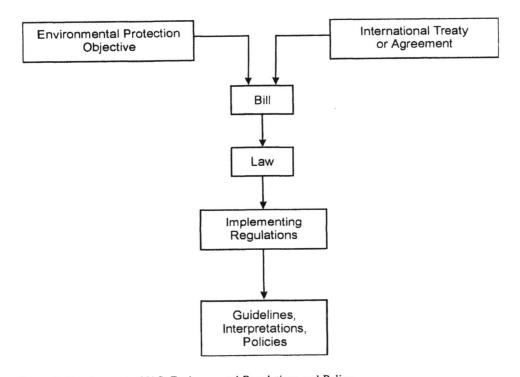

Figure 2 Development of U.S. Environmental Regulations and Policy

Unlike international environmental law in which individuals are not recognized (although recent decisions by the International Court of Justice indicate this is changing), regulatory action by the EPA requires public notification accomplished through advanced publication of the action in the Federal Register. The action will call for public comment and EPA may sponsor workshops or presentations in various cities around the country regarding proposed regulations as another means to inform the public and solicit comment and participation.

Federal regulations may require states to take action, including development of state laws resulting in two and sometimes three layers of command and control type requirements that are lengthy and complicated. The third layer is a local municipality or authority. Under this framework, laws have been developed that establish nationally acceptable pollution levels such as National Ambient Air Quality Standards (NAAQS) under the Clean Air Act (CAA) and Maximum Contaminant Levels (MCLs) under the Safe Drinking Water Act (SDWA), both designed to protect human health.

Standards have also been established that specify required technology and pollution control reductions. Much has been accomplished to reduce and control pollution through these mechanisms. However, the very nature of the system precludes flexibility both in compliance and enforcement of the regulation. In our fast-changing world, today's businesses must be allowed to determine the best ways to achieve compliance or better yet methods to meet environmental objectives. The current system is fragmented and duplicative. Not only are the regulations organized independent from one another, so are the offices and, consequently, the people tasked with implementing these programs. Federal and state acts provide for substantial civil and criminal penalties creating litigation issues that promote a focus on compliance even if not accomplishing improvements to environmental quality. As reported by Davies and Mazurek, pollution levels in the United States remain high when compared to other industrialized nations. Substantial improvements to air quality have been realized under the current system for a number of air pollutants; however, some still remain elusive and may present a risk to human health and the environment.

Would an integrated approach such as that adopted by numerous countries (Table 1) provide a more efficient system both in terms of cost and environmental protection? It may promote technological advances in pollution control, provide the necessary focus on pollution prevention and minimization, and provide the necessary program integration to eliminate duplication and facilitate environmental protection. In spite of the regulatory structure, businesses have begun to realize that an Environmental Management System improves environmental performance while at the same time improving efficiency. Although the need for regulatory requirements still exists, national policies should promote and support integrated environmental management programs adopted by businesses.

Table 1: Integrated Pollution Control

European Union	Council Directive 96/61/EC, 24 September 1996
The Netherlands	National Environmental Policy Plan, 1989 and Environmental Management Act of 1993
Norway	Norwegian Pollution Control Act
Sweden	Environmental Protection Act
United Kingdom	Environmental Protection Act of 1990
United States/Mexico	Integrated Border Environmental Plan (1992)

REFERENCES

Pollution Control in the United States, Evaluating the System, J. Clarence Davies and Jan Mazurek, Resources for the Future, 1998.

Direct Effect of European Law and the Regulation of Dangerous Substances, Christopher J. M. Smith, Gordon and Breach Publishers, 1995.

Environmental Change and International Law, edited by Edith Brown Weiss, United Nations University Press, 1992.

International Environmental Law and Policy, David Hunter, James Salzman, Durwood Zaelke, Foundation Press, 1998.

Environmental Management Systems, Jay G. Martin and Gerald J. Edgley, Governments Institutes, 1998.

International Environmental Auditing, David D. Nelson, Government Institutes, 1998.

Precautionary Legal Duties and Principles of Modern International Environmental Law, Harold Hohmann, Graham & Trotman/Martinus Nijhoff, 1998.

Environmental Management in European Companies, Success Stories and Evaluation, edited by Jobst Conrad, Gordon and Breach Science Publishers, 1998.

Pollution Control in the United States, Evaluating the System, J. Clarence Davies and Jan Mazurek, Resources for the Future, 1998.

Public Policies for Environmental Protection, Editor, Paul R. Portney, Resources for the Future, 1992.

Environmental Strategies Handbook, A Guide to Effective Policies & Practices, Rao V. Kolluru, McGraw-Hill, Inc. 1994.

4 Key International Agreements

CONTENTS

GENERAL AGREEMENT ON TARIFFS AND TRADE (GATT)

Issues relating to trade and/or the environment have been addressed in key multilateral, bilateral, or regional agreements. National laws designed to protect human health and the environment have been determined to present barriers to trade such as restrictions on importation of goods or tariffs. The World Trade Organization (WTO), which replaced the General Agreement on Tariffs and Trade (GATT) in 1995, is a major multilateral agreement on international trade. WTO is a comprehensive series of agreements and obligations composed of the original 1947 agreement and accompanying legal instruments, the Uruguay Round Protocol to the GATT, and other Uruguay Round Agreements. Environmental exceptions to GATT allow nations to adopt programs "necessary to protect human, animal or plant life or health" or relating to the "conservation of exhaustible natural resources if such measures are made effective in conjunction with restrictions on domestic production or consumption."

GATT contains three main principles, all designed to reduce trade barriers.

Article I – General Most-Favoured-Nation Treatment

"1. With respect to customs duties and charges of any kind imposed on or in connection with importation or exportation or imposed on the international transfer of payments for imports or exports, and with respect to the method of levying such duties and charges, and with respect to all rules and formalities in connection with importation and exportation, and with respect to all matters referred to in paragraphs 2 and 4 of Article III, any advantage, favour, privilege or immunity granted by any contracting party to any product originating in or destined for any other country shall be accorded immediately and unconditionally to the like product originating in or destined for the territories of all other contracting parties..."

This article evens the playing field by requiring all countries to be treated the same. No favoritism or discrimination is allowed.

Article III – National Treatment on Internal Taxation and Regulation

"1. The contracting parties recognize that internal taxes and other internal charges, and laws, regulations, and requirements affecting the internal sale, offering for sale, purchase, transportation, distribution or use of products, and internal quantitative regulations requiring the mixture, processing or use of products in specified amounts or proportions, should not be applied to imported or domestic products so as to afford protection to domestic production.

2. The products of the territory of any contracting party imported into the territory of any other contracting party shall not be subject, directly or indirectly, to internal taxes or other internal charges of any kind in excess of those applied, directly or indirectly, to like domestic products. Moreover, no contracting party shall otherwise apply internal taxes or other internal charges to imported or domestic products in a manner contrary to the principles set forth in paragraph 1.

3. The products of the territory of any contracting party imported into the territory of any other contracting party shall be accorded treatment not less favourable than that accorded to like products of national origin in respect of all laws, regulations and requirements affecting their internal sale, offering for sale, purchase, transportation, distribution or use. The provisions of this paragraph shall not prevent the application of differential internal transportation charges which are based exclusively on the economic operation of the means of transport and not on the nationality of the product."

As described in these three paragraphs, domestic and foreign products must be treated the same, without discrimination or domestic programs designed to protect their own products.

Article XI – General Elimination of Quantitative Restrictions

"No prohibitions or restrictions other than duties, taxes or other charges, whether made effective through quotas, import or export licenses or other measures, shall be instituted or maintained by any contracting party on the importation of any product of the territory of any other contracting party or on the exportation or sale for export of any product destined for the territory of any other contracting party."

NORTH AMERICAN FREE TRADE AGREEMENT (NAFTA)

NAFTA is designed to eliminate trade barriers between the three North American countries party to this agreement: Canada, the United States, and Mexico. NAFTA was signed by the three parties on October 7, 1992, and signed into law by President Clinton on December 8, 1993. This landmark agreement contains environmental protection provisions in the main body of the agreement but added a "side agreement" called the *Supplemental Agreement of the Environment* that deals specifically with environmental issues. The majority of environmental provisions are contained in Articles 904, 905 and 907 of Chapter 9 (Standards-Related Measures) of the agreement. This chapter is a component of Part Three, *Technical Barriers to Trade*. Article 904 allows parties to implement standards-related measures to protect the environment. If a party does not comply with such measures, the issuing party can ban importation of goods or the provision of service. Paragraph 2 of Articles 904 and 905 allow the establishment of appropriate levels of protection based on international standards. Article 905 also allows parties to establish more stringent provisions.

The key environmental provisions of NAFTA are described below.

"Article 904: Basic Rights and Obligations

Right to Take Standards-Related Measures

1. Each Party may, in accordance with this Agreement, adopt, maintain or apply any standards-related measure, including any such measure relating to safety, the protection

of human, animal or plant life or health, the environment or consumers, and any measure to ensure its enforcement or implementation. Such measures include those to prohibit the importation of a good of another Party or the provision of a service by a service provider of another Party that fails to comply with the applicable requirements of those measures or to complete the Party's approval procedures.

Right to Establish Level of Protection

2. Notwithstanding any other provision of this Chapter, each Party may, in pursuing its legitimate objectives of safety or the protection of human, animal or plant life or health, the environment or consumers, establish the levels of protection that it considers appropriate in accordance with Article 907(2)."

"Article 905: Use of International Standards

1. Each Party shall use, as a basis for its standards-related measures, relevant international standards or international standards whose completion is imminent, except where such standards would be an ineffective or inappropriate means to fulfill its legitimate objectives, for example because of fundamental climatic, geographical, technological or infrastructural factors, scientific justification or the level of protection that the Party considers appropriate...

3. Nothing in paragraph 1 shall be construed to prevent a Party, in pursuing its legitimate objectives, from adopting, maintaining or applying any standards-related measure that results in a higher level of protection than would be achieved if the measure were based on the relevant international standard."

"Article 907: Assessment of Risk

1. A Party may, in pursuing its legitimate objectives, conduct an assessment of risk. In conducting an assessment, a Party may take into account, among other factors relating to a good or service:
   (a) available scientific evidence or technical information;
   (b) intended end uses;
   (c) processes or production, operating, inspection, sampling or testing methods; or
   (d) environmental conditions."

NAFTA Preamble

The Government of Canada, the Government of the United Mexican States and the Government of the United States of America, resolved to:
   STRENGTHEN the special bonds of friendship and cooperation among their nations;
   CONTRIBUTE to the harmonious development and expansion of world trade and provide a catalyst to broader international cooperation;
   CREATE an expanded and secure market for the goods and services produced in their territories;
   REDUCE distortions to trade;
   ESTABLISH clear and mutually advantageous rules governing their trade;
   ENSURE a predictable commercial framework for business planning and investment;
   BUILD on their respective rights and obligations under the General Agreement on Tariffs and Trade and other multilateral and bilateral instruments of cooperation;
   ENHANCE the competitiveness of their firms in global markets;

FOSTER creativity and innovation, and promote trade in goods and services that are the subject of intellectual property rights;

CREATE new employment opportunities and improve working conditions and living standards in their respective territories;

UNDERTAKE each of the preceding in a manner consistent with environmental protection and conservation;

PRESERVE their flexibility to safeguard the public welfare;

PROMOTE sustainable development;

STRENGTHEN the development and enforcement of environmental laws and regulations; and

PROTECT, enhance and enforce basic workers' rights;

HAVE AGREED as follows: ...

Objectives of the North American Free Trade Agreement (NAFTA)

Article 101: Establishment of the Free Trade Area

The Parties to this Agreement, consistent with Article XXIV of the General Agreement on Tariffs and Trade, hereby establish a free trade area.

Article 102: Objectives

1. The objectives of this Agreement, as elaborated more specifically through its principles and rules, including national treatment, most-favored-nation treatment and transparency are to:
   (a) eliminate barriers to trade in, and facilitate the cross border movement of, goods and services between the territories of the Parties;
   (b) promote conditions of fair competition in the free trade area;
   (c) increase substantially investment opportunities in their territories;
   (d) provide adequate and effective protection and enforcement of intellectual property rights in each Party's territory;
   (e) create effective procedures for the implementation and application of this Agreement, and for its joint administration and the resolution of disputes; and
   (f) establish a framework for further trilateral, regional and multilateral cooperation to expand and enhance the benefits of this Agreement.
2. The Parties shall interpret and apply the provisions of this Agreement in the light of its objectives set out in paragraph 1 and in accordance with applicable rules of international law.

Article 103: Relation to Other Agreements

1. The Parties affirm their existing rights and obligations with respect to each other under the General Agreement on Tariffs and Trade and other agreements to which such Parties are party.
2. In the event of any inconsistency between the provisions of this Agreement and such other agreements, the provisions of this Agreement shall prevail to the extent of the inconsistency, except as otherwise provided in this Agreement.

Article 104: Relation to Environmental and Conservation Agreements

1. In the event of any inconsistency between this Agreement and the specific trade obligations set out in:
   (a) Convention on the International Trade in Endangered Species of Wild Fauna and Flora, done at Washington, March 3, 1973;
   (b) the Montreal Protocol on Substances that Deplete the Ozone Layer, done at Montreal, September 16, 1987, as amended June 29, 1990;
   (c) Basel Convention on the Control of Transboundary Movements of Hazardous Wastes and Their Disposal, done at Basel, March 22, 1989, upon its entry into force for Canada, Mexico and the United States; or
   (d) the agreements set out in Annex 104.1, such obligations shall prevail to the extent of the inconsistency, provided that where a Party has a choice among equally effective and reasonably available means of complying with such obligations, the Party chooses the alternative that is the least inconsistent with the other provisions of this Agreement.
2. The Parties may agree in writing to modify Annex 104.1 to include any amendment to the agreements listed in paragraph 1, and any other environmental or conservation agreement.

Article 105: Extent of Obligations

The Parties shall ensure that all necessary measures are taken in order to give effect to the provisions of this Agreement, including their observance, except as otherwise provided in this Agreement, by state and provincial governments.

Annex 104

Bilateral and Other Environmental and Conservation Agreements

1. The Agreement Between the Government of Canada and the Government of the United States of America Concerning the Transboundary Movement of Hazardous Waste, signed at Ottawa, October 28, 1986.
2. The Agreement between the United States of America and the United Mexican States on Cooperation for the Protection and Improvement of the Environment in the Border Area, signed at La Paz, Baja California Sur, August 14, 1983.

SUMMARY OF THE AGREEMENT ON ENVIRONMENTAL COOPERATION (AS AGREED UPON BY THE THREE PARTIES)

On August 13, 1993, Canadian Minister for International Trade, Thomas Hockin, Mexican Secretary of Trade and Industrial Development, Jaime Serra, and United States Trade Representative, Mickey Kantor completed negotiations on a proposed North American Agreement on Environmental Cooperation.

By strengthening environmental cooperation and the effective enforcement of domestic environmental laws and regulations, the environmental agreement will support the achievement of the economic, trade, and environmental goals and objectives of the NAFTA. The two agreements will work in a complementary manner to promote sustainable development, to create jobs, and to make the region more competitive.

The following description does not itself constitute an agreement between the three countries and is not intended as an interpretation of the text.

Preamble and Objectives

The Preamble sets out the goals, principles, and aspirations on which the Agreement is based. It recognizes a tradition of cooperation on the environment, and expresses a commitment to support and build on international environmental agreements and on existing institutions. The Objectives of the Agreement include the promotion of sustainable development, cooperation on the conservation, protection and enhancement of the environment, and the effective enforcement of and compliance with domestic environmental laws. The Agreement promotes transparency and public participation in the development and improvement of environmental laws and policies.

Obligations

While affirming the right of each Party to establish its own levels of protection, policies, and priorities, the Agreement requires that each Party ensure that its laws provide for high levels of environmental protection and strive to continue to improve those laws. This agreement also protects the rights of states and provinces to set high levels of protection, consistent with the NAFTA.

To achieve high levels of environmental protection and compliance, each Party agrees to effectively enforce its environmental law through appropriate government actions such as appointment and training of inspectors; monitoring compliance and examining suspected violations of law; seeking voluntary compliance agreements; and, using legal proceedings to sanction, or to seek appropriate remedies for, violations of its environmental law. The Agreement does not empower one Party's authorities to undertake environmental law enforcement activities within the territory of another Party.

Each Party undertakes, with respect to its territory, to

- report on the state of the environment.
- develop environmental emergency preparedness measures.
- promote environmental education, scientific research, and technological development.
- assess, as appropriate, environmental impacts.
- promote the use of economic instruments for the efficient achievement of environmental goals.

Each Party will notify the other Parties of a decision to ban or severely restrict a pesticide or chemical and will consider banning the export to another Party of toxic substances, the use of which is banned within its own territory.

Parties agree to ensure that their procedures for the enforcement of environmental law are fair, open, and equitable. Each Party undertakes to ensure appropriate public access to procedures for the enforcement of their environmental law. Such access includes the right to

- request action for the enforcement of domestic environmental law.
- sue another person under that Party's jurisdiction for damages.

Commission for Environmental Cooperation

The Agreement establishes a Commission for Environmental Cooperation, comprising a governing Council, a central Secretariat, and a Joint Public Advisory Committee.

The Council

The Council, the governing institution of the Commission, will be composed of cabinet-level officers or equivalent representatives of the Parties. It will oversee the implementation of the Agreement, serve as a forum for discussion of environmental matters, promote and facilitate cooperation, oversee the Secretariat, and address questions and disputes that may arise regarding the interpretation or application of the Agreement. The Council has key responsibilities related to the agreement's dispute settlement provisions concerning persistent patterns of failure by any Party to enforce its environmental laws.

The Council will strengthen cooperation on the development and continuing improvement of environmental laws and regulations, by

- promoting the exchange of information on criteria and methodologies used in establishing domestic environmental standards.
- developing recommendations on greater compatibility of environmental standards without reducing the level of environmental protection.

The Council will cooperate with the Free Trade Commission to achieve the environmental goals and objectives of the NAFTA, by

- contributing to the prevention or resolution of environment-related trade disputes.
- maintaining a list of experts who could provide information or technical advice to NAFTA institutions.

The Council will consider and develop recommendations with respect to assessing the environmental impact of proposed projects likely to cause significant adverse transboundary effects. It will also consider and may develop recommendations on

- public access to information, including information on hazardous materials and activities.
- appropriate limits for specific pollutants, taking into account differences in ecosystems.
- reciprocal access to rights and remedies for damage or injury resulting from transboundary pollution.

The Council will meet at least once a year. There will be public meetings at all regular sessions.

The Secretariat

The Agreement establishes a central Secretariat responsible for providing technical, administrative, and operational support to the Council and to committees and groups established by the Council. The Secretariat will prepare an annual budget and program, including proposed cooperative activities. The Secretariat will also prepare reports on matters within the scope of the annual program.

The Secretariat will consider submissions from any person or nongovernmental organization or association alleging a Party's failure to effectively enforce its environmental law. Provided the submission meets criteria set out in the Agreement, the Secretariat may propose that a factual record be developed. In developing this record, the Secretariat can seek information from a variety of sources, including submissions from interested persons and information developed by independent experts.

The size and location of the Secretariat will be determined by the Parties.

The Joint Public Advisory Committee

The Joint Public Advisory Committee will include five members of the public from each country. It will meet at least once a year, concurrent with the regular session of the Council. The Joint Committee will advise the Council and provide technical, scientific, or other information to the

Secretariat. It will also provide input to the annual program and budget of the Council as well as the annual and other reports.

Consultations

A Party may request consultations regarding any matter that affects the operation of the Agreement. Should the consultations fail to resolve the matter, any Party may call a meeting of the Council. In seeking a resolution, the Council may consult technical advisors or create working groups or expert groups and make recommendations.

Resolution of Disputes

If the Council cannot resolve a dispute involving a Party's alleged persistent pattern of failure to effectively enforce an environmental law relating to a situation involving the production of goods or services traded between parties, any party may request an arbitral panel. A panel will be established on a two-thirds vote of the Council.

Panelists will normally be chosen from a previously agreed roster of experts, including experts on environmental matters. With the approval of the disputing Parties, a panel may seek information and technical advice from any person or body that it deems appropriate. The report of the panel will be made publicly available five days after it is transmitted to the Parties.

If a panel makes a finding that a Party has engaged in a persistent pattern of failure to effectively enforce its environmental law, the Parties may, within 60 days, agree on a mutually satisfactory action plan to remedy the nonenforcement.

If there is no agreed action plan, then between 60 and 120 days after the final panel report, the panel may be reconvened to evaluate an action plan proposed by the Party complained against or to set out an action plan in its stead. The panel would also make a determination on the imposition of monetary enforcement assessments on the Party complained against.

The panel may be reconvened at any time to determine if an action plan is being duly implemented. If it is not being fully implemented, the panel is to impose a monetary enforcement assessment on the Party complained against.

In the event that a Party complained against fails to pay a monetary enforcement assessment or continues in its failure to enforce its environmental law, the Party is liable to ongoing enforcement actions. In the case of Canada, the Commission, on the request of a complaining Party, collects the monetary enforcement assessment and enforces an action plan in summary proceedings before a Canadian court of competent jurisdiction. In the case of Mexico and the United States, the complaining Party or Parties may suspend NAFTA benefits based on the amount of the assessment.

TREATY OF ROME, 1952, SINGLE EUROPEAN ACT, 1987, AND THE MAASTRICHT TREATY, 1992

The European Union (EU) is a community of 15 countries and 372 million citizens. The EU, or European Economic Community (EEC) as it was formerly known, was formed in 1957 with the signing of the Treaty of Rome by Belgium, Germany, France, Italy, Luxemborg, and The Netherlands. This treaty and others to follow provided a basis for economic and social growth but did not contain provisions for environmental protection. The Treaty of Rome does, however, contain two key articles that allowed the European Commission to issue directives and a provision allowing the Community to take action when no specific provision was available or appropriate. Although not specific to environmental issues, these two articles provided the necessary provisions to establish legislation for environmental protection.

Article 100 of the Treaty of Rome states

"The Council, acting unanimously on a proposal from the Commission, issue directives for the approximation of such provisions laid down by law, regulations or administrative action in Member States as directly affect the establishment or functioning of the common market."

Article 235 of the Treaty of Rome reads

"If action by the Commission should prove necessary to attain, in the course of the operation of the common market, one of the objectives of the Community and this Treaty has not provided necessary powers, the Council shall, acting on a proposal from the Commission and after consulting the European Parliament, take appropriate measures."

The Single European Act amended the Treaty of Rome and for the first time included three articles addressing the environment and provided

- More power to the European Parliament allowing
- The European Commission to evaluate health, safety and environmental proposals from a "high level of protection"
- Qualified majority voting instead of unanimously on environmental proposals.

Environmental provisions of the Single European Act

Article 130 r

"1. Action by the Community relating to the environment shall have the following objectives:
   (i)   to preserve, protect and improve the quality of the environment;
   (ii) to contribute toward protecting human health;
   (iii) to ensure a prudent and rational utilization of natural resources.
2. Action by the Community relating to the environment shall be based on the principle that preventative action should be taken, that environmental damage should as a priority be rectified at source, and that the polluter should pay. Environmental protection requirements shall be a component of the Community's other policies.
3. In preparing its action relating to the environment, the Community shall take account of:
   (i)   available scientific and technical data;
   (ii) environmental conditions in the various regions of the Community;
   (iii) the potential benefits and costs of action or lack of action
   (iv) the economic and social development of the Community as a whole and the balanced development of its regions.
4. The Community shall take action relating to the environment to the extent to which the objectives referred to in paragraph 1 can be attained better at Community level than at the level of the individual Member State. Without prejudice to certain measures of a community nature, the Member States shall finance and implement other measures.
5. Within their respective spheres of competence, the Community and the Member States shall cooperate with third countries and with relevant international organizations. The arrangements for Community cooperation may be the subject of agreement between the Community and the third parties concerned, which shall be negotiated and concluded in accordance with Article 228.

The previous paragraph shall be without prejudice to Member States' competence to negotiate in international bodies and to conclude international agreements."

Article 130s

"The Council, acting unanimously on a proposal from the Commission and after consulting the European Parliament and the Economic and Socila Committee, shall decide what action is to be taken by the Community.

The Council shall, under the condition laid down in the preceding subparagraph, define those matters on which decisions are to be taken by a qualified majority."

Article 130t

"The protective measures adopted in common pursuant to Articles shall not prevent any Member State from maintaining or introducing more stringent protective measures compatible with this treaty."

The 1992 Maastricht Treaty formally established the concept of sustainable development in European Union law. As stated in the Treaty objectives, "Sustainable and non-inflationary growth respecting the environment." Articles 130 r, s, and t were also amended requiring environmental protection factors be integrated into other Community legislation, providing qualified majority voting on environmental proposals, and a requirement that Member States implementing more stringent requirements notify the Commission.

DECLARATION OF THE UNITED NATIONS CONFERENCE ON THE HUMAN ENVIRONMENT (1972)

This meeting produced some of the most important principles and guidelines on environmental protection issues to that time or since. The participants agreed on 26 guiding principles and 106 recommendations for environmental management as part of the Stockholm Action Plan. This conference was sponsored by the United Nations and marked a new era of modern international environmental law.

TEXT OF THE DECLARATION OF THE UNITED NATIONS CONFERENCE ON THE HUMAN ENVIRONMENT (1972)

The United Nations Conference on the Human Environment,
    Having met at Stockholm from 5 to 16 June 1972,
    Having considered the need for a common outlook and for common principles to inspire and guide the peoples of the world in the preservation and enhancement of the human environment,
    Proclaims that:

1. Man is both creature and moulder of his environment, which gives him physical sustenance and affords him the opportunity for intellectual, moral, social and spiritual growth. In the long and tortuous evolution of the human race on this planet a stage has been reached when, through the rapid acceleration of science and technology, man has acquired the power to transform his environment in countless ways and on an unprecedented scale. Both aspects of man's environment, the natural and the man-made, are essential to his well-being and to the enjoyment of basic human rights—even the right to life itself.
2. The protection and improvement of the human environment is a major issue which affects the well-being of peoples and economic development throughout the world; it is the urgent desire of the peoples of the whole world and the duty of all Governments.
3. Man has constantly to sum up experience and go on discovering, inventing, creating and advancing. In our time, man's capability to transform his surroundings, if used wisely,

can bring to all peoples the benefits of development and the opportunity to enhance the quality of life. Wrongly or heedlessly applied, the same power can do incalculable harm to human beings and the human environment. We see around us growing evidence of man-made harm in many regions of the earth: dangerous levels of pollution in water, air, earth and living beings; major and undesirable disturbances to the ecological balance of the biosphere; destruction and depletion of irreplaceable resources; and gross deficiencies, harmful to the physical, mental and social health of man, in the man-made environment, particularly in the living and working environment.

4. In the developing countries most of the environmental problems are caused by under-development. Millions continue to live far below the minimum levels required for a decent human existence, deprived of adequate food and clothing, shelter and education, health and sanitation. Therefore, the developing countries must direct their efforts to development, bearing in mind their priorities and the need to safeguard and improve the environment. For the same purpose, the industrialized countries should make efforts to reduce the gap between themselves and the developing countries. In the industrialized countries, environmental problems are generally related to industrialization and technological development.

5. The natural growth of population continuously presents problems for the preservation of the environment, and adequate policies and measures should be adopted, as appropriate, to face these problems. Of all things in the world, people are the most precious. It is the people that propel social progress, create social wealth, develop science and technology and, through their hard work, continuously transform the human environment. Along with social progress and the advance of production, science and technology, the capability of man to improve the environment increases with each passing day.

6. A point has been reached in history when we must shape our actions throughout the world with a more prudent care for their environmental consequences. Through ignorance or indifference we can do massive and irreversible harm to the earthly environment on which our life and well-being depend. Conversely, through fuller knowledge and wiser action, we can achieve for ourselves and our posterity a better life in an environment more in keeping with human needs and hopes. There are broad vistas for the enhancement of environmental quality and the creation of a good life. What is needed is an enthusiastic but calm state of mind and intense but orderly work. For the purpose of attaining freedom in the world of nature, man must use knowledge to build, in collaboration with nature, a better environment. To defend and improve the human environment for present and future generations has become an imperative goal for mankind—a goal to be pursued together with, and in harmony with, the established and fundamental goals of peace and of worldwide economic and social development.

7. To achieve this environmental goal will demand the acceptance of responsibility by citizens and communities and by enterprises and institutions at every level; all sharing equitably in common efforts. Individuals in all walks of life as well as organizations in many fields, by their values and the sum of their actions, will shape the world environment of the future. Local and national governments will bear the greatest burden for large-scale environmental policy and action within their jurisdictions. International co-operation is also needed in order to raise resources to support the developing countries in carrying out their responsibilities in this field. A growing class of environmental problems, because they are regional or global in extent or because they affect the common international realm, will require extensive co-operation among nations and action by international organizations in the common interest. The Conference calls upon Governments and peoples to exert common efforts for the preservation and improvement of the human environment, for the benefit of all the people and for their posterity.

II. Principles

States the common conviction that:

Principle 1

Man has the fundamental right to freedom, equality and adequate conditions of life, in an environment of a quality that permits a life of dignity and well-being, and he bears a solemn responsibility to protect and improve the environment for present and future generations. In this respect, policies promoting or perpetuating apartheid, racial segregation, discrimination, colonial and other forms of oppression and foreign domination stand condemned and must be eliminated.

Principle 2

The natural resources of the earth, including the air, water, land, flora and fauna and especially representative samples of natural ecosystems, must be safeguarded for the benefit of present and future generations through careful planning or management, as appropriate.

Principle 3

The capacity of the earth to produce vital renewable resources must be maintained and, wherever practicable, restored or improved.

Principle 4

Man has a special responsibility to safeguard and wisely manage the heritage of wildlife and its habitat, which are now gravely imperilled by a combination of adverse factors. Nature conservation, including wildlife, must therefore receive importance in planning for economic development.

Principle 5

The non-renewable resources of the earth must be employed in such a way as to guard against the danger of their future exhaustion and to ensure that benefits from such employment are shared by all mankind.

Principle 6

The discharge of toxic substances or of other substances and the release of heat, in such quantities or concentrations as to exceed the capacity of the environment to render them harmless, must be halted in order to ensure that serious or irreversible damage is not inflicted upon ecosystems. The just struggle of the peoples of all countries against pollution should be supported.

Principle 7

States shall take all possible steps to prevent pollution of the seas by substances that are liable to create hazards to human health, to harm living resources and marine life, to damage amenities or to interfere with other legitimate uses of the sea.

Principle 8

Economic and social development is essential for ensuring a favourable living and working environment for man and for creating conditions on earth that are necessary for the improvement of the quality of life.

Principle 9

Environmental deficiencies generated by the conditions of under-development and natural disasters pose grave problems and can best be remedied by accelerated development through the transfer of substantial quantities of financial and technological assistance as a supplement to the domestic effort of the developing countries and such timely assistance as may be required.

Principle 10

For the developing countries, stability of prices and adequate earnings for primary commodities and raw materials are essential to environmental management since economic factors as well as ecological processes must be taken into account.

Principle 11

The environmental policies of all States should enhance and not adversely affect the present or future development potential of developing countries, nor should they hamper the attainment of better living conditions for all, and appropriate steps should be taken by States and international organizations with a view to reaching agreement on meeting the possible national and international economic consequences resulting from the application of environmental measures.

Principle 12

Resources should be made available to preserve and improve the environment, taking into account the circumstances and particular requirements of developing countries and any costs which may emanate from their incorporating environmental safeguards into their development planning and the need for making available to them, upon their request, additional international technical and financial assistance for this purpose.

Principle 13

In order to achieve a more rational management of resources and thus to improve the environment, States should adopt an integrated and coordinated approach to their development planning so as to ensure that development is compatible with the need to protect and improve environment for the benefit of their population.

Principle 14

Rational planning constitutes an essential tool for reconciling any conflict between the needs of development and the need to protect and improve the environment.

Principle 15

Planning must be applied to human settlements and urbanization with a view to avoiding adverse effects on the environment and obtaining maximum social, economic and environmental benefits for all. In this respect, projects which are designed for colonialist and racist domination must be abandoned.

Principle 16

Demographic policies which are without prejudice to basic human rights and which are deemed appropriate by Governments concerned should be applied in those regions where the rate of population growth or excessive population concentrations are likely to have adverse effects on the environment of the human environment and impede development.

Principle 17

Appropriate national institutions must be entrusted with the task of planning, managing or controlling the environmental resources of States with a view to enhancing environmental quality.

Principle 18

Science and technology, as part of their contribution to economic and social development, must be applied to the identification, avoidance and control of environmental risks and the solution of environmental problems and for the common good of mankind.

Principle 19

Education in environmental matters, for the younger generation as well as adults, giving due consideration to the underprivileged, is essential in order to broaden the basis for an enlightened opinion and responsible conduct by individuals, enterprises and communities in protecting and improving the environment in its full human dimension. It is also essential that mass media of communications avoid contributing to the deterioration of the environment, but, on the contrary, disseminate information of an educational nature on the need to protect and improve the environment in order to enable man to develop in every respect.

Principle 20

Scientific research and development in the context of environmental problems, both national and multinational, must be promoted in all countries, especially the developing countries. In this connection, the free flow of up-to-date scientific information and transfer of experience must be supported and assisted, to facilitate the solution of environmental problems; environmental technologies should be made available to developing countries on terms which would encourage their wide dissemination without constituting an economic burden on the developing countries.

Principle 21

States have, in accordance with the Charter of the United Nations and the principles of international law, the sovereign right to exploit their own resources pursuant to their own environmental policies, and the responsibility to ensure that activities within their jurisdiction or control do not cause damage to the environment of other States or of areas beyond the limits of national jurisdiction.

Principle 22

States shall cooperate to develop further the international law regarding liability and compensation for the victims of pollution and other environmental damage caused by activities within the jurisdiction or control of such States to areas beyond their jurisdiction.

Principle 23

Without prejudice to such criteria as may be agreed upon by the international community, or to standards which will have to be determined nationally, it will be essential in all cases to consider the systems of values prevailing in each country, and the extent of the applicability of standards which are valid for the most advanced countries but which may be inappropriate and of unwarranted social cost for the developing countries.

Principle 24

International matters concerning the protection and improvement of the environment should be handled in a co-operative spirit by all countries, big and small, on an equal footing. Co-operation through multilateral or bilateral arrangements or other appropriate means is essential to effectively control, prevent, reduce and eliminate adverse environmental effects resulting from activities conducted in all spheres, in such a way that due account is taken of the sovereignty and interests of all States.

Principle 25

States shall ensure that international organizations play a coordinated, efficient and dynamic role for the protection and improvement of the environment.

Principle 26

Man and his environment must be spared the effects of nuclear weapons and all other means of mass destruction. States must strive to reach prompt agreement, in the relevant international organs, on the elimination and complete destruction of such weapons.

21st plenary meeting 16 June 1972

THE RIO DECLARATION ON ENVIRONMENT AND DEVELOPMENT (1992)

The Rio Earth Summit expanded upon and reaffirmed principles of the Stockholm Declaration and established binding conventions on Biodiversity and Climate Change and Agenda 21, guidelines for sustainable development in the twenty-first century. Further, the Rio Earth Summit established forestry principles, legal instruments for various existing treaties, and the Commission on Sustainable Development.

TEXT OF THE RIO DECLARATION

Preamble
    The United Nations Conference on Environment and Development,
    Having met at Rio de Janeiro from 3 to 14 June 1992,

Reaffirming the Declaration of the United Nations Conference on the Human Environment, adopted at Stockholm on 16 June 1972, and seeking to build upon it,

With the goal of establishing a new and equitable global partnership through the creation of new levels of cooperation among States, key sectors of societies and people,

Working towards international agreements which respect the interests of all and protect the integrity of the global environmental and developmental system,

Recognizing the integral and interdependent nature of the Earth, our home,

Proclaims that:

Principle 1

Human beings are at the centre of concerns for sustainable development. They are entitled to a healthy and productive life in harmony with nature.

Principle 2

States have, in accordance with the Charter of the United Nations and the principles of international law, the sovereign right to exploit their own resources pursuant to their own environmental and developmental policies, and the responsibility to ensure that activities within their jurisdiction or control do not cause damage to the environment of other States or of areas beyond the limits of national jurisdiction.

Principle 3

The right to development must be fulfilled so as to equitably meet developmental and environmental needs of present and future generations.

Principle 4

In order to achieve sustainable development, environmental protection shall constitute an integral part of the development process and cannot be considered in isolation from it.

Principle 5

All States and all people shall cooperate in the essential task of eradicating poverty as an indispensable requirement for sustainable development, in order to decrease the disparities in standards of living and better meet the needs of the majority of the people of the world.

Principle 6

The special situation and needs of developing countries, particularly the least developed and those most environmentally vulnerable, shall be given special priority. International actions in the field of environment and development should also address the interests and needs of all countries.

Principle 7

States shall cooperate in a spirit of global partnership to conserve, protect and restore the health and integrity of the Earth's ecosystem. In view of the different contributions to global environmental degradation, States have common but differentiated responsibilities. The developed countries acknowledge the responsibility that they bear in the international pursuit of sustainable development

in view of the pressures their societies place on the global environment and of the technologies and financial resources they command.

Principle 8

To achieve sustainable development and a higher quality of life for all people, States should reduce and eliminate unsustainable patterns of production and consumption and promote appropriate demographic policies.

Principle 9

States should cooperate to strengthen endogenous capacity-building for sustainable development by improving scientific understanding through exchanges of scientific and technological knowledge, and by enhancing the development, adaptation, diffusion and transfer of technologies, including new and innovative technologies.

Principle 10

Environmental issues are best handled with the participation of all concerned citizens, at the relevant level. At the national level, each individual shall have appropriate access to information concerning the environment that is held by public authorities, including information on hazardous materials and activities in their communities, and the opportunity to participate in decision-making processes. States shall facilitate and encourage public awareness and participation by making information widely available. Effective access to judicial and administrative proceedings, including redress and remedy, shall be provided.

Principle 11

States shall enact effective environmental legislation. Environmental standards, management objectives and priorities should reflect the environmental and developmental context to which they apply. Standards applied by some countries may be inappropriate and of unwarranted economic and social cost to other countries, in particular developing countries.

Principle 12

States should cooperate to promote a supportive and open international economic system that would lead to economic growth and sustainable development in all countries, to better address the problems of environmental degradation. Trade policy measures for environmental purposes should not constitute a means of arbitrary or unjustifiable discrimination or a disguised restriction on international trade. Unilateral actions to deal with environmental challenges outside the jurisdiction of the importing country should be avoided. Environmental measures addressing transboundary or global environmental problems should, as far as possible, be based on an international consensus.

Principle 13

States shall develop national law regarding liability and compensation for the victims of pollution and other environmental damage. States shall also cooperate in an expeditious and more determined manner to develop further international law regarding liability and compensation for adverse effects of environmental damage caused by activities within their jurisdiction or control to areas beyond their jurisdiction.

Principle 14

States should effectively cooperate to discourage or prevent the relocation and transfer to other States of any activities and substances that cause severe environmental degradation or are found to be harmful to human health.

Principle 15

In order to protect the environment, the precautionary approach shall be widely applied by States according to their capabilities. Where there are threats of serious or irreversible damage, lack of full scientific certainty shall not be used as a reason for postponing cost-effective measures to prevent environmental degradation.

Principle 16

National authorities should endeavour to promote the internalization of environmental costs and the use of economic instruments, taking into account the approach that the polluter should, in principle, bear the cost of pollution, with due regard to the public interest and without distorting international trade and investment.

Principle 17

Environmental impact assessment, as a national instrument, shall be undertaken for proposed activities that are likely to have a significant adverse impact on the environment and are subject to a decision of a competent national authority.

Principle 18

States shall immediately notify other States of any natural disasters or other emergencies that are likely to produce sudden harmful effects on the environment of those States. Every effort shall be made by the international community to help States so afflicted.

Principle 19

States shall provide prior and timely notification and relevant information to potentially affected States on activities that may have a significant adverse transboundary environmental effect and shall consult with those States at an early stage and in good faith.

Principle 20

Women have a vital role in environmental management and development. Their full participation is therefore essential to achieve sustainable development.

Principle 21

The creativity, ideals and courage of the youth of the world should be mobilized to forge a global partnership in order to achieve sustainable development and ensure a better future for all.

Principle 22

Indigenous people and their communities, and other local communities, have a vital role in environmental management and development because of their knowledge and traditional practices. States should recognize and duly support their identity, culture and interests and enable their effective participation in the achievement of sustainable development.

Principle 23

The environment and natural resources of people under oppression, domination and occupation shall be protected.

Principle 24

Warfare is inherently destructive of sustainable development. States shall therefore respect international law providing protection for the environment in times of armed conflict and cooperate in its further development, as necessary.

Principle 25

Peace, development and environmental protection are interdependent and indivisible.

Principle 26

States shall resolve all their environmental disputes peacefully and by appropriate means in accordance with the Charter of the United Nations.

Principle 27

States and people shall cooperate in good faith and in a spirit of partnership in the fulfilment of the principles embodied in this Declaration and in the further development of international law in the field of sustainable development.

WORLD CHARTER FOR NATURE (1982)

The World Charter for Nature was born from the work of 34 developing nations. The Charter is a code of conduct for the treatment of nature and was adopted by the United Nations General Assembly in 1982. The vote was 111 for, one against (United States), 18 abstentions. It recognizes the importance of ecosystems and their protection and incorporates precautionary principles.

TEXT OF THE WORLD CHARTER FOR NATURE

The General Assembly,

Reaffirming the fundamental purposes of the United Nations, in particular the maintenance of international peace and security, the development of friendly relations among nations and the achievement of international cooperation in solving international problems of an economic, social, cultural, technical, intellectual or humanitarian character.

Aware that:

(a) Mankind is a part of nature and life depends on the uninterrupted functioning of natural systems which ensure the supply of energy and nutrients.
(b) Civilization is rooted in nature, which has shaped human culture and influenced all artistic and scientific achievements, and living in harmony with nature gives man the best opportunities for the development of his creativity, and for rest and recreation.

Convinced that:

(a) Every form of life is unique, warranting respect regardless of its worth to man, and, to accord other organisms such recognition, man must be guided by a moral code of action.
(b) Man can alter nature and exhaust natural resources by his action or its consequences and, therefore, must fully recognize the urgency of maintaining the stability and quality of nature and of conserving natural resources.

Persuaded that:

(a) Lasting benefits from nature depend upon the maintenance of essential ecological processes and life support systems, and upon the diversity of life forms, which are jeopardized through excessive exploitation and habitat destruction by man.
(b) The degradation of natural systems owing to excessive consumption and misuse of natural resources, as well as to failure to establish an appropriate economic order among peoples and among States, leads to the breakdown of the economic, social and political framework of civilization.
(c) Competition for scarce resources creates conflicts, whereas the conservation of nature and natural resources contributes to justice and the maintenance of peace and cannot be achieved until mankind learns to live in peace and to forsake war and armaments.

Reaffirming that man must acquire the knowledge to maintain and enhance his ability to use natural resources in a manner which ensures the preservation of the species and ecosystems for the benefit of present and future generations.

Firmly convinced of the need for appropriate measures, at the national and international, individual and collective, and private and public levels, to protect nature and promote international co-operation in this field.

Adopts, to these ends, the present World Charter for Nature, which proclaims the following principles of conservation by which all human conduct affecting nature is to be guided and judged.

I. General Principles

1. Nature shall be respected and its essential processes shall not be impaired.
2. The genetic viability on the earth shall not be compromised; the population levels of all life forms, wild and domesticated, must be at least sufficient for their survival, and to this end necessary habitat shall be safeguarded.
3. All areas of the earth, both land and sea, shall be subject to these principles of conservation; special protection shall be given to unique areas, to representative samples of all the different types of ecosystems and to the habitat of rare or endangered species.
4. Ecosystems and organisms, as well as the land, marine and atmospheric resources that are utilized by man, shall be managed to achieve and maintain optimum sustainable

productivity, but not in such a way as to endanger the integrity of those other ecosystems or species with which they coexist.

5. Nature shall be secured against degradation caused by warfare or other hostile activities.

II. Functions

6. In the decision-making process it shall be recognized that man's needs can be met only by ensuring the proper functioning of natural systems and by respecting the principles set forth in the present Charter.

7. In the planning and implementation of social and economic development activities, due account shall be taken of the fact that the conservation of nature is an integral part of those activities.

8. In formulating long-term plans for economic development, population growth and the improvement of standards of living, due account shall be taken of the long-term capacity of natural systems to ensure the subsistence and settlement of the populations concerned, recognizing that this capacity may be enhanced through science and technology.

9. The allocation of areas of the earth to various uses shall be planned and due account shall be taken of the physical constraints, the biological productivity and diversity and the natural beauty of the areas concerned.

10. Natural resources shall not be wasted, but used with a restraint appropriate to the principles set forth in the present Charter, in accordance with the following rules:

    (a) Living resources shall not be utilized in excess of their natural capacity for regeneration;

    (b) The productivity of soils shall be maintained or enhanced through measures which safeguard their long-term fertility and the process of organic decomposition, and prevent erosion and all other forms of degradation;

    (c) Resources, including water, which are not consumed as they are used shall be reused or recycled;

    (d) Non-renewable resources which are consumed as they are used shall be exploited with restraint, taking into account their abundance, their rational possibilities of converting them for consumption, and the compatibility of their exploitation with the functioning of natural systems.

11. Activities which might have an impact on nature shall be controlled, and the best available technologies that minimize significant risks to nature or other adverse effects shall be used; in particular:

    (a) Activities which are likely to cause irreversible damage to nature shall be avoided;

    (b) Activities which are likely to pose a significant risk to nature shall be preceded by an exhaustive examination; their proponents shall demonstrate that expected benefits outweigh potential damage to nature, and where potential adverse effects are not fully understood, the activities should not proceed;

    (c) Activities which may disturb nature shall be preceded by assessment of their consequences, and environmental impact studies of development projects shall be conducted sufficiently in advance, and if they are to be undertaken, such activities shall be planned and carried out so as to minimize potential adverse effects;

    (d) Agriculture, grazing, forestry and fisheries practices shall be adapted to the natural characteristics and constraints of given areas;

    (e) Areas degraded by human activities shall be rehabilitated for purposes in accord with their natural potential and compatible with the well-being of affected populations.

12. Discharge of pollutants into natural systems shall be avoided and:
    (a) Where this is not feasible, such pollutants shall be treated at the source, using the best practicable means available;
    (b) Special precautions shall be taken to prevent discharge of radioactive or toxic wastes.
13. Measures intended to prevent, control or limit natural disasters, infestations and diseases shall be specifically directed to the causes of these scourges and shall avoid adverse side-effects on nature.

III. Implementation

14. The principles set forth in the present Charter shall be reflected in the law and practice of each State, as well as at the international level.
15. Knowledge of nature shall be broadly disseminated by all possible means, particularly by ecological education as an integral part of general education.
16. All planning shall include, among its essential elements, the formulation of strategies for the conservation of nature, the establishment of inventories of ecosystems and assessments of the effects on nature of proposed policies and activities; all of these elements shall be disclosed to the public by appropriate means in time to permit effective consultation and participation.
17. Funds, programmes and administrative structures necessary to achieve the objective of the conservation of nature shall be provided.
18. Constant efforts shall be made to increase knowledge of nature by scientific research and to disseminate such knowledge unimpeded by restrictions of any kind.
19. The status of natural processes, ecosystems and species shall be closely monitored to enable early detection of degradation or threat, ensure timely intervention and facilitate the evaluation of conservation policies and methods.
20. Military activities damaging to nature shall be avoided.
21. States and, to the extent they are able, other public authorities, international organizations, individuals, groups and corporations shall:
    (a) cooperate in the task of conserving nature through common activities and other relevant actions, including information exchange and consultations;
    (b) Establish standards for products and other manufacturing processes that may have adverse effects on nature, as well as agreed methodologies for assessing these effects;
    (c) Implement the applicable international legal provisions for the conservation of nature and the protection of the environment;
    (d) Ensure that activities within their jurisdictions or control do not cause damage to the natural systems located within other States or in the areas beyond the limits of national jurisdiction;
    (e) Safeguard and conserve nature in areas beyond national jurisdiction.
22. Taking fully into account the sovereignty of States over their natural resources, each State shall give effect to the provisions of the present Charter through its competent organs and in co-operation with other States.
23. All persons, in accordance with their national legislation, shall have the opportunity to participate, individually or with others, in the formulation of decisions of direct concern to their environment, and shall have access to means of redress when their environment has suffered damage or degradation.
24. Each person has a duty to act in accordance with the provisions of the present Charter, acting individually, in association with others or through participation in the political process, each person shall strive to ensure that the objectives and requirements of the present Charter are met.

REFERENCES

Trade and the Environment, Law, Economics and Policy, edited by Durwood Zaelke, Paul Orbuch, Robert F. Housman, Center for International Environmental Law, 1993.

Pollution Control in the United States, Evaluating the System, J. Clarence Davies and Jan Mazurek, Resources for the Future, 1998.

Direct Effect of European Law and the Regulation of Dangerous Substances, Christopher J. M. Smith, Gordon and Breach Publishers, 1995.

Environmental Change and International Law, edited by Edith Brown Weiss, United Nations University Press, 1992.

International Environmental Law and Policy, David Hunter, James Salzman, Durwood Zaelke, Foundation Press, 1998.

Environmental Management Systems, Jay G. Martin and Gerald J. Edgley, Government Institutes, 1998.

International Environmental Auditing, David D. Nelson, Government Institutes, 1998.

Precautionary Legal Duties and Principles of Modern International Environmental Law, Harold Hohmann, Graham & Trotman/Martinus Nijhoff, 1998.

Environmental Management in European Companies, Success Stories and Evaluation, edited by Jobst Conrad, Gordon and Breach Science Publishers, 1998.

Pollution Control in the United States, Evaluating the System, J. Clarence Davies and Jan Mazurek, Resources for the Future, 1998.

Public Policies for Environmental Protection, Editor, Paul R. Portney, Resources for the Future, 1992.

Environmental Strategies Handbook, A Guide to Effective Policies & Practices, Rao V. Kolluru, McGraw-Hill, Inc., 1994.

5 Media-Specific Conventions, Agreements and Standards

CONTENTS

INTRODUCTION

Important resources such as air, water, and soil have been the subject of agreements and disagreements throughout history. With regard to resource protection, regulations have been developed, implemented, and frequently amended by countries around the world since the 1960s and 1970s. Entire acts or treaties have been dedicated to air and water pollution and hazardous waste management. However, prior to these laws, international measures regarding protection of the environment and natural resources were mainly designed to maximize resource exploitation. According to the principle of territorial sovereignty, states have the right to use their territory extending to the country's geographical boundary as they wish, including resource exploitation. State sovereignty has been reaffirmed in the following international agreements:

- Article III of the International Law Association's Principles of Law Governing the Uses of the Waters of International Rivers
- Article 6 of the UNESCO World Heritage Convention
- Article 15 of the Biodiversity Convention
- Principle 21 of the Stockholm Declaration
- Principle 2 of the Rio Declaration on Environment and Development

Principle 2, Rio Declaration on Environment and Development (1992) reads

"States have, in accordance with the Charter of the United Nations and the principles of international law, the sovereign right to exploit their own resources pursuant to their own environmental and developmental policies, and the responsibility to ensure that activities within their jurisdiction or control do not cause damage to the environment of other states or of areas beyond the limits of national jurisdiction."

Principle 21 of the Stockholm Declaration is identical to that of the Rio Declaration excerpt above except that the Rio Declaration has included "developmental policies" with environmental policies. This clearly indicates the relationship the Rio Declaration has with balancing environmental and developmental activities. Notice that the activity must not exceed their boundaries. Following this period of resource exploitation, national and international law began to focus on medium-specific environmental laws. In the 1970s, Germany, United States, France, Italy, Japan, The Netherlands, and the United Kingdom established laws to protect air, water, and land resources. These media-specific laws, especially those from the United States, have been used as models for environmental laws by other countries. The adoption of media-specific regulations has continued although the shift to prevention, management, and planning had begun. This shift was supported by numerous international agreements focusing on sustainability. In support of the regulatory framework, individual countries have developed technical and guidance documents to facilitate compliance. Also, standards organizations such as ASTM (U.S.), ANSI (U.S.), and DIN (Germany) have developed hundreds of detailed technical standards covering test methods, practices, and procedures relating to all aspects of the environment. For example, ASTM has developed a standard for the performance of environmental site assessments. DIN has developed standard methods for the examination of water and wastewater. In addition, data series, manuals, monographs, and special technical publications have been developed. Further, EU directives covering air and water pollution and hazardous waste have been established. This chapter lists media-specific regulations for various countries and the EU, reviews some of the media-specific international agreements, and lists media-specific standards from organizations.

MEDIA-SPECIFIC REGULATIONS FOR SELECTED COUNTRIES

AUSTRALIA

- National Environmental Protection Measures (Implementation) Act 1998
- Hazardous Waste, Regulation of Exports and Imports, 1989 (Amendment 1996)
- Ozone Protection Act 1989
- Environmental Protection (Sea Dumping) Act 1981
- Protection of the Sea Act, Prevention of Pollution from Ships, 1983
- Dangerous Goods Act 1975 and 1985
- Environmentally Hazardous Chemicals Act 1985
- Environmental Effects Act 1978
- Environmental Protection Act 1970
- Water Act 1958
- Endangered Species Protection Act 1992
- National Parks and Wildlife Conservation Act 1975
- Great Barrier Reef Marine Part Act 1975

AUSTRIA

Agreement on Ambient Air Quality Standards, entered in 1987 between the federal government and the nine states.

- Air Pollution Law for Boiler Facilities, 1988
- Chlorinated Hydrocarbon Facilities Regulation, promulgated 1990
- Constitutional Federal Law on Comprehensive Environmental Protection, 1984
- General Regulation on Wastewater Discharges, promulgated 1991
- Law on Old Hazardous Waste Sites, 1989
- Law on Ozone, 1992
- Law on Smog Alarm, 1989
- Law on Waste Material, 1990
- Regulation on Asbestos, promulgated 1990
- Regulation on the Prohibition of Halons, promulgated 1990
- Regulation Prohibiting Fluorochlorohydrocarbons as a Propellant, promulgated 1990

- Regulation Prohibiting Fully Halogenated Fluorochlorohydrocarbons, promulgated 1990
- Waste Manifest Regulation, 1991
- Waste Oil Regulation, 1987
- Water Law (last amendments 1990)

BELARUS

Applicable former USSR laws include laws addressing air, water, land, and minerals.

BERMUDA

- Agriculture Regulations, 1967 (Soil Erosion)
- The Clean Air Act, 1991
- Hamilton Sewerage Act, 1917
- Marine and Ports Authority Regulations, 1967 (Dumping)
- The Merchant Shipping Order, 1975 (Oil Pollution)
- The Merchant Shipping Order, 1988 (Prevention of Oil Pollution)
- The Prevention of Oil Pollution Act, 1971
- The Prevention of Oil Pollution, Enforcement of Convention Order, 1980
- The Prevention of Oil Pollution, Shipping Casualties Order, 1980
- Protected Waters Act (Castle Harbour), 1951
- Public Health Regulations, 1951 (Water Storage)
- Waste and Litter Control Act, 1987
- Water Resources Act, 1975
- Water Resources Regulations, Appeals to the Minister, 1976
- Water Resources Regulations, 1976 (Well Diggers)

CANADA (FEDERAL)

- Arctic Waters Pollution Prevention Act, 1985
- Canada Water Act, 1985
- Northern Inland Waters Act, 1985
- Territorial Lands Act, 1985

Note: Each province (Alberta, British Columbia, Manitoba, New Brunswick, Newfoundland, Northwest Territories, Nova Scotia, Ontario, Prince Edward Island, Quebec, Saskatchewan, and the Yukon) has also established pertinent environmental regulations.

CHINA

- Air Pollution Prevention and Control Law, 1987
- Marine Environmental Protection Law, 1982
- Regulation Concerning Control of Waste Dumping at Sea, promulgated 1986
- Regulations Concerning Environmental Protection in Offshore Oil Exploration and Exploitation, promulgated 1983
- Regulations Concerning Prevention of Land-Originating Pollutants from Damaging the Marine Environment Promulgated 1990
- Regulations Concerning Prevention of Pollution from Coastal Construction Projects Damaging the Marine Environment, promulgated 1990
- Regulations Concerning Pollution Prevention and Control of Sea Areas by Vessels, promulgated 1983

- Regulations Concerning Water and Soil Conservation, promulgated 1982
- Rules for the Implementation of the Air Pollution Prevention and Control Law, promulgated 1991
- Rules for the Implementation of the Water Pollution Prevention and Control Law, promulgated 1989
- Water Law, 1988
- Water Pollution Prevention and Control Law, 1984

CZECH REPUBLIC

- Act on the State Administration of Waste Management
- Act on Charges for the Deposit of Waste
- Air Quality Monitoring and Emissions Reductions
- Clean Air Act
- Clean Water Act
- Restrictions on Import and Export of Waste
- The Law of the Czech National Council on the Administration of Water Management
- The Waste Act

FRANCE

- Air (three acts starting in 1961)
- Waste (nine acts starting in 1975)
- Water (seven acts starting in 1964)

GERMANY

- Act on the Protection against Dangerous Substances, 1980
- Act on the Prevention of Harmful Effects on the Environment Caused by Air Pollution, Noise, Vibration and Similar Phenomena – Federal Emission Control Act, last amended 1993
- Federal Water Act, 1960
- Federal Waste Water Tax Act, 1987 (Amended 1990)
- Waste Avoidance and Waste Management Act, 1986
- Regulation on the Avoidance of Packaging Waste, 1991

INDIA

- Constitutional Provisions for Environmental Protection
- The Environment Protect Act
- The Water Act (Prevention and Control of Pollution)
- The Air Act (Prevention and Control of Pollution)

ITALY

- Regulation to Prevent Air Pollution, 1967
- Decree of the President of Ministries March 28, 1983 Relating to Limits of Concentrations of each Pollutant Present in Air, 1983

- Regulation to Preserve Waters from Pollution, 1976
- Presidential Decree September 10, 1982 Number 915, Implementation of EEC Directives Numbers 75/442 and 76/403 Relating to Solid Waste Discharge, 1982

JAMAICA

- Air Pollution
- Water Resources and Pollution (Water Allocation, Protection of Supplies, Groundwater, Watershed Protection)
- Public Health

JAPAN

- Air Pollution Control Law, 1968
- Basic Law for Environmental Pollution Control, 1971
- Law Concerning Protection of Ozone Layer by Regulating Special Substances, 1988
- Law for Promotion of Utilization of Recyclable Resources, 1991
- Law for Special Measures Concerning the Total Emission Reduction of Nitrogen Oxides from Automobiles in Specified Areas, 1992
- Law for the Control of Export, Import, etc. of Specified Hazardous Waste and Other Waste, 1992
- Law for the Punishment of Crimes Relating to Environmental Pollution which Adversely Affects the Health of Persons, 1970
- Marine Pollution Prevention Law, 1970
- Pollution Related Health Damage Compensation Law, 1973
- Waste Management Law, 1970
- Water Pollution Control Law, 1970

MEXICO

Treaties become national law following the adoption of specific implementing legislation. Mexico is party to more than 40 international conventions. Ecological Technical Standards addressing maximum allowed limits of pollutants for air and water discharges, pollutant monitoring methods, and technical requirements have been developed. Many of the ecological technical standards are industry specific. Examples of ecological technical standards include the following:

- Manifest for Occasional Generators of PCBs Residues Generated by Electrical Equipment
- Standards on Requirements for Operation of Controlled Confinement Facilities for Hazardous Waste, 1989
- Ecological Criteria on Water Quality, 1989
- Ecological Technical Standard Establishing Maximum Water Pollution Levels for the Coffee Industry, 1991
- Ecological Technical Standard for the Maximum Opacity Levels of Smoke Generated by Diesel Motors, 1991
- Ecological Technical Standard Relating to Particles, Carbon Monoxide, Sulphur Dioxide, and Nitrogen Oxides from Fixed Source Fuel Combustion, 1988

THE NETHERLANDS

- Air Pollution Act, 1970
- Chemical Waste Act, 1976
- Groundwater Act, 1981
- Hazardous Substances Act, 1963
- Sea Water Pollution Act, 1975
- Soil Protection Act, 1986
- Waste Act, 1977
- Water Management Act, 1989
- Water Suppliers Act, 1957

POLAND

Poland has, among others, environmental laws covering air and water pollution, hazardous waste, toxic substances, and medical waste.

RUSSIAN FEDERATION

Environmental laws of the former USSR involving air and water pollution, wildlife, forests, land use, mineral resources, and environmental impact assessment are still applicable. New environmental laws include

- The Law of Environmental Protection
- Law on the Sanitary-Epidemiological Well Being of the Population

SINGAPORE

- The Clean Air Act of 1986 (Singapore Statutes)
- The Clean Air Act Regulations of 1972 (Standards)
- The Clean Air Order of 1973 (Prohibition on the Use of Open Fires)
- The Merchant Ship Act of 1985 (Oil Pollution)
- The Poisons Rules of 1988 (Hazardous Substances)
- The Environmental Public Health Regulations of 1988 (Toxic Industrial Waste)
- The Water Pollution Control and Drainage Act of 1985
- The Trade Effluent Regulations of 1976
- Prevention of Pollution of the Sea of 1985 (Singapore Statutes)
- Prevention of Pollution of the Sea Regulations of 1976
- The Merchant Shipping Act (Oil Pollution) 1985
- The Surface Water Drainage Regulations of 1976

SOUTH KOREA

- Air Environment Preservation Act, 1991
- Act Concerning Transnational Movement and Disposal of Wastes (based on Basel Convention)
- Marine Pollution Preservation Act, 1987
- Oil Pollution Damages Compensation Act
- Public Water Surface Control Act, 1966
- Public Water Surface Reclamation Act, 1986
- Sewage Act, 1982

- Synthetic Resin Waste Processing Business Act, 1980
- Toxic Chemicals Control Act, 1991
- Waste Control Act, 1987
- Waterworks Act, 1966
- Water Quality Environment Preservation Act, 1991

UKRAINE

Environmental protection law in the Ukraine includes provisions for environmental protection management, monitoring and recording rules, air regulations, and legislation in effect from the former USSR, including water law.

UNITED KINGDOM

- Clean Air Acts (1956 and 1968)
- Health and Safety Regulations (Emissions into the Atmosphere), 1983
- Clean Air Regulations (Emission of Grit and Dust from Furnaces), 1971
- Control of Pollution Act, 1974
- Control of Pollution Act, 1989
- Control of Pollution Regulations (Special Waste), 1980
- Transfrontier Shipment of Hazardous Waste Regulations, 1988
- Controlled Waste Regulations (Registration Legislation of Carriers and Seizure of Vehicles), 1991
- Environmental Protection Regulations, 1991 (Duty of Care)
- Controlled Waste Regulations, 1992
- Prevention of Oil Pollution Act, 1971
- Merchant Shipping Act (Oil Pollution), 1971
- Merchant Shipping Regulations (Prevention of Oil Pollution), 1983
- Water Resources Act, 1991
- Control of Pollution Act, 1974 (Water)
- Water Industry Act, 1991
- Trade Effluents Regulations (Prescribed Processes and Substances), 1989
- Sewerage Act, 1968 (Scotland)

AIR POLLUTION

Few transboundary air pollution cases exist in international environmental law but one provides an excellent case study. Beginning in 1896 and into the 1930s, a smelter operated near Trail, B.C. Canada. Damage was caused to areas within the U.S. when air pollutants from the facility migrated across the border. On behalf of local citizens, the U.S. government, with concurrence of the Canadian government, brought the case to the International Joint Commission (IJC). The IJC is a bilateral commission originally established to address water quality issues but is utilized for other environmental issues. In this case, the IJC issued a report that included assessment of damages. This judgment demonstrates an important component of modern environmental law; the acts of one state cannot be injurious to another.

The Trail Smelter case involved, among others, sulphur dioxide (SO_2) emissions. SO_2 and nitrogen oxides (NO_x) are precursers to acid rain. Acid rain is one transboundary air pollution problem to be addressed by international agreement. SO_2 and NO_x are industrial pollutants found in industrial regions around the world especially North America and Europe. In North America, acid deposition is most prevalent between the United States and Canada. Negotiations have been ongoing between the two countries since the 1970s resulting in the 1991 U.S.-Canada Bilateral Air

Quality Agreement. In Europe, large amounts of pollutants, namely SO_2 and NO_X, are transported from one country to another. Monitoring of these pollutants and estimates of pollution deposition have indicated which countries are net exporters and which are net importers. Net exporters include the United Kingdom, the former East Germany, Spain, and Italy. Austria, Switzerland, Norway, and Sweden received the majority of SO_2 from foreign sources. To address this problem, the 1979 Convention on Long-Range Transboundary Air Pollution was signed following negotiations convened by the United Nations Economic Commission for Europe.

U.S.-Canada Bilateral Air Quality Agreement, 1991

This agreement involved the control of SO_2 and NO_X emissions and specified reductions of both pollutants by each country. Technology-based standards were called to reduce emissions of NO_X. Further, the agreement included provisions to facilitate cooperation between the two nations, evaluation of activities that may result in transboundary air pollution, and notification of such proposed activities. Also, an Air Quality Committee was established with the responsibility of helping implement the agreement as well as reporting progress.

Convention on Long-Range Transboundary Air Pollution, 1979

This convention was designed to address the problems of pollutant and acid deposition on countries outside the country of origin and entered into force March 16, 1983. For purposes of the convention, "air pollution means the introduction by man, directly or indirectly, of substances or energy into the air resulting in deleterious effects of such a nature as to endanger human health, harm living resources and ecosystems and material property and impair or interfere with amenities and other legitimate uses of the environment, and "air pollutants" shall be construed accordingly." This definition covers the traditional problems caused by pollutant deposition, including human health and environmental effects. However, the definition also includes harm to ecosystems. From a regulatory perspective in the United States, ecosystem protection is a relatively new concept. This treaty was one of the first of the pollution-specific international treaties to address ecosystem protection. The Long-Range Transboundary Air Pollution treaty is composed of 18 articles and according to Article 11..." The Executive Secretary of the Economic Commission for Europe shall carry out, for the Executive Body, the following secretariat functions:

   (a) to convene and prepare the meetings of the Executive Body;
   (b) to transmit to the Contracting Parties reports and other information received in accordance with the provisions of the present Convention;
   (c) to discharge the functions assigned by the Executive Body."

Although somewhat general, the treaty establishes the framework for cooperation, information exchange, and monitoring. The program for evaluating air pollutants in Europe (EMEP) has been an important source of information, and this treaty calls for expansion of the program. The full text of this convention has been included at the end of this chapter.

Other Agreements (actions by the UNEP and EU)

   • Protocol to the 1979 Convention on Long-Range Transboundary Air Pollution Concerning the Control of Emissions of Nitrogen Oxides or their Transboundary Fluxes (NO_X Protocol), 1986
   • Protocol to the 1979 Convention on Long-Range Transboundary Air Pollution Concerning the Control of Emissions of VOCs or their Transboundary Fluxes, 1992

- Protocol on the Reduction of Sulphur Emissions or their Transboundary Fluxes by at Least 30%, 1988
- Protocol on Further Reduction of Sulphur Emissions, 1994

In addition to these treaties, the European Union has established Directives geared toward air pollution. As discussed in Chapter 6, the European Union establishes legislation that must be adopted and implemented by member states. Twenty-one EU directives, including amendments, address air quality, air pollution from both stationary and mobile sources, and ozone depletion including:

- 70/220/EEC, Air pollution by gases from positive ignition engines of motor vehicles, 1970
- 75/716/EEC, The sulphur content of certain liquid fuels, 1975
- 80/779/EEC, Air quality limit values and the guide values for sulphur dioxide and suspended particulates, 1989
- 93/12/EEC, Sulphur content of certain liquid fuels, 1993
- 85/203/EEC, Air quality standards for nitrogen dioxide, 1985
- 88/609/EEC, Prevention and reduction of environmental pollution by asbestos, 1987
- 88/609/EEC, The limitation of emissions of certain pollutants into the air from large combustion plants, 1988
- 89/369/EEC, Air pollution from new municipal waste incineration plants, 1989
- 89/429/EEC, Air pollution from existing municipal waste incineration plants, 1989
- 89/349/EEC, Recommendations on the reduction of chlorofluorocarbons by the aerosol industry, 1989
- 93/C56/05, Emissions from motor vehicles, 1993
- 594/91, Substances that deplete the ozone layer, 1991
- 92/3952, Concerning ozone depletion, 1992
- 92/72/EEC, Ozone pollution of the atmosphere, 1992
- 93/361/EEC, Long-range transboundary nitrogen oxide pollution, 1992
- COM(93) 277EC, Proposal for a Council Directive relating to measures to be taken against air pollution
- COM(87) 706EC, 1994 Methods for measurement of air pollution Part 13

BRITISH STANDARDS INSTITUTION (BSI)

- BS 1747: PART 12BSI1: 1993 Methods for Measurement of Air Pollution Part 11: Determination of a Black Smoke
- BS 1747: PART 11 BSI1: 1990 Methods for the Measurement of Air Pollution Part 10: Determination of the Mass
- BS 1747: PART 10 BSI1: 1987 Methods for the Measurement of Air Pollution Part 9: Determination of the Mass
- BS 1747: PART 9 BSI1: 1986 Methods for the Measurement of Air Pollution Part 8: Determination of the Mass
- BS 1747: PART 8 BSI1: 1983 Methods for the Measurement of Air Pollution Part 7: Determination of Mass Concentration
- BS 1747: PART 7 BSI1: 1983 Methods for the Measurement of Air Pollution Part 6: Sampling Equipment Used
- BS 1747: PART 13BSI1: 1972 Methods for the Measurement of Air Pollution Part 5: Directional Dust Gauges
- BS 1747: PART 5BSI1: 1969 Methods for the Measurement of Air Pollution Part 3: Determination of Sulphur

- BS 1747: PART 3BSI1: 1969 Methods for the Measurement of Air Pollution Part 2: Determination of Concentration
- BS 1747: PART 2BSI1: 1969 Methods for the Measurement of Air Pollution Part 1: Deposit Gauges

AMERICAN SOCIETY FOR TESTING AND MATERIALS (ASTM)

A compilation of air sampling and monitoring standards has been developed by ASTM. The standards have been published with methodologies for water and waste sampling. The compilation contains 138 standards. In addition, Volume 11.03 of the annual ASTM standards books covers atmospheric analysis (as well as occupational safety and health and protective clothing) and contains 160 standards. The section on Sampling and Analysis of Atmospheres includes 83 tests and practices on standard procedures for assessing ambient and workplace atmospheres and measuring asbestos and fluoride content, lead, sulfur-dioxide, and particulates. Also, ASTM has produced several special technical publications and manuals developed by experts and practitioners around the world that have been peer reviewed. The following publications cover testing techniques and their application:

- Manual 15 – Radon: Prevalence, Measurements, Health Risks and Control, Editor, Dr. N.L. Nagda
- STP 653 – Air Quality Meteorology and Atmospheric Ozone
- STP 721 – Sampling and Analysis of Toxic Organics in the Atmosphere
- STP 732 – Health Effects of Synthetic Silica Particulates
- STP 786 – Toxic Materials in the Atmosphere, Sampling and Analysis
- STP – Inhalation Toxicology of Air Pollution: Clinical Research Considerations
- STP 957 – Sampling and Calibration for Atmospheric Measurements
- STP 1002 – Design and Protocol for Monitoring Indoor Air Quality
- STP 1024 – Susceptibility to Inhaled Pollutants
- STP 1052 – Monitoring Methods for Toxics in the Atmosphere
- STP 1082 – Characterization and Toxicity of Smoke
- STP 1181 – Alternatives to Chlorofluorocarbon Fluids in the Cleaning of Oxygen and Aerospace Systems and Components
- STP 1205 – Modeling for Indoor Air Quality and Exposure
- STP 1226 – Lead in Paint, Soil and Dust: Health Risks, Exposure Studies, Control Measures, Measurement Methods, and Quality Assurance, Editors, Beard and Iske
- STP 1287 Characterizing Sources of Indoor Air Pollution and Related Sink Effects, Editor, B.A. Tichenor

WATER RESOURCES

Water, as an important resource, has been the subject of many debates, legal actions, national and international laws, international tension, and even acts of violence. International law of waterways is particularly important considering that more than two hundred river basins are multinational, supporting almost forty percent of the world's population. Not surprisingly, international cooperation regarding water issues dates as far back as Roman times. Early agreements were designed to protect economic interests of navigation, fishing, agriculture, and industry. Principles of ecology and proper management and protection from pollution were considered in later agreements and legislation. More than two hundred international agreements regulating shared water resources and hundreds of standards exist worldwide. Further, countries have developed national legislation to protect water resources from degradation. However, even though many developing countries such as Africa and Asia have important multinational river basins, relatively few agreements have been established in these areas. In the United States, water laws date back to the turn of the century and have evolved into laws designed to protect waterways from sources of pollution from vessels, end of pipe discharges, and nonpoint sources of stormwater and drinking water. The water environment includes the oceans and marine environment, freshwater, wetlands, coastal zones, groundwater, and other sources of drinking water. Pollution sources include:

- Dumping of waste from airplanes and ships such as oil, grease, solvents, and garbage from vessels dumped or discharged from bilge or ballast water or ash from waste incineration from ships.
- Land-based pollution sources, including point source discharges or end of pipe discharges that may include organic and inorganic chemical compounds, sewage, oil and grease, pH altering waste, or heat.
- Land-based nonpoint source pollution from stormwater (precipitation induced) runoff containing fertilizers, pesticides, oil, petroleum products, heavy metals, sediment and biological and chemical oxygen demand waste (BOD,COD). Also, in times of high flow as during a rain storm, treatment systems that combine the storm sewer system with the sanitary sewer cannot treat the large flow and allow untreated waste to by-pass the treatment system directly flowing into waterways.

  - Dredge and fill operations near or in waterways.
  - Oil drilling and exploration.
  - Deep sea bed mining.
  - Oil and chemical spills.
  - Contaminant deposition carried by the atmosphere.

Other environmental problems include overfishing, destruction of nontarget species caught during fishing operations, destruction or loss of habitat, and the introduction of non-native species into a waterway.

PROTECTION OF FRESH WATER RESOURCES

The International Law Association

The International Law Association adopted the resolution "Principles of Law Governing the Uses of the Waters of International Rivers" in 1956. Principles III and IV of the resolution established state obligations to protect water resources from pollution that may damage another state and reads:

"III. While each State has sovereign control over the international rivers within its own boundaries, the State must exercise this control with due consideration for its effects upon other riparian States.

IV. A State is responsible, under international law, for public or private acts producing change in the existing regime of a river to the injury of another State, which it could have prevented by reasonable diligence."

These twelve principles were based on Recommendations on Fresh Water Pollution Control in Europe, 1965,

"a) Control of water pollution forms and integral part of water resource and water utilization policies.
b) Water pollution control constitutes a fundamental governmental responsibility and requires systematic international collaboration.
c) It also requires the cooperation of the local communities and of all users of water."

From these recommendations came Recommendation 436 on Fresh Water Pollution Control in Europe, 1965. The preamble states:

"a) Control of water pollution forms an integral part of water resource and water utilization policies.
 b) Water pollution control constitutes a fundamental government responsibility and requires systematic international collaboration.
 c) It also requires cooperation of the local communities and of all users of water."

Although not considered part of traditional international environmental law, the European Water Charter was established by the Council of Europe in 1968 based on Recommendation 436 on Fresh Water Pollution Control in Europe, 1965 and included twelve principles.

"I.   There is no life without water. It is a treasure indispensable to all human activity.
II.   Fresh water resources are not inexhaustible. It is essential to conserve, control, and whenever possible, to increase them. The population explosion and the rapidly expanding needs of modern industry and agriculture are making increasing demands on water resources. It will be impossible to meet these demands and to achieve rising standards of living, unless each one of us regards water as a precious commodity to be preserved and used wisely.
III.   To pollute water is to harm man and other living creatures which are dependent on water... surface and underground waters should be preserved from pollution. Water in nature is a medium containing beneficial organisms which help to keep it clean. If we pollute the water, we risk destroying those organisms...and perhaps modifying the living medium infavourably and irrevocably.
IV.   The quality of water must be maintained at levels suitable for the use to be made of it and, in particular, must meet appropriate public health standards.
V.   When used water is returned to a common source it must not impair the further uses, both public and private, to which the common source will be put.
VI.   The maintenance of an adequate vegetation cover, preferable forest land, is imperative for the conservation of water resources.
VII.   Water resources must be assessed. It is essential to know surface and underground water sources, bearing in mind the water cycle, the quality of water and its utilisation. Assessment, in this context, involves the survey, recording and appraisal of water resources.
VIII. The wise husbandry of water resources must be planned by the appropriate authorities...Furthermore, maintenance of quality and quantity calls for development and improvement of utilisation, recycling and purification techniques.
IX.   Conservation of water calls for intensified scientific research, training of specialists and public information services... Means of providing information should be increased and international exchange facilitated...
X.   Water is a common heritage, the value of which must be recognized by all. Everyone has the duty to use water carefully and economically. Each human being is a consumer and user of water and is therefore responsible to other users. To use water thoughtlessly is to misuse our natural heritage.
XI.   The management of water resources should be based on their natural basins rather than on political and administrative boundaries. ...Within a drainage basin, all uses of surface and underground waters are interdependent and should be managed bearing in mind their interrelationship.
XII.   Water knows no frontiers; as a common resource it demands international cooperation.

More currently, the UN Convention on the Law of the Non-Navigational Uses of International Watercourses, 1997, includes articles addressing:

- Equitable and reasonable utilization of watercourses
- Participation in the use, development, and protection of international watercourses
- An obligation not to cause significant harm to watercourses
- Regular exchange of data and information between watercourse states
- Notification concerning planned measures with possible adverse effects
- Protection and preservation of ecosystems
- Prevention, reduction, and control of pollution
- Introduction of alien or new species
- Protection and preservation of the marine environment

Unlike earlier agreements, this convention specifically calls for ecosystem protection (Article 20) and pollution prevention (Article 21). Article 21 calls for watercourse states to "...prevent, reduce and control the pollution of international watercourses that may cause significant harm to other watercourse states or to their environment, including harm to human health or safety, to the use of the waters for any beneficial use or to the living resources of the watercourse." The focus on prevention is especially important. However, the term "significant harm" is difficult to define and subject to interpretation. For example, the National Environmental Policy Act (NEPA) of 1969 is a broadly written U.S. statute that uses similar terms. Virtually every word in the law has since been defined in case law. To make the debate more complicated, current theories suggest that certain compounds called "endocrine disrupters" interfere with normal growth and can lead to disease in both animals and humans. The theory proposes that effects may be realized with relatively small doses of exposure which may completely change how we view environmental contamination and exposure protection. The theory may have significant impact on the determination of the threshold of significant harm and, consequently, national and international environmental law.

PROTECTION OF THE OCEANS AND THE MARINE ENVIRONMENT

Three conventions were held; the first commenced in 1958 and attempted to resolve conflicts between maritime states interested in freedom of the seas for navigation and coastal states promoting tight national jurisdiction over adjacent waters. The three conventions were organized by the UN and have created the Law of the Sea. The first meeting of the United Nations Conference on the Law of the Sea (UNCLOS I) actually produced four conventions:

1) Convention of the Territorial Sea and Contiguous Zone
2) Convention of the High Seas
3) Convention of Fishing and Conservation of Living Resources of the High Seas
4) Convention on the Continental Shelf

Upon signing of UNCLOS III in 1982 that went into force in 1994, a global law of the sea with the goal of protecting and preserving the marine environment had been established for the first time. UNCLOS includes an obligation requiring states to protect and preserve the marine environment and to monitor the risks or effects of pollution. It also established jurisdictional zones based on distance offshore including the Exclusive Economic Zone (EEZ). The EEZ allows states to exploit resources as well as pass legislation to protect those resources. The goal of these provisions was to provide responsibility and thus ensure protection and conservation of marine resources.

MARine POLlution (MARPOL, 1973 and 1978) is another key international agreement designed to protect the marine environment. Specifically, the agreement addresses discharges from ships. The MARPOL Secretariat is the International Maritime Organization (IMO). The IMO is an agency created by the UN in 1958 to regulate the shipping industry. MARPOL contains mandatory discharge standards for oil, construction, design, equipment and manning specifications such as segregated ballast tanks, and navigation standards to protect sensitive areas. States must adopt laws to prevent, reduce, and control pollution from ships. Annex I (in force) is a key mandatory annex to MARPOL, which addresses pollution from oil tankers and other vessels. Design standards require that tankers over 150 tons (and all ships over 400 tons) receive an International Oil Pollution Prevention Certificate issued by the flag state. Before a certificate is issued, the vessel must be inspected. The inspection includes a survey of the vessels structure, equipment, fittings, arrangements, and material. Also, Annex I requires an Oil Record Book be maintained covering ship operations such as ballasting or cleaning of fuel tanks, discharges of dirty ballast or cleaning water, disposal of oil residues, discharge of bilge water, and the loading and unloading of oil cargo.

The London Convention (1972) covers "…any deliberate disposal at sea of wastes or other matter from vessels, aircraft, platforms, or other man-made structures at sea." Covered wastes include sewage sludge, dredged materials, radioactive wastes, construction, and demolition debris. To avoid overlap, certain types of wastes have been excluded because they are addressed by other agreements. This convention prohibits the incineration of waste at sea to prevent ash being deposited into the ocean and impacting marine life. Parties to the 1996 Protocol to the Convention on the Prevention of Marine Pollution by Dumping of Wastes and Other Matter (The London Convention) must establish an authority to issue permits for dumping at sea. The use of a permitting system to control ocean dumping is a familiar regulatory tool used throughout the world to address pollution problems especially air and water pollution. In the United States, the regulation of air and water releases and the disposal of hazardous wastes are controlled using permits issued to the polluter or disposal facility by a regulatory authority, usually a state. The permit establishes criteria above and beyond the regulatory requirements, that must be met by the permit holder. Some permits, such as air permits, are specifically designed for the types and amounts of pollutants from an individual site. Stormwater permits are issued under a general permit in which the conditions are the same for all. Similarly, in the United Kingdom, water discharges from industries must be authorized and maximum concentrations and quantity limits are established.

THE EUROPEAN UNION

Water resources have been a subject of EU programs and directives since the early 1970s. Regulated waters include inland waters, groundwater, drinking water, estuarine waters, wetlands, fisheries, and water for home use. EU directives have been a key instrument to address water pollution problems. Directives provide a framework for member states to establish maximum concentrations and maximum quantities of substances in a discharge. Also, member states must provide the EU Commission with information regarding the discharges they have authorized, an inventory of discharges, and monitoring results. One problem with the implementation of EU directives by member states is the difficulty of identifying sources of pollution. Federal authorities do not have the knowledge of local industry to make this determination.   The United Kingdom uses a variety of regulatory bodies to address water resources, including local water companies that supply drinking water and treat wastewater. The U.K. has vested authority in private water companies under the Water Industry Act of 1991. One function they perform is identification and control of discharges to waterways. Further, the U.K. has designated specific uses for waterways under the Surface Waters Classification Regulations, 1989, No. 2286 and passed the following legislation and regulatory instruments:

- Water Industry Act, 1991
- Water Resources Act, 1991
- Water Act, 1989
- Environmental Protection Act, 1990

- Control of Pollution Act, 1974
- Circular 7/89 (regarding authorization to discharge)
- Statutory Instrument, No. 1156, The Trade Effluents (Prescribed Processes and Substances) Regulations, 1989
- Statutory Instrument No. 1057, The Surface Water (River Ecosystems) Classification Regulations, 1994
- Statutory Instrument No. 337, The Surface Water (Dangerous Substances) (Classification) Regulations, 1992
- Water Quality Objectives for Waters Classified in Accordance with the Surface Water (Dangerous Substances) (Classification) Regulations 1989 and relating to the Council Directive on Discharges of Dangerous Substances
- The Sewarage Undertakers (Information) Direction, 1991

These programs and legislation in the U.K. have been implemented to protect water quality as well as meet their obligation to EU directives (Figure 3), specifically, 76/464/EEC Pollution caused by certain dangerous substances discharged into the aquatic environment, 1976. This directive was initially proposed in 1974 and official Journal notification was made in 1976. The directive provides a framework for water pollution control and covers discharges to the sewer system and to aquatic environments. The program phased in requirements for release authorization, pollution reduction, and program implementation from 1978 to 1986. The goal was to reduce or eliminate discharges of dangerous substances into waterways by establishing a list of substances for which discharge limits are established. Directive 86/280/EEC contains limit values and environmental quality objectives for specific compounds. Limit values are specific to a type of industry and control of discharges by environmental quality objectives applies to all discharges.

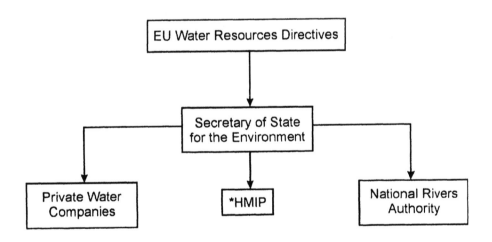

*Her Majesty's Inspectorate of Pollution

Figure 3 Implementation of EU Directives in the U.K.

EUROPEAN UNION DIRECTIVES

- 75/440/EEC, The quality required of surface water intended for the abstraction of drinking water, 1975
- 76/160/EEC, The quality of bathing water, 1976
- 76/464/EEC, Pollution caused by certain dangerous substances discharged into the aquatic environment, 1976
- 78/176/EEC, Waste from titanium dioxide industry, 1978
- 78/659/EEC, The quality of fresh water needing protection or improvement in order to support fresh water life, 1978
- 79/869/EEC, Methods of measurement of surface water intended for the extraction of drinking water, 1979
- 79/923/EEC, The quality required of shellfish waters, 1979
- 80/68/EEC, The protection of groundwater against pollution caused by certain dangerous substances, 1980
- 80/778/EEC, The quality of water intended for human consumption, 1980
- 82/176/EEC, Mercury discharges by the chlor-alkali electolysis industry, 1982
- 83/513/EEC, Cadmium discharges, 1983
- 86/280/EEC, Discharges of certain dangerous substances included in I to directive 76/463/EEC, 1986
- 91/271/EEC, Directive on municipal wastewater treatment, 1991
- 91/676/EEC, Pollution caused by nitrates from agriculture source, 1991
- 92/C59/02, Council Resolution on the future community groundwater policy, 1992
- 92/446/EEC, Questionnaires relating to water pollution, 1992

AMERICAN SOCIETY OF TESTING AND MATERIALS (ASTM)

Volume 11.01 Water (I)

This volume contains 117 standards and is the first of two volumes (I and II) relating to the assessment of water. Specifically, Volume 11.01 includes standard procedures for assessing water and is divided into four sections:

1) Terminology, Reagents, and the Reporting of Results
2) Sampling and Flow Measurement
3) General Properties of Water (20 tests, practices, and standard techniques for water analysis)
4) Inorganic Constituents (70 tests and practices for determining inorganic constituents, such as D 3559 Test Method for Lead in Water)

Volume 11.02 Water (II)

Volume 11.02 contains 167 standards and is divided into six sections:

1) Organic Constituents (contains 40 standard procedures on general analysis methods and tests for specific constituents and waterborne oils)
   D 2036 Test Methods for Cyanides in Water (revised)
2) Radioactivity (20 procedures for measuring radioactivity and specific radionuclides)
3) Saline and Brackish Waters, Seawaters, and Brines (over 10 tests for determining specific constituents such as barium, iodide and bromide, and chloride ions)
4) Microbiological Examination (14 standard tests and practices)

5) Water-Formed Deposits (20 standard tests and guides)
6) Water-Treatment Materials (30 tests on chemicals, particulate ion-exchange materials, membrane filters, and reverse osmosis devices)

SPECIFIC ASTM WATER STANDARDS (PARTIAL LIST)

- D5259 ASTM: Standard Test Method for Isolation and Enumeration of Enterococci from Water
- D5259 ASTM: Standard Guide for Monitoring Aqueous Nutrients in Watersheds
- D6146 ASTM: Standard Test Method for Isolation and Enumeration of Escherichia Coli in Water
- D5392 ASTM: Standard Guide for Monitoring Aqueous Nutrients in Watersheds
- D6145 ASTM: Standard Guide for Monitoring Sediment in Watersheds
- D5392 ASTM: Standard Guide for Monitoring Sediment in Watersheds
- E1609 ASTM: Standard Guide for Planning and Implementing Water Monitoring
- D5851 ASTM: Standard Practices for Extraction of Trace Elements from Sediments
- D3974 ASTM: Dual Check Valve Type Backflow Preventers
- E679ASTM: Specification and Performance of On-Site Instrumentation for Continuously Monitoring
- 4516 ASTM: Standard Guide for Sampling Groundwater Monitoring Wells
- D4448 ASTM: Standard Guide for Water Analysis for Reverse Osmosis Application
- D4195 ASTM: Standard Guide for Measuring Matric Potential in the Vadose Zone Using Tensiometer
- D3404 ASTM: Standard Test Method for Organochlorine Pesticides In Water

Additional Standards

- MIL-V-24509 AASSIST: Standard Test Method for Isolation and Enumeration of Enterococci from Water
- ANSI/ASSE 1024-1998 ASSE: Protection Against Pollution of Potable Water in Drinking Water Installations
- PREN 1717CEN: Standard Practice for Determination of Odor and Taste Thresholds by a Forced-Choice
- N42.18-1980IEEE: Standard Practice for Use of Reversed-Phase High Performance Liquid Chromatography
- M0026PEWEF: Methods of Test for Drinking and Mineral Water
- 707SASO: Ultraviolet Photometer for Monitoring of Water Pollution
- DIN 38405 P14 German Standard Methods for the Examination of Water, Waste Water and Sludge; Anions (Group D); Determination of Cyanides in Drinking Water, and in Groundwater and Surface Water with Low Pollution Levels (D 14)
- DIN 38405 P19 (E) German Standard Methods for the Examination of Water, Waste Water and Sludge; Anions (Group D); Determination of Fluoride, Chloride, Nitrite, (Ortho) Phosphate, Bromide, Nitrate and Sulfate Anions in Water with a Low Pollution Level by Ion Exchange
- DIN 6625 P1 Steel Tanks Erected on Site, for the Above Ground Storage of Hazardous Flammable Water Polluting Liquids of Class A III and of Non-Flammable Water Polluting Liquids; Requirements and Testing

WASTE

The United Nations Environment Program (UNEP) estimates that more than 400 million tons of hazardous waste are annually generated worldwide. Although tremendous benefits have been realized through industrialization and technological advances, a legacy of chemical waste has been left behind. In spite of efforts to reduce or eliminate the generation of hazardous waste, it continues to be produced in large amounts. Protection of human health and the environment depends on proper management and disposal of these wastes. National legislation has been developed by countries around the world to ensure that hazardous waste is handled and disposed of properly. To address movement of waste from one country to another, the United States and other countries, the EU, UNEP, and OECD have developed laws or programs. Some countries, including Nigeria and Cameroon, have banned the importation of hazardous waste as a result of scams designed to send waste to unsuspecting countries. There is an economic incentive for developing countries to accept hazardous waste for both the receiving country and the generator. However, many of these countries do not have adequate facilities to properly handle or dispose of the waste. To address this problem, UNEP, of which 110 nations are a party, convened the Basel Convention.

THE BASEL CONVENTION (UNEP) ON THE CONTROL OF TRANSBOUNDARY MOVEMENTS OF HAZARDOUS WASTE AND THEIR DISPOSAL, 1989

One hundred and sixteen countries participated in two years of negotiation. The United States has signed but not ratified the convention. To do so would require amendments to the hazardous waste laws. The convention establishes a global notification system and controls the movement of hazardous waste to less developed countries. The classification of a waste stream as hazardous waste can be a complicated process. In the United States, the Resource Conservation and Recovery Act (RCRA) (Chapter 12) defines a hazardous waste. The classification of a waste stream is dependent on the method in which the waste was generated and the waste's characteristics. Four lists and four characteristics are used to classify the waste. The Basel Convention defines a hazardous waste by listing specific categories of wastes from specific industries, similar to the "K" or specific source wastes under RCRA as well as the characteristics of the waste. The waste streams are identified alphanumerically. The letter "Y" symbolizes the category of waste and the letter "H" the characteristic wastes.   Finally, the waste is hazardous waste if national laws decreed that

- the waste originated,
- the waste is to be transported through, or
- the waste is to be disposed.

The Basel Convention definition includes medical waste. The convention excludes those wastes covered by other treaties as is the case with radioactive wastes and waste from ships. The general obligations of the convention include

- The state receiving the waste must consent in writing to the importation of the waste. They can also conditionally consent or deny permission. Transit states must also grant written consent.
- Provisions to prevent the exportation of waste to countries that have banned its importation.
- States must be notified of the transboundary movement of hazardous waste and the effects on human health and the environment.
- Requires proper packaging, labeling, and transport.
- Requires the use of a "movement document" (manifest) that all parties are required to sign.

- Places restrictions on exportation. The following conditions must be met before exportation is allowed:

  - State of export does not have adequate storage or disposal facilities
  - The waste is a raw material for the recycling or recovery industry

- Parties to the convention can implement more stringent requirements.

Reviewing the conditions that must be met before exportation of hazardous is allowed, notice how they discourage exportation altogether and promote recycling and recovery. Other provisions require the disposal facility upon receipt and disposal of the waste to notify the exporter and the "competent authority of the state." Also, an exporter of hazardous waste must have a contract with a proper disposal company.

ORGANIZATION FOR ECONOMIC COOPERATION AND DEVELOPMENT (OECD)

The OECD has developed numerous provisions for movement of hazardous waste from one country to another especially as it relates to OECD members.   The recommendations and nonbinding provisions are similar to the requirements contained in the Basel Convention.

- Practical Information for the Implementation of the OECD Control system: Transfrontier Movements of Wastes Destined for Recovery Operations, 1992 and 1997
- Decision and Recommendation on Transfrontier Movements of Hazardous Waste, 1984 (requires importing countries be notified)
- Council Decision and Recommendation on Exports of Hazardous Wastes from the OECD Area, 1986

  - Requires consent from importing country and transit countries
  - Must provide information on the origin, nature composition, and quantity of waste
  - Must provide information on environmental risks presented during transport
  - Importing country must have adequate disposal facilities

U.S. BILATERAL AGREEMENTS

- Bilateral Agreement Between the United States and Canada Concerning the Transboundary Movement of Hazardous Waste, 1986
- Bilateral Agreement of Cooperation Between the United States of America and the United Mexican States Regarding the Transboundary Shipments of Hazardous Waste and Hazardous Substances, 1986

EU DIRECTIVES

- 75/439/EEC, Directive on the disposal of waste oils, 1975
- 75/442/EEC, Framework directive on waste, 1975
- 76/403/EEC, Directive on the disposal of PCBs and PCTs, 1976
- 84/631/EEC, The transfrontier shipment of hazardous waste, 1984
- 93/259, Supervision and control of waste shipments, 1993
- 86/278/EEC, The use of sewage sludge, 1986
- 90/C251/3, Proposed directive on civil liability for waste, 1989
- 89/428/EEC, The reduction and eventual elimination of pollution caused by waste from the titanium dioxide industry, 1989

- 91/689/EEC, Hazardous waste, 1991
- COM(93)275, Amended proposal on landfill disposal of waste
- 91/C299/05, Disposal of PCBs and PCTs, 1991
- 91/C192/04, Amended proposal of civil liability for damage caused by waste
- 92/3, Euratom shipment of radioactive waste, 1993
- 93/C190/05, Incineration of hazardous waste, 1993

ASTM STANDARDS

- E1605 ASTM: Standard Practice for Processing Mixtures of Lime, Fly Ash, and Heavy Metal Wastes
- E887 ASTM: Standard Test Method for Ash in the Analysis Sample of Refuse-Derived Fuel
- E790 ASTM: Standard Test Method for Forms of Chlorine in Refuse-Derived Fuel
- E776 ASTM: Standard Test Methods for Total Sulfur in the Analysis Sample of Refuse-Derived Fuel
- E1294 ASTM: Standard Test Method for Total Moisture in a Refuse-Derived Fuel Laboratory Sample
- E949 ASTM: Standard Test Methods for Analyses of Metals in Refuse-Derived Fuel by Atomic Absorption
- E828ASTM: Standard Test Method for Calculating Refuse-Derived Fuel Analysis Data
- E791ASTM: Standard Practice for Sampling and Handling of Fuels for Volatility Measurement
- E1294 ASTM: Standard Test Method for Total Moisture in a Refuse-Derived Fuel Laboratory Sample
- E791 ASTM: Standard Practice for Sampling and Handling of Fuels for Volatility Measurement
- D5841 ASTM: Standard Practice for Sampling Single or Multilayered Liquids
- ASTM 11.04: Pesticides, Resource Recovery; Hazardous Substances and Oil Spill Responses; Waste Disposal; Biological Effects
- ASTM D 5177: Waste Acceptance at Hazardous Waste Incinerators
- ASTM D 5530: Test Method for Total Moisture of Hazardous Waste Fuel by Karl Fischer Titrimetry
- ASTM D 5839: Trace Element Analysis of Hazardous Waste Fuel by Energy-Dispersive X-Ray Fluorescence Spectrometry
- ASTM D 6050: Standard Test Method for Determination of Insoluble Solids in Organic Liquid Hazardous Waste
- ASTM D 6052: Standard Test for Preparation and Elemental Analysis of Liquid Hazardous Waste by Energy-Dispersive X-Ray Fluorescence Spectrometry
- ASTM PS 85: Standard Provisional Guide for Expedited Site Characterization of Hazardous Waste Contaminated Sites
- ASTM STP 1240: Stabilization and Solidification of Hazardous, Radioactive and Mixed Wastes, third Volume

ADDITIONAL STANDARDS

- ASME QHO 1: Qualifications and Certification of Hazardous Waste Incinerator Operators
- ANSI/AIIM TR20-1994AIIM: Hazardous Waste Site Remediation: Assessment & Characterization

- AASH: Measurement of BTEX Emission Fluxes from Refinery Wastewater Impoundments
- Publication 4518 API: Evaluation of Analytical Methods for Measuring Appendix IX Constituents in Groundwater
- Publication 4499 API: Land Treatment Safe and Efficient Disposal of Petroleum Waste
- Publication 4388 API: Generation and Management of Residual Materials
- Publication 4518 API: Evaluation of Analytical Methods for Measuring Appendix IX Constituents in Groundwater
- Publication 333A PI: Test Method for Determining the Anaerobic Biodegradability of Plastic Materials
- Z218.0-93 CSA: Standard Test Method for Pore Size Characteristics of Membrane Filters
- EPA 530/SW-86-044: Technical Resource Document for the Storage and Treatment of Hazardous Waste in Tank Systems
- EPA 530/SW-87-016: A Catalog of Hazardous and Solid Waste Publications
- EPA 625/6-86/012: Hazardous Waste Incineration

INTERNATIONAL TREATIES RELATING TO THE ENVIRONMENT

AIR POLLUTION AND THE OZONE LAYER

Convention on Long-Range Transboundary Air Pollution
Geneva, 13 November 1979

The Parties to the present Convention,

Determined to promote relations and cooperation in the field of environmental protection,

Aware of the significance of the activities of the United Nations Economic Commission for Europe in strengthening such relations and cooperation, particularly in the field of air pollution including long-range transport of air pollutants,

Recognizing the contribution of the Economic Commission for Europe to the multilateral implementation of the pertinent provisions of the Final Act of the Conference on Security and Cooperation in Europe,

Cognizant of the references in the chapter on environment of the Final Act of the Conference on Security and Cooperation in Europe calling for cooperation to control air pollution and its effects, including long-range transport of air pollutants, and to the development through international cooperation of an extensive programme for the monitoring and evaluation of long-range transport of air pollutants, starting with sulphur dioxide and with possible extension to other pollutants,

Considering the pertinent provisions of the Declaration of the United Nations Conference on the Human Environment, and in particular principle 21, which expresses the common conviction that States have, in accordance with the Charter of the United Nations and the principles of international law, the sovereign right to exploit their own resources pursuant to their own environmental policies, and the responsibility to ensure that activities within their jurisdiction or control do not cause damage to the environment of other States or of areas beyond the limits of national jurisdiction,

Recognizing the existence of possible adverse effects, in the short and long term, of air pollution including transboundary air pollution,

Concerned that a rise in the level of emissions of air pollutants within the region as forecast may increase such adverse effects,

Recognizing the need to study the implications of the long-range transport of air pollutants and the need to seek solutions for the problems identified,

Affirming their willingness to reinforce active international cooperation to develop appropriate national policies and by means of exchange of information, consultation, research and monitoring, to coordinate national action for combating air pollution including long-range transboundary air pollution,

Have agreed as follows:

Article 1
Definitions

For the purposes of the present Convention:

    (a) "air pollution" means the introduction by man, directly or indirectly, of substances or energy into the air resulting in deleterious effects of such a nature as to endanger human health, harm living resources and ecosystems and material property and impair or interfere with amenities and other legitimate uses of the environment, and "air pollutants" shall be construed accordingly;

    (b) "long-range transboundary air pollution" means air pollution whose physical origin is situated wholly or in part within the area under the national jurisdiction of one State and which has adverse effects in the area under the jurisdiction of another State at such a distance that it is not generally possible to distinguish the contribution of individual emission sources or groups of sources.

Article 2
Fundamental Principles

The Contracting Parties, taking due account of the facts and problems involved, are determined to protect man and his environment against air pollution and shall endeavour to limit and, as far as possible, gradually reduce and prevent air pollution including long-range transboundary air pollution.

Article 3

The Contracting Parties, within the framework of the present Convention, shall by means of exchanges of information, consultation, research and monitoring, develop without undue delay policies and strategies which shall serve as a means of combating the discharge of air pollutants, taking into account efforts already made at national and international levels.

Article 4

The Contracting Parties shall exchange information on and review their policies, scientific activities and technical measures aimed at combating, as far as possible, the discharge of air pollutants which may have adverse effects, thereby contributing to the reduction of air pollution including long-range transboundary air pollution.

Article 5

Consultations shall be held, upon request, at an early stage between, on the one hand, Contracting Parties which are actually affected by or exposed to a significant risk of long-range transboundary air pollution and, on the other hand, Contracting Parties within which and subject to whose jurisdiction a significant contribution to long-range transboundary air pollution originates, or could originate, in connexion with activities carried on or contemplated therein.

Article 6
Air Quality Management

Taking into account articles 2 to 5, the ongoing research, exchange of information and monitoring and the results thereof, the cost and effectiveness of local and other remedies and, in order to combat air pollution, in particular that originating from new or rebuilt installations, each Contracting Party undertakes to develop the best policies and strategies including air quality management systems and, as part of them, control measures compatible with balanced development, in particular by using the best available technology which is economically feasible and low- and non-waste technology.

Article 7
Research and Development

The Contracting Parties, as appropriate to their needs, shall initiate and cooperate in the conduct of research into and/or development of:

(a) existing and proposed technologies for reducing emissions of sulphur compounds and other major air pollutants, including technical and economic feasibility, and environmental consequences;
(b) instrumentation and other techniques for monitoring and measuring emission rates and ambient concentrations of air pollutants;
(c) improved models for a better understanding of the transmission of long-range transboundary air pollutants;
(d) the effects of sulphur compounds and other major air pollutants on human health and the environment, including agriculture, forestry, materials, aquatic and other natural ecosystems and visibility, with a view to establishing a scientific basis for dose/effect relationships designed to protect the environment;
(e) the economic, social and environmental assessment of alternative measures for attaining environmental objectives including the reduction of long-range transboundary air pollution;
(f) education and training programmes related to the environmental aspects of pollution by sulphur compounds and other major air pollutants.

Article 8
Exchange of Information

The Contracting Parties, within the framework of the Executive Body referred to in Article 10 and bilaterally, shall, in their common interests, exchange available information on:

(a) data on emissions at periods of time to be agreed upon, of agreed air pollutants, starting with sulphur dioxide, coming from grid-units of agreed size; or on the fluxes of agreed air pollutants, starting with sulphur dioxide, across national borders, at distances and at periods of time to be agreed upon;
(b) major changes in national policies and in general industrial development, and their potential impact, which would be likely to cause significant changes in long-range transboundary air pollution;
(c) control technologies for reducing air pollution relevant to long-range transboundary air pollution;
(d) the projected cost of the emission control of sulphur compounds and other major air pollutants on a national scale;

(e) meteorological and physicochemical data relating to the processes during transmission;

(f) physico-chemical and biological data relating to the effects of long-range transboundary air pollution and the extent of the damage which these data indicate can be attributed to long-range transboundary air pollution;

(g) national, subregional and regional policies and strategies for the control of sulphur compounds and other major air pollutants.

Article 9
Implementation and Further Development of the Cooperative Programme for the Monitoring and Evaluation of the Long-Range Transmission of Air Pollutants in Europe

The Contracting Parties stress the need for the implementation of the existing "Cooperative Programme for the Monitoring and Evaluation of the Long-Range Transmission of Air Pollutants in Europe" (hereinafter referred to as EMEP) and, with regard to the further development of this programme, agree to emphasize

(a) the desirability of Contracting Parties joining in and fully implementing EMEP which, as a first step, is based on the monitoring of sulphur dioxide and related substances;

(b) the need to use comparable or standardized procedures for monitoring whenever possible;

(c) the desirability of basing the monitoring programme on the framework of both national and international programmes. The establishment of monitoring stations and the collection of data shall be carried out under the national jurisdiction of the country in which the monitoring stations are located;

(d) the desirability of establishing a framework for a cooperative environmental monitoring programme, based on and taking into account present and future national, subregional, regional and other international programmes;

(e) the need to exchange data on emissions at periods of time to be agreed upon, of agreed air pollutants, starting with sulphur dioxide, coming from grid-units of agreed size; or on the fluxes of agreed air pollutants, starting with sulphur dioxide, across national borders, at distances and at periods of time to be agreed upon. The method, including the model, used to determine the fluxes, as well as the method, including the model, used to determine the transmission of air pollutants based on the emissions per grid-unit, shall be made available and periodically reviewed, in order to improve the methods and the models;

(f) their willingness to continue the exchange and periodic updating of national data on total emissions of agreed air pollutants, starting with sulphur dioxide;

(g) the need to provide meteorological and physico-chemical data relating to processes during transmission;

(h) the need to monitor chemical components in other media such as water, soil and vegetation, as well as a similar monitoring programme to record effects on health and environment;

(i) the desirability of extending the national EMEP networks to make them operational for control and surveillance purposes.

Article 10
Executive Body

1. The representatives of the Contracting Parties shall, within the framework of the Senior Advisers to ECE Governments on Environmental Problems, constitute the Executive Body of the present Convention, and shall meet at least annually in that capacity.

2. The Executive Body shall:
    (a) review the implementation of the present Convention;
    (b) establish, as appropriate, working groups to consider matters related to the implementation and development of the present Convention and to this end to prepare appropriate studies and other documentation and to submit recommendations to be considered by the Executive Body;
    (c) fulfill such other functions as may be appropriate under the provisions of the present Convention.
3. The Executive Body shall utilize the Steering Body for the EMEP to play an integral part in the operation of the present Convention, in particular with regard to data collection and scientific cooperation.
4. The Executive Body, in discharging its functions, shall, when it deems appropriate, also make use of information from other relevant international organizations.

Article 11
Secretariat

The Executive Secretary of the Economic Commission for Europe shall carry out, for the Executive Body, the following secretariat functions:

(a) to convene and prepare the meetings of the Executive Body;
(b) to transmit to the Contracting Parties reports and other information received in accordance with the provisions of the present Convention;
(c) to discharge the functions assigned by the Executive Body.

Article 12
Amendments to the Convention

1. Any Contracting Party may propose amendments to the present Convention.
2. The text of proposed amendments shall be submitted in writing to the Executive Secretary of the Economic Commission for Europe, who shall communicate them to all Contracting Parties. The Executive Body shall discuss proposed amendments at its next annual meeting provided that such proposals have been circulated by the Executive Secretary of the Economic Commission for Europe to the Contracting Parties at least ninety days in advance.
3. An amendment to the present Convention shall be adopted by consensus of the representatives of the Contracting Parties, and shall enter into force for the Contracting Parties which have accepted it on the ninetieth day after the date on which two-thirds of the Contracting Parties have deposited their instruments of acceptance with the depositary. Thereafter, the amendment shall enter into force for any other Contracting Party on the ninetieth day after the date on which that Contracting Party deposits its instrument of acceptance of the amendment.

Article 13
Settlements of Disputes

If a dispute arises between two or more Contracting Parties to the present Convention as to the interpretation or application of the Convention, they shall seek a solution by negotiation or by any other method of dispute settlement acceptable to the parties to the dispute.

Article 14
Signature

1. The present Convention shall be open for signature at the United Nations Office at Geneva from 13 to 16 November 1979 on the occasion of the High-level Meeting within the framework of the Economic Commission for Europe on the Protection of the Environment, by the member States of the Economic Commission for Europe as well as States having consultative status with the Economic Commission for Europe, pursuant to paragraph 8 of Economic and Social Council resolution 36 (IV) of 28 March 1947, and by regional economic integration organizations, constituted by sovereign States members of the Economic Commission for Europe, which have competence in respect of the negotiation, conclusion and application of international agreements in matters covered by the present Convention.

2. In matters within their competence, such regional economic integration organizations shall, on their own behalf, exercise the rights and fulfil the responsibilities which the present Convention attributes to their member States. In such cases, the member States of these organizations shall not be entitled to exercise such rights individually.

Article 15
Ratification, Acceptance, Approval and Accession

1. The present Convention shall be subject to ratification, acceptance or approval.
2. The present Convention shall be open for accession from 17 November 1979 by the States and organizations referred to in Article 14, paragraph 1.
3. The instruments of ratification, acceptance, approval or accession shall be deposited with the Secretary-General of the United Nations, who will perform the functions of the depositary.

Article 16
Entry into Force

1. The present Convention shall enter into force on the ninetieth day after the date of deposit of the twenty-fourth instrument of ratification, acceptance, approval or accession.
2. For each Contracting Party which ratifies, accepts or approves the present Convention or accedes thereto after the deposit of the twenty-fourth instrument of ratification, acceptance, approval or accession, the Convention shall enter into force on the ninetieth day after the date of deposit by such Contracting Party of its instrument of ratification, acceptance, approval or accession.

Article 17
Withdrawal

At any time after five years from the date on which the present Convention has come into force with respect to a Contracting Party, that Contracting Party may withdraw from the Convention by giving written notification to the depositary. Any such withdrawal shall take effect on the ninetieth day after the date of its receipt by the depositary.

Article 18
Authentic Texts

The original of the present Convention, of which the English, French and Russian texts are equally authentic, shall be deposited with the Secretary-General of the United Nations.

In witness whereof the undersigned, being duly authorized thereto, have signed the present Convention. Done at Geneva, this thirteenth day of November, one thousand nine hundred and seventy-nine.

Protocol on Substances that Deplete the Ozone Layer
Montreal, 16 September 1987

The Parties to this Protocol,

Being Parties to the Vienna Convention for the Protection of the Ozone Layer,

Mindful of their obligation under that Convention to take appropriate measures to protect human health and the environment against adverse effects resulting or likely to result from human activities which modify or are likely to modify the ozone layer,

Recognizing that world-wide emissions of certain substances can significantly deplete and otherwise modify the ozone layer in a manner that is likely to result in adverse effects on human health and the environment,

Conscious of the potential climatic effects of emissions of these substances,

Aware that measures taken to protect the ozone layer from depletion should be based on relevant scientific knowledge, taking into account technical and economic considerations,

Determined to protect the ozone layer by taking precautionary measures to control equitably total global emissions of substances that deplete it, with the ultimate objective of their elimination on the basis of developments in scientific knowledge, taking into account technical and economic considerations,

Acknowledging that special provision is required to meet the needs of developing countries for these substances,

Noting the precautionary measures for controlling emissions of certain chlorofluorocarbons that have already been taken at national and regional levels,

Considering the importance of promoting international cooperation in the research and development of science and technology relating to the control and reduction of emissions of substances that deplete the ozone layer, bearing in mind in particular the needs of developing countries,

Have agreed as follows:

Article 1
Definitions

For the purposes of this Protocol:

1. "Convention" means the Vienna Convention for the Protection of the Ozone Layer, adopted on 22 March 1985.
2. "Parties" means, unless the text otherwise indicates, Parties to this Protocol.
3. "Secretariat" means the secretariat of the Convention.
4. "Controlled substance" means a substance listed in Annex A to this Protocol, whether existing alone or in a mixture. It excludes, however, any such substance or mixture which is in a manufactured product other than a container used for the transportation or storage of the substance listed.
5. "Production" means the amount of controlled substances produced minus the amount destroyed by technologies to be approved by the Parties.

6. "Consumption" means production plus imports minus exports of controlled substances.
7. "Calculated levels" of production, imports, exports and consumption means levels determined in accordance with Article 3.
8. "Industrial rationalization" means the transfer of all or a portion of the calculated level of production of one Party to another, for the purpose of achieving economic efficiencies or responding to anticipated shortfalls in supply as a result of plant closures.

Article 2
Control Measures

1. Each Party shall ensure that for the twelve-month period commencing on the first day of the seventh month following the date of the entry into force of this Protocol, and in each twelve-month period thereafter, its calculated level of consumption of the controlled substances in Group I of Annex A does not exceed its calculated level of consumption in 1986. By the end of the same period, each Party producing one or more of these substances shall ensure that its calculated level of production of the substances does not exceed its calculated level of production in 1986, except that such level may have increased by no more than ten per cent based on the 1986 level. Such increase shall be permitted only so as to satisfy the basic domestic needs of the Parties operating under Article 5 and for the purposes of industrial rationalization between Parties.

2. Each Party shall ensure that for the twelve-month period commencing on the first day of the thirty-seventh month following the date of the entry into force of this Protocol, and in each twelve-month period thereafter, its calculated level of consumption of the controlled substances listed in Group II of Annex A does not exceed its calculated level of consumption in 1986. Each Party producing one or more of these substances shall ensure that its calculated level of production of the substances does not exceed its calculated level of production in 1986, except that such level may have increased by no more than ten per cent based on the 1986 level. Such increase shall be permitted only so as to satisfy the basic domestic needs of the Parties operating under Article 5 and for the purposes of industrial rationalization between Parties. The mechanisms for implementing these measures shall be decided by the Parties at their first meeting following the first scientific review.

3. Each Party shall ensure that for the period 1 July 1993 to 30 June 1994 and in each twelve-month period thereafter, its calculated level of consumption of the controlled substances in Group I of Annex A does not exceed, annually, eighty per cent of its calculated level of consumption in 1986. Each Party producing one or more of these substances shall, for the same periods, ensure that its calculated level of production of the substances does not exceed, annually, eighty per cent of its calculated level of production in 1986. However, in order to satisfy the basic domestic needs of the Parties operating under Article 5 and for the purposes of industrial rationalization between Parties, its calculated level of production may exceed that limit by up to ten per cent of its calculated level of production in 1986.

4. Each Party shall ensure that for the period 1 July 1998 to 30 June 1999, and in each twelve-month period thereafter, its calculated level of consumption of the controlled substances in Group I of Annex A does not exceed, annually, fifty per cent of its calculated level of consumption in 1986. Each Party producing one or more of these substances shall, for the same periods, ensure that its calculated level of production of the substances does not exceed, annually, fifty per cent of its calculated level of production in 1986. However, in order to satisfy the basic domestic needs of the Parties operating under Article 5 and for the purposes of industrial rationalization between Parties, its calculated level of production may exceed that limit by up to fifteen per cent of its

calculated level of production in 1986. This paragraph will apply unless the Parties decide otherwise at a meeting by a two-thirds majority of Parties present and voting, representing at least two-thirds of the total calculated level of consumption of these substances of the Parties. This decision shall be considered and made in the light of the assessments referred to in Article 6.

5. Any Party whose calculated level of production in 1986 of the controlled substances in Group I of Annex A was less than twenty-five kilotonnes may, for the purposes of industrial rationalization, transfer to or receive from any other Party, production in excess of the limits set out in paragraphs 1, 3 and 4 provided that the total combined calculated levels of production of the Parties concerned does not exceed the production limits set out in this Article. Any transfer of such production shall be notified to the secretariat, no later than the time of the transfer.

6. Any Party not operating under Article 5, that has facilities for the production of controlled substances under construction, or contracted for, prior to 16 September 1987, and provided for in national legislation prior to 1 January 1987, may add the production from such facilities to its 1986 production of such substances for the purposes of determining its calculated level of production for 1986, provided that such facilities are completed by 31 December 1990 and that such production does not raise that Party's annual calculated level of consumption of the controlled substances above 0.5 kilograms per capita.

7. Any transfer of production pursuant to paragraph 5 or any addition of production pursuant to paragraph 6 shall be notified to the secretariat, no later than the time of the transfer or addition.

8. (a) Any Parties which are Member States of a regional economic integration organization as defined in Article 1 (6) of the Convention may agree that they shall jointly fulfill their obligations respecting consumption under this Article provided that their total combined calculated level of consumption does not exceed the levels required by this Article.

   (b) The Parties to any such agreement shall inform the secretariat of the terms of the agreement before the date of the reduction in consumption with which the agreement is concerned.

   (c) Such agreement will become operative only if all Member States of the regional economic integration organization and the organization concerned are Parties to the Protocol and have notified the secretariat of their manner of implementation.

9. (a) Based on the assessments made pursuant to Article 6, the Parties may decide whether:
   (i) adjustments to the ozone depleting potentials specified in Annex A should be made and, if so, what the adjustments should be; and
   (ii) further adjustments and reductions of production or consumption of the controlled substances from 1986 levels should be undertaken and, if so, what the scope, amount and timing of any such adjustments and reductions should be.

   (b) Proposals for such adjustments shall be communicated to the Parties by the secretariat at least six months before the meeting of the Parties at which they are proposed for adoption.

   (c) In taking such decisions, the Parties shall make every effort to reach agreement by consensus. If all efforts at consensus have been exhausted, and no agreement reached, such decisions shall, as a last resort, be adopted by a two-thirds majority vote of the Parties present and voting representing at least fifty per cent of the total consumption of the controlled substances of the Parties.

   (d) The decisions, which shall be binding on all Parties, shall forthwith be communicated to the Parties by the Depositary. Unless otherwise provided in the decisions, they shall enter into force on the expiry of six months from the date of the circulation of the communication by the Depositary.

10. (a) Based on the assessments made pursuant to Article 6 of this Protocol and in accordance with the procedure set out in Article 9 of the Convention, the Parties may decide:

    (i) whether any substances, and if so which, should be added to or removed from any annex to this Protocol; and

    (ii) the mechanism, scope and timing of the control measures that should apply to those substances;

  (b) Any such decision shall become effective, provided that it has been accepted by a two-thirds majority vote of the Parties present and voting.

11. Notwithstanding the provisions contained in this Article, Parties may take more stringent measures than those required by this Article.

Article 3
Calculation of Control Levels

For the purposes of Articles 2 and 5, each Party shall, for each Group of substances in Annex A, determine its calculated levels of:

(a) production by:
    (i) multiplying its annual production of each controlled substance by the ozone depleting potential specified in respect of it in Annex A; and
    (ii) adding together, for each such Group, the resulting figures;

(b) imports and exports, respectively, by following, mutatis mutandis, the procedure set out in subparagraph (a); and

(c) consumption by adding together its calculated levels of production and imports and subtracting its calculated level of exports as determined in accordance with subparagraphs (a) and (b). However, beginning on 1 January 1993, any export of controlled substances to non-Parties shall not be subtracted in calculating the consumption level of the exporting Party.

Article 4
Control of Trade with Non-Parties

1. Within one year of the entry into force of this Protocol, each Party shall ban the import of controlled substances from any State not party to this Protocol.

2. Beginning on 1 January 1993, no Party operating under paragraph 1 of Article 5 may export any controlled substance to any State not party to this Protocol.

3. Within three years of the date of the entry into force of this Protocol, the Parties shall, following the procedures in Article 10 of the Convention, elaborate in an annex a list of products containing controlled substances. Parties that have not objected to the annex in accordance with those procedures shall ban, within one year of the annex having become effective, the import of those products from any State not party to this Protocol.

4. Within five years of the entry into force of this Protocol, the Parties shall determine the feasibility of banning or restricting, from States not party to this Protocol, the import of products produced with, but not containing, controlled substances. If determined feasible, the Parties shall, following the procedures in Article 10 of the Convention, elaborate in an annex a list of such products. Parties that have not objected to it in accordance with those procedures shall ban or restrict, within one year of the annex having become effective, the import of those products from any State not party to this Protocol.

5. Each Party shall discourage the export, to any State not party to this Protocol, of technology for producing and for utilizing controlled substances.
6. Each Party shall refrain from providing new subsidies, aid, credits, guarantees or insurance programmes for the export to States not party to this Protocol of products, equipment, plants or technology that would facilitate the production of controlled substances.
7. Paragraphs 5 and 6 shall not apply to products, equipment, plants or technology that improve the containment, recovery, recycling or destruction of controlled substances, promote the development of alternative substances, or otherwise contribute to the reduction of emissions of controlled substances.
8. Notwithstanding the provisions of this Article, imports referred to in paragraphs 1, 3 and 4 may be permitted from any State not party to this Protocol if that State is determined, by a meeting of the Parties, to be in full compliance with Article 2 and this Article, and has submitted data to that effect as specified in Article 7.

Article 5
Special Situation of Developing Countries

1. Any Party that is a developing country and whose annual calculated level of consumption of the controlled substances is less than 0.3 kilograms per capita on the date of the entry into force of the Protocol for it, or any time thereafter within ten years of the date of entry into force of the Protocol shall, in order to meet its basic domestic needs, be entitled to delay its compliance with the control measures set out in paragraphs 1 to 4 of Article 2 by ten years after that specified in those paragraphs. However, such Party shall not exceed an annual calculated level of consumption of 0.3 kilograms per capita. Any such Party shall be entitled to use either the average of its annual calculated level of consumption for the period 1995 to 1997 inclusive or a calculated level of consumption of 0.3 kilograms per capita, whichever is the lower, as the basis for its compliance with the control measures.
2. The Parties undertake to facilitate access to environmentally safe alternative substances and technology for Parties that are developing countries and assist them to make expeditious use of such alternatives.
3. The Parties undertake to facilitate bilaterally or multilaterally the provision of subsidies, aid, credits, guarantees or insurance programmes to Parties that are developing countries for the use of alternative technology and for substitute products.

Article 6
Assessment and Review of Control Measures

Beginning in 1990, and at least every four years thereafter, the Parties shall assess the control measures provided for in Article 2 on the basis of available scientific, environmental, technical and economic information. At least one year before each assessment, the Parties shall convene appropriate panels of experts qualified in the fields mentioned and determine the composition and terms of reference of any such panels. Within one year of being convened, the panels will report their conclusions, through the secretariat, to the Parties.

Article 7
Reporting of Data

1. Each Party shall provide to the secretariat, within three months of becoming a Party, statistical data on its production, imports and exports of each of the controlled substances

for the year 1986, or the best possible estimates of such data where actual data are not available.

2. Each Party shall provide statistical data to the secretariat on its annual production (with separate data on amounts destroyed by technologies to be approved by the Parties), imports, and exports to Parties and non-Parties, respectively, of such substances for the year during which it becomes a Party and for each year thereafter. It shall forward the data no later than nine months after the end of the year to which the data relate.

Article 8
Non-Compliance

The Parties, at their first meeting, shall consider and approve procedures and institutional mechanisms for determining non-compliance with the provisions of this Protocol and for treatment of Parties found to be in non-compliance.

Article 9
Research, Development, Public Awareness and Exchange of Information

1. The Parties shall cooperate, consistent with their national laws, regulations and practices and taking into account in particular the needs of developing countries, in promoting, directly or through competent international bodies, research, development and exchange of information on:
   (a) best technologies for improving the containment, recovery, recycling or destruction of controlled substances or otherwise reducing their emissions;
   (b) possible alternatives to controlled substances, to products containing such substances, and to products manufactured with them; and
   (c) costs and benefits of relevant control strategies.
2. The Parties, individually, jointly or through competent international bodies, shall cooperate in promoting public awareness of the environmental effects of the emissions of controlled substances and other substances that deplete the ozone layer.
3. Within two years of the entry into force of this Protocol and every two years thereafter, each Party shall submit to the secretariat a summary of the activities it has conducted pursuant to this Article.

Article 10
Technical Assistance

1. The Parties shall, in the context of the provisions of Article 4 of the Convention, and taking into account in particular the needs of developing countries, cooperate in promoting technical assistance to facilitate participation in and implementation of this Protocol.
2. Any Party or Signatory to this Protocol may submit a request to the secretariat for technical assistance for the purposes of implementing or participating in the Protocol.
3. The Parties, at their first meeting, shall begin deliberations on the means of fulfilling the obligations set out in Article 9, and paragraphs 1 and 2 of this Article, including the preparation of workplans. Such workplans shall pay special attention to the needs and circumstances of the developing countries. States and regional economic integration organizations not party to the Protocol should be encouraged to participate in activities specified in such workplans.

Article 11
Meetings of the Parties

1. The Parties shall hold meetings at regular intervals. The secretariat shall convene the first meeting of the Parties not later than one year after the date of the entry into force of this Protocol and in conjunction with a meeting of the Conference of the Parties to the Convention, if a meeting of the latter is scheduled within that period.

2. Subsequent ordinary meetings of the Parties shall be held, unless the Parties otherwise decide, in conjunction with meetings of the Conference of the Parties to the Convention. Extraordinary meetings of the Parties shall be held at such other times as may be deemed necessary by a meeting of the Parties, or at the written request of any Party, provided that, within six months of such a request being communicated to them by the secretariat, it is supported by at least one third of the Parties.

3. The Parties, at their first meeting, shall:
   (a) adopt by consensus rules of procedure for their meetings;
   (b) adopt by consensus the financial rules referred to in paragraph 2 of Article 13;
   (c) establish the panels and determine the terms of reference referred to in Article 6;
   (d) consider and approve the procedures and institutional mechanisms specified in Article 8; and
   (e) begin preparation of workplans pursuant to paragraph 3 of Article 10.

4. The functions of the meetings of the Parties shall be to:
   (a) review the implementation of this Protocol;
   (b) decide on any adjustments or reductions referred to in paragraph 9 of Article 2;
   (c) decide on any addition to, insertion in or removal from any annex of substances and on related control measures in accordance with paragraph 10 of Article 2;
   (d) establish, where necessary, guidelines or procedures for reporting of information as provided for in Article 7 and paragraph 3 of Article 9;
   (e) review requests for technical assistance submitted pursuant to paragraph 2 of Article 10;
   (f) review reports prepared by the secretariat pursuant to subparagraph (c) of Article 12;
   (g) assess, in accordance with Article 6, the control measures provided for in Article 2;
   (h) consider and adopt, as required, proposals for amendment of this Protocol or any annex and for any new annex;
   (i) consider and adopt the budget for implementing this Protocol; and
   (j) consider and undertake any additional action that may be required for the achievement of the purposes of this Protocol.

5. The United Nations, its specialized agencies and the International Atomic Energy Agency, as well as any State not party to this Protocol, may be represented at meetings of the Parties as observers. Any body or agency, whether national or international, governmental or non-governmental, qualified in fields relating to the protection of the ozone layer which has informed the secretariat of its wish to be represented at a meeting of the Parties as an observer may be admitted unless at least one third of the Parties present object. The admission and participation of observers shall be subject to the rules of procedure adopted by the Parties.

Article 12
Secretariat

For the purposes of this Protocol, the secretariat shall:

(a) arrange for and service meetings of the Parties as provided for in Article 11;
(b) receive and make available, upon request by a Party, data provided pursuant to Article 7;
(c) prepare and distribute regularly to the Parties reports based on information received pursuant to Articles 7 and 9;
(d) notify the Parties of any request for technical assistance received pursuant to Article 10 so as to facilitate the provision of such assistance;
(e) encourage non-Parties to attend the meetings of the Parties as observers and to act in accordance with the provisions of this Protocol;
(f) provide, as appropriate, the information and requests referred to in subparagraphs (c) and (d) to such non-party observers; and
(g) perform such other functions for the achievement of the purposes of this Protocol as may be assigned to it by the Parties.

Article 13
Financial Provisions

1. The funds required for the operation of this Protocol, including those for the functioning of the secretariat related to this Protocol shall be charged exclusively against contributions from the Parties.
2. The Parties, at their first meeting, shall adopt by consensus financial rules for the operation of this Protocol.

Article 14
Relationship of this Protocol to the Convention

Except as otherwise provided in this Protocol, the provisions of the Convention relating to its protocols shall apply to this Protocol.

Article 15
Signature

This Protocol shall be open for signature by States and by regional economic integration organizations in Montreal on 16 September 1987, in Ottawa from 17 September 1987 to 16 January 1988, and at United Nations Headquarters in New York from 17 January 1988 to 15 September 1988.

Article 16
Entry into Force

1. This Protocol shall enter into force on 1 January 1989, provided that at least eleven instruments of ratification, acceptance, approval of the Protocol or accession thereto have been deposited by States or regional economic integration organizations representing at least two-thirds of 1986 estimated global consumption of the controlled substances, and the provisions of paragraph 1 of Article 17 of the Convention have been fulfilled. In the event that these conditions have not been fulfilled by that date, the Protocol shall enter

into force on the ninetieth day following the date on which the conditions have been fulfilled.

2. For the purposes of paragraph 1, any such instrument deposited by a regional economic integration organization shall not be counted as additional to those deposited by member States of such organization.

3. After the entry into force of this Protocol, any State or regional economic integration organization shall become a Party to it on the ninetieth day following the date of deposit of its instrument of ratification, acceptance, approval or accession.

Article 17
Parties Joining after Entry into Force

Subject to Article 5, any State or regional economic integration organization which becomes a Party to this Protocol after the date of its entry into force, shall fulfill forthwith the sum of the obligations under Article 2, as well as under Article 4, that apply at that date to the States and regional economic integration organizations that became Parties on the date the Protocol entered into force.

Article 18
Reservations

No reservations may be made to this Protocol.

Article 19
Withdrawal

For the purposes of this Protocol, the provisions of Article 19 of the Convention relating to withdrawal shall apply, except with respect to Parties referred to in paragraph 1 of Article 5. Any such Party may withdraw from this Protocol by giving written notification to the Depositary at any time after four years of assuming the obligations specified in paragraphs 1 to 4 of Article 2. Any such withdrawal shall take effect upon expiry of one year after the date of its receipt by the Depositary, or on such later date as may be specified in the notification of the withdrawal.

Article 20
Authentic Texts

The original of this Protocol, of which the Arabic, Chinese, English, French, Russian and Spanish texts are equally authentic shall be deposited with the Secretary-General of the United Nations.

In witness whereof the undersigned, being duly authorized to that effect, have signed this Protocol.

Done at Montreal this sixteenth day of September, one thousand nine hundred and eighty-seven.

Annex A

Controlled Substances

Hazardous Wastes
Basel Convention on the Control of Transboundary Movements of Hazardous Wastes and Their Disposal (1989)

Entered into force 5 May, 1992

Preamble

The Parties to this Convention,

Aware of the risk of damage to human health and the environment caused by hazardous wastes and other wastes and the transboundary movement thereof,

Mindful of the growing threat to human health and the environment posed by the increased generation and complexity, and transboundary movement of hazardous wastes and other wastes,

Mindful also that the most effective way of protecting human health and the environment from the dangers posed by such wastes is the reduction of their generation to a minimum in terms of quantity and/or hazard potential,

Convinced that States should take necessary measures to ensure that the management of hazardous wastes and other wastes including their transboundary movement and disposal is consistent with the protection of human health and the environment whatever the place of their disposal,

Noting that States should ensure that the generator should carry out duties with regard to the transport and disposal of hazardous wastes and other wastes in a manner that is consistent with the protection of the environment, whatever the place of disposal,

Fully recognizing that any State has the sovereign right to ban the entry or disposal of foreign hazardous wastes and other wastes in its territory,

Recognizing also the increasing desire for the prohibition of transboundary movements of hazardous wastes and their disposal in other States, especially developing countries,

Convinced that hazardous wastes and other wastes should, as far as is compatible with environmentally sound and efficient management, be disposed of in the State where they were generated,

Aware also that transboundary movements of such wastes from the State of their generation to any other State should be permitted only when conducted under conditions which do not endanger human health and the environment, and under conditions in conformity with the provisions of this Convention,

Considering that enhanced control of transboundary movement of hazardous wastes and other wastes will act as an incentive for their environmentally sound management and for the reduction of the volume of such transboundary movement,

Convinced that States should take measures for the proper exchange of information on and control of the transboundary movement of hazardous wastes and other wastes from and to those States,

Noting that a number of international and regional agreements have addressed the issue of protection and preservation of the environment with regard to the transit of dangerous goods,

Taking into account the Declaration of the United Nations Conference on the Human Environment (Stockholm, 1972), the Cairo Guidelines and Principles for the Environmentally Sound Management of Hazardous Wastes adopted by the Governing Council of the United Nations Environment Programme (UNEP) by decision 14/30 of 17 June 1987, the Recommendations of the United Nations Committee of Experts on the Transport of Dangerous Goods (formulated in 1957 and updated biennially), relevant recommendations, declarations, instruments and regulations adopted within the United Nations system and the work and studies done within other international and regional organizations,

Mindful of the spirit, principles, aims and functions of the World Charter for Nature adopted by the General Assembly of the United Nations at its thirty-seventh session (1982) as the rule of ethics in respect of the protection of the human environment and the conservation of natural resources,

Affirming that States are responsible for the fulfilment of their international obligations concerning the protection of human health and protection and preservation of the environment, and are liable in accordance with international law,

Recognizing that in the case of a material breach of the provisions of this Convention or any protocol thereto the relevant international law of treaties shall apply,

Aware of the need to continue the development and implementation of environmentally sound low-waste technologies, recycling options, good house-keeping and management systems with a view to reducing to a minimum the generation of hazardous wastes and other wastes,

Aware also of the growing international concern about the need for stringent control of transboundary movement of hazardous wastes and other wastes, and of the need as far as possible to reduce such movement to a minimum,

Concerned about the problem of illegal transboundary traffic in hazardous wastes and other wastes,

Taking into account also the limited capabilities of the developing countries to manage hazardous wastes and other wastes,

Recognizing the need to promote the transfer of technology for the sound management of hazardous wastes and other wastes produced locally, particularly to the developing countries in accordance with the spirit of the Cairo Guidelines and decision 14/16 of the Governing Council of UNEP on Promotion of the transfer of environmental protection technology,

Recognizing also that hazardous wastes and other wastes should be transported in accordance with relevant international conventions and recommendations,

Convinced also that the transboundary movement of hazardous wastes and other wastes should be permitted only when the transport and the ultimate disposal of such wastes is environmentally sound, and

Determined to protect, by strict control, human health and the environment against the adverse effects which may result from the generation and management of hazardous wastes and other wastes,

Have Agreed as Follows:

Article 1
Scope of the Convention

1. The following wastes that are subject to transboundary movement shall be "hazardous wastes" for the purposes of this Convention:
   (a) Wastes that belong to any category contained in Annex I, unless they do not possess any of the characteristics contained in Annex III; and
   (b) Wastes that are not covered under paragraph (a) but are defined as, or are considered to be, hazardous wastes by the domestic legislation of the Party of export, import or transit.
2. Wastes that belong to any category contained in Annex II that are subject to transboundary movement shall be "other wastes" for the purposes of this Convention.
3. Wastes which, as a result of being radioactive, are subject to other international control systems, including international instruments, applying specifically to radioactive materials, are excluded from the scope of this Convention.
4. Wastes which derive from the normal operations of a ship, the discharge of which is covered by another international instrument, are excluded from the scope of this Convention.

Article 2
Definitions

For the purposes of this Convention:

1. "Wastes" are substances or objects which are disposed of or are intended to be disposed of or are required to be disposed of by the provisions of national law;
2. "Management" means the collection, transport and disposal of hazardous wastes or other wastes, including after-care of disposal sites;
3. "Transboundary movement" means any movement of hazardous wastes or other wastes from an area under the national jurisdiction of one State to or through an area under the national jurisdiction of another State or to or through an area not under the national jurisdiction of any State, provided at least two States are involved in the movement;
4. "Disposal" means any operation specified in Annex IV to this Convention;
5. "Approved site or facility" means a site or facility for the disposal of hazardous wastes or other wastes which is authorized or permitted to operate for this purpose by a relevant authority of the State where the site or facility is located;
6. "Competent authority" means one governmental authority designated by a Party to be responsible, within such geographical areas as the Party may think tit, for receiving the notification of a transboundary movement of hazardous wastes or other wastes, and any information related to it, and for responding to such a notification, as provided in Article 6;
7. "Focal point" means the entity of a Party referred to in Article 5 responsible for receiving and submitting information as provided for in Articles 13 and 16;
8. "Environmentally sound management of hazardous wastes or other wastes" means taking all practicable steps to ensure that hazardous wastes or other wastes are managed in a manner which will protect human health and the environment against the adverse effects which may result from such wastes;
9. "Area under the national jurisdiction of a State" means any land, marine area or airspace within which a State exercises administrative and regulatory responsibility in accordance with international law in regard to the protection of human health or the environment;
10. "State of export" means a Party from which a transboundary movement of hazardous wastes or other wastes is planned to be initiated or is initiated;
11. "State of import" means a Party to which a transboundary movement of hazardous wastes or other wastes is planned or takes place for the purpose of disposal therein or for the purpose of loading prior to disposal in an area not under the national jurisdiction of any State;
12. "State of transit" means any State, other than the State of export or import, through which a movement of hazardous wastes or other wastes is planned or takes place;
13. "States concerned" means Parties which are States of export or import, or transit States, whether or not Parties;
14. "Person" means any natural or legal person;
15. "Exporter" means any person under the jurisdiction of the State of export who arranges for hazardous wastes or other wastes to be exported;
16. "Importer" means any person under the jurisdiction of the State of import who arranges for hazardous wastes or other wastes to be imported;
17. "Carrier" means any person who carries out the transport of hazardous wastes or other wastes;
18. "Generator" means any person whose activity produces hazardous wastes or other wastes or, if that person is not known, the person who is in possession and/or control of those wastes;

19. "Disposer" means any person to whom hazardous wastes or other wastes are shipped and who carries out the disposal of such wastes;
20. "Political and/or economic integration organization" means an organization constituted by sovereign States to which its member States have transferred competence in respect to matters governed by this Convention and which has been duly authorized, in accordance with its internal procedures, to sign, ratify, accept, approve, formally confirm or accede to it;
21. "Illegal traffic" means any transboundary movement of hazardous wastes or other wastes as specified in Article 9.

Article 3
National Definitions of Hazardous Wastes

1. Each Party shall, within six months of becoming a Party to this Convention, inform the Secretariat of the convention of the wastes, other than those listed in Annexes I and II, considered or defined as hazardous under its national legislation and of any requirements concerning transboundary movement procedures applicable to such wastes.
2. Each Party shall subsequently inform the Secretariat of any significant changes to the information it has provided pursuant to paragraph 1.
3. The Secretariat shall forthwith inform all Parties of the information it has received pursuant to paragraphs 1 and 2.
4. Parties shall be responsible for making the information transmitted to them by the Secretariat under paragraph 3 available to their exporters.

Article 4
General Obligations

1. (a) Parties exercising their right to prohibit the import of hazardous wastes or other wastes for disposal shall inform the other Parties of their decision pursuant to Article 13.
   (b) Parties shall prohibit or shall not permit the export of hazardous wastes and other wastes to the Parties which have prohibited the import of such wastes, when notified pursuant to subparagraph (a) above.
   (c) Parties shall prohibit or shall not permit the export of hazardous wastes and other wastes if the State of import does not consent in writing to the specific import, in the case where that State of import has not prohibited the import of such wastes.
2. Each Party shall take the appropriate measures to:
   (a) Ensure that the generation of hazardous wastes and other wastes within it is reduced to a minimum, taking into account social, technological and economic aspects;
   (b) Ensure the availability of adequate disposal facilities, for the environmentally sound management of hazardous wastes and other wastes, that shall be located, to the extent possible, within it, whatever the place of their disposal;
   (c) Ensure that persons involved in the management of hazardous wastes or other wastes within it take such steps as are necessary to prevent pollution due to hazardous wastes and other wastes arising from such management and, if such pollution occurs, to minimize the consequences thereof for human health and the environment;
   (d) Ensure that the transboundary movement of hazardous wastes and other wastes is reduced to the minimum consistent with the environmentally sound and efficient management of such wastes, and is conducted in a manner which will protect human health and the environment against the adverse effects which may result from such movement;

(e) Not allow the export of hazardous wastes or other wastes to a State or group of States belonging to an economic and/or political integration organization that are Parties, particularly developing countries, which have prohibited by their legislation all imports, or if it has reason to believe that the wastes in question will not be managed in an environmentally sound manner, according to criteria to be decided on by the Parties at their first meeting.

(f) Require that information about a proposed transboundary movement of hazardous wastes and other wastes be provided to the States concerned, according to Annex V A, to state clearly the effects of the proposed movement on human health and the environment;

(g) Prevent the import of hazardous wastes and other wastes if it has reason to believe that the wastes in question will not be managed in an environmentally sound manner;

(h) Cooperate in activities with other Parties and interested organizations, directly and through the Secretariat, including the dissemination of information on the transboundary movement of hazardous wastes and other wastes, in order to improve the environmentally sound management of such wastes and to achieve the prevention of illegal traffic.

3. The Parties consider that illegal traffic in hazardous wastes or other wastes is criminal.

4. Each Party shall take appropriate legal, administrative and other measures to implement and enforce the provisions of this Convention, including measures to prevent and punish conduct in contravention of the Convention.

5. A Party shall not permit hazardous wastes or other wastes to be exported to a non-Party or to be imported from a non-Party.

6. The Parties agree not to allow the export of hazardous wastes or other wastes for disposal within the area south of 60 degrees South latitude, whether or not such wastes are subject to transboundary movement.

7. Furthermore, each Party shall:

(a) Prohibit all persons under its national jurisdiction from transporting or disposing of hazardous wastes or other wastes unless such persons are authorized or allowed to perform such types of operations;

(b) Require that hazardous wastes and other wastes that are to be the subject of a transboundary movement be packaged, labelled, and transported in conformity with generally accepted and recognized international rules and standards in the field of packaging, labelling, and transport, and that due account is taken of relevant internationally recognized practices;

(c) Require that hazardous wastes and other wastes be accompanied by a movement document from the point at which a transboundary movement commences to the point of disposal.

8. Each Party shall require that hazardous wastes or other wastes, to be exported, are managed in an environmentally sound manner in the State of import or elsewhere. Technical guidelines for the environmentally sound management of wastes subject to this Convention shall be decided by the Parties at their first meeting.

9. Parties shall take the appropriate measures to ensure that the transboundary movement of hazardous wastes and other wastes only be allowed if:

(a) The State of export does not have the technical capacity and the necessary facilities, capacity or suitable disposal sites in order to dispose of the wastes in question in an environmentally sound and efficient manner; or

(b) The wastes in question are required as a raw material for recycling or recovery industries in the State of import; or

(c) The transboundary movement in question is in accordance with other criteria to be decided by the Parties, provided those criteria do not differ from the objectives of this Convention.

10. The obligation under this Convention of States in which hazardous wastes and other wastes are generated to require that those wastes are managed in an environmentally sound manner may not under any circumstances be transferred to the States of import or transit.

11. Nothing in this Convention shall prevent a Party from imposing additional requirements that are consistent with the provisions of this Convention, and are in accordance with the rules of international law, in order better to protect human health and the environment.

12. Nothing in this Convention shall affect in any way the sovereignty of States over their territorial sea established in accordance with international law, and the sovereign rights and the jurisdiction which States have in their exclusive economic zones and their continental shelves in accordance with international law, and the exercise by ships and aircraft of all States of navigational rights and freedoms as provided for in international law and as reflected in relevant international instruments.

13. Parties shall undertake to review periodically the possibilities for the reduction of the amount and/or the pollution potential of hazardous wastes and other wastes which are exported to other States, in particular to developing countries.

Article 5
Designation of Competent Authorities and Focal Point

To facilitate the implementation of this Convention, the Parties shall:

1. Designate or establish one or more competent authorities and one focal point. One competent authority shall be designated to receive the notification in case of a State of transit.

2. Inform the Secretariat, within three months of the date of the entry into force of this Convention for them, which agencies they have designated as their focal point and their competent authorities.

3. Inform the Secretariat, within one month of the date of decision, of any changes regarding the designation made by them under paragraph 2 above.

Article 6
Transboundary Movement between Parties

1. The State of export shall notify, or shall require the generator or exporter to notify, in writing, through the channel of the competent authority of the State of export, the competent authority of the States concerned of any proposed transboundary movement of hazardous wastes or other wastes. Such notification shall contain the declarations and information specified in Annex V A, written in a language acceptable to the State of import. Only one notification needs to be sent to each State concerned.

2. The State of import shall respond to the notifier in writing, consenting to the movement with or without conditions, denying permission for the movement, or requesting additional information. A copy of the final response of the State of import shall be sent to the competent authorities of the States concerned which are Parties.

3. The State of export shall not allow the generator or exporter to commence the transboundary movement until it has received written confirmation that:
   (a) The notifier has received the written consent of the State of import; and

    (b) The notifier has received from the State of import confirmation of the existence of a contract between the exporter and the disposer specifying environmentally sound management of the wastes in question.

4. Each State of transit which is a Party shall promptly acknowledge to the notifier receipt of the notification. It may subsequently respond to the notifier in writing, within 60 days, consenting to the movement with or without conditions, denying permission for the movement, or requesting additional information. The State of export shall not allow the transboundary movement to commence until it has received the written consent of the State of transit. However, if at any time a Party decides not to require prior written consent, either generally or under specific conditions, for transit transboundary movements of hazardous wastes or other wastes, or modifies its requirements in this respect, it shall forthwith inform the other Parties of its decision pursuant to Article 13. In this latter case, if no response is received by the State of export within 60 days of the receipt of a given notification by the State of transit, the State of export may allow the export to proceed through the State of transit.

5. In the case of a transboundary movement of wastes where the wastes are legally defined as or considered to be hazardous wastes only:

    (a) By the State of export, the requirements of paragraph 9 of this Article that apply to the importer or disposer and the State of import shall apply mutatis mutandis to the exporter and State of export, respectively;

    (b) By the State of import, or by the States of import and transit which are Parties, the requirements of paragraphs 1, 3, 4 and 6 of this Article that apply to the exporter and State of export shall apply mutatis mutandis to the importer or disposer and State of import, respectively; or

    (c) By any State of transit which is a Party, the provisions of paragraph 4 shall apply to such State.

6. The State of export may, subject to the written consent of the States concerned, allow the generator or the exporter to use a general notification where hazardous wastes or other wastes having the same physical and chemical characteristics are shipped regularly to the same disposer via the same customs office of exit of the State of export via the same customs office of entry of the State of import, and, in the case of transit, via the same customs office of entry and exit of the State or States of transit.

7. The States concerned may make their written consent to the use of the general notification referred to in paragraph 6 subject to the supply of certain information, such as the exact quantities or periodical lists of hazardous wastes or other wastes to be shipped.

8. The general notification and written consent referred to in paragraphs 6 and 7 may cover multiple shipments of hazardous wastes or other wastes during a maximum period of 12 months.

9. The Parties shall require that each person who takes charge of a transboundary movement of hazardous wastes or other wastes sign the movement document either upon delivery or receipt of the wastes in question. They shall also require that the disposer inform both the exporter and the competent authority of the State of export of receipt by the disposer of the wastes in question and, in due course, of the completion of disposal as specified in the notification. If no such information is received within the State of export, the competent authority of the State of export or the exporter shall so notify the State of import.

10. The notification and response required by this Article shall be transmitted to the competent authority of the Parties concerned or to such governmental authority as may be appropriate in the case of non-Parties.

11. Any transboundary movement of hazardous wastes or other wastes shall be covered by insurance, bond or other guarantee as may be required by the State of import or any State of transit which is a Party.

Article 7
Transboundary Movement from a Party through States which are not Parties

Paragraph 2 of Article 6 of the Convention shall apply mutatis mutandis to transboundary movement of hazardous wastes or other wastes from a Party through a State or States which are not Parties.

Article 8
Duty to Reimport

When a transboundary movement of hazardous wastes or other wastes to which the consent of the States concerned has been given, subject to the provisions of this Convention, cannot be completed in accordance with the terms of the contract, the State of export shall ensure that the wastes in question are taken back into the State of export, by the exporter, if alternative arrangements cannot be made for their disposal in an environmentally sound manner, within 90 days from the time that the importing State informed the State of export and the Secretariat, or such other period of time as the States concerned agree. To this end, the State of export and any Party of transit shall not oppose, hinder or prevent the return of those wastes to the State of export.

Article 9
Illegal Traffic

1. For the purpose of this Convention, any transboundary movement of hazardous wastes or other wastes:
   (a) without notification pursuant to the provisions of this Convention to all States concerned; or
   (b) without the consent pursuant to the provisions of this Convention of a State concerned; or
   (c) with consent obtained from States concerned through falsification, misrepresentation or fraud; or
   (d) that does not conform in a material way with the documents; or
   (e) that results in deliberate disposal (e.g., dumping) of hazardous wastes or other wastes in contravention of this Convention and of general principles of international law, shall be deemed to be illegal traffic.

2. In case of a transboundary movement of hazardous wastes or other wastes deemed to be illegal traffic as the result of conduct on the part of the exporter or generator, the State of export shall ensure that the wastes in question are:
   (a) taken back by the exporter or the generator or, if necessary, by itself into the State of export, or, if impracticable,
   (b) are otherwise disposed of in accordance with the provisions of this Convention, within 30 days from the time the State of export has been informed about the illegal traffic or such other period of time as States concerned may agree. To this end the Parties concerned shall not oppose, hinder or prevent the return of those wastes to the State of export.

3. In the case of a transboundary movement of hazardous wastes or other wastes deemed to be illegal traffic as the result of conduct on the part of the importer or disposer, the State of import shall ensure that the wastes in question are disposed of in an environmentally sound manner by the importer or disposer or, if necessary, by itself within 30 days from the time the illegal traffic has come to the attention of the State of import or such other period of time as the States concerned may agree. To this end, the Parties concerned shall cooperate, as necessary, in the disposal of the wastes in an environmentally sound manner.

4. In cases where the responsibility for the illegal traffic cannot be assigned either to the exporter or generator or to the importer or disposer, the Parties concerned or other Parties, as appropriate, shall ensure, through co-operation, that the wastes in question are disposed of as soon as possible in an environmentally sound manner either in the State of export or the State of import or elsewhere as appropriate.

5. Each Party shall introduce appropriate national/domestic legislation to prevent and punish illegal traffic. The Parties shall cooperate with a view to achieving the objects of this Article.

Article 10
International Co-operation

1. The Parties shall cooperate with each other in order to improve and achieve environmentally sound management of hazardous wastes and other wastes.

2. To this end, the Parties shall:
   (a) Upon request, make available information, whether on a bilateral or multilateral basis, with a view to promoting the environmentally sound management of hazardous wastes and other wastes, including harmonization of technical standards and practices for the adequate management of hazardous wastes and other wastes;
   (b) Cooperate in monitoring the effects of the management of hazardous wastes on human health and the environment;
   (c) Cooperate, subject to their national laws, regulations and policies, in the development and implementation of new environmentally sound low-waste technologies and the improvement of existing technologies with a view to eliminating, as far as practicable, the generation of hazardous wastes and other wastes and achieving more effective and efficient methods of ensuring their management in an environmentally sound manner, including the study of the economic, social and environmental effects of the adoption of such new or improved technologies;
   (d) Cooperate actively, subject to their national laws, regulations and policies, in the transfer of technology and management systems related to the environmentally sound management of hazardous wastes and other wastes. They shall also cooperate in developing the technical capacity among Parties, especially those which may need and request technical assistance in this field;
   (e) Cooperate in developing appropriate technical guidelines and/or codes of practice.

3. The Parties shall employ appropriate means to cooperate in order to assist developing countries in the implementation of subparagraphs a, b, c and d of paragraph 2 of Article 4.

4. Taking into account the needs of developing countries, co-operation between Parties and the competent international organizations is encouraged to promote, inter alia, public awareness, the development of sound management of hazardous wastes and other wastes and the adoption of new low-waste technologies.

Article 11
Bilateral, Multilateral and Regional Agreements

1. Notwithstanding the provisions of Article 4 paragraph 5, Parties may enter into bilateral, multilateral, or regional agreements or arrangements regarding transboundary movement of hazardous wastes or other wastes with Parties or non-Parties provided that such agreements or arrangements do not derogate from the environmentally sound management of hazardous wastes and other wastes as required by this Convention. These agreements or arrangements shall stipulate provisions which are not less environmentally sound than those provided for by this Convention in particular taking into account the interests of developing countries.

2. Parties shall notify the Secretariat of any bilateral, multilateral or regional agreements or arrangements referred to in paragraph 1 and those which they have entered into prior to the entry into force of this Convention for them, for the purpose of controlling transboundary movements of hazardous wastes and other wastes which take place entirely among the Parties to such agreements. The provisions of this Convention shall not affect transboundary movements which take place pursuant to such agreements provided that such agreements are compatible with the environmentally sound management of hazardous wastes and other wastes as required by this Convention.

Article 12
Consultations on Liability

The Parties shall cooperate with a view to adopting, as soon as practicable, a protocol setting out appropriate rules and procedures in the field of liability and compensation for damage resulting from the transboundary movement and disposal of hazardous wastes and other wastes.

Article 13
Transmission of Information

1. The Parties shall, whenever it comes to their knowledge, ensure that, in the case of an accident occurring during the transboundary movement of hazardous wastes or other wastes or their disposal, which are likely to present risks to human health and the environment in other States, those states are immediately informed.

2. The Parties shall inform each other, through the Secretariat, of:
   (a) Changes regarding the designation of competent authorities and/or focal points, pursuant to Article 5;
   (b) Changes in their national definition of hazardous wastes, pursuant to Article 3; and, as soon as possible,
   (c) Decisions made by them not to consent totally or partially to the import of hazardous wastes or other wastes for disposal within the area under their national jurisdiction;
   (d) Decisions taken by them to limit or ban the export of hazardous wastes or other wastes;
   (e) Any other information required pursuant to paragraph 4 of this Article.

3. The Parties, consistent with national laws and regulations, shall transmit, through the Secretariat, to the Conference of the Parties established under Article 15, before the end of each calendar year, a report on the previous calendar year, containing the following information:
   (a) Competent authorities and focal points that have been designated by them pursuant to Article 5;

  (b) Information regarding transboundary movements of hazardous wastes or other wastes in which they have been involved, including:
  - (i) The amount of hazardous wastes and other wastes exported, their category, characteristics, destination, any transit country and disposal method as stated on the response to notification;
  - (ii) The amount of hazardous wastes and other wastes imported, their category, characteristics, origin, and disposal methods;
  - (iii) Disposals which did not proceed as intended;
  - (iv) Efforts to achieve a reduction of the amount of hazardous wastes or other wastes subject to transboundary movement;
  (c) Information on the measures adopted by them in implementation of this Convention;
  (d) Information on available qualified statistics which have been compiled by them on the effects on human health and the environment of the generation, transportation and disposal of hazardous wastes or other wastes;
  (e) Information concerning bilateral, multilateral and regional agreements and arrangements entered into pursuant to Article 11 of this Convention;
  (f) Information on accidents occurring during the transboundary movement and disposal of hazardous wastes and other wastes and on the measures undertaken to deal with them;
  (g) Information on disposal options operated within the area of the international jurisdiction;
  (h) Information on measures undertaken for development of technologies for the reduction and/or elimination of production of hazardous wastes and other wastes; and
  (i) Such other matters as the Conference of the Parties shall deem relevant.
4. The Parties, consistent with national laws and regulations, shall ensure that copies of each notification concerning any given transboundary movement of hazardous wastes or other wastes, and the response to it, are sent to the Secretariat when a Party considers that its environment may be affected by that transboundary movement has requested that this should be done.

Article 14
Financial Aspects

1. The Parties agree that, according to the specific needs of different regions and subregions, regional or subregional centres for training and technology transfers regarding the management of hazardous wastes and other wastes and the minimization of their generation should be established. The Parties shall decide on the establishment of appropriate funding mechanisms of a voluntary nature.
2. The Parties shall consider the establishment of a revolving fund to assist on an interim basis in case of emergency situations to minimize damage from accidents arising from transboundary movements of hazardous wastes and other wastes or during the disposal of those wastes.

Article 15
Conference of the Parties

1. A Conference of the Parties is hereby established. The first meeting of the Conference of the Parties shall be convened by the Executive Director of UNEP not later than one year after the entry into force of this Convention. Thereafter, ordinary meetings of the

Conference of the Parties shall be held at regular intervals to be determined by the Conference at its first meeting.

2. Extraordinary meetings of the Conference of the Parties shall be held at such other times as may be deemed necessary by the Conference, or at the written request of any Party, provided that, within six months of the request being communicated to them by the Secretariat, it is supported by at least one third of the Parties.

3. The Conference of the Parties shall by consensus agree upon and adopt rules of procedure for itself and for any subsidiary body it may establish, as well as financial rules to determine in particular the financial participation of the Parties under this Convention.

4. The Parties at their first meeting shall consider any additional measures needed to assist them in fulfilling their responsibilities with respect to the protection and the preservation of the marine environment in the context of this Convention.

5. The Conference of the Parties shall keep under continuous review and evaluation the effective implementation of this Convention, and, in addition, shall:
   (a) Promote the harmonization of appropriate policies, strategies and measures for minimizing harm to human health and the environment by hazardous wastes and other wastes;
   (b) Consider and adopt, as required, amendments to this Convention and its annexes, taking into consideration, inter alia, available scientific, technical, economic and environmental information;
   (c) Consider and undertake any additional action that may be required for the achievement of the purposes of this Convention in the light of experience gained in its operation and in the operation of the agreements and arrangements envisaged in Article 11;
   (d) Consider and adopt protocols as required; and
   (e) Establish such subsidiary bodies as are deemed necessary for the implementation of this Convention.

6. The United Nations, its specialized agencies, as well as any State not party to this Convention, may be represented as observers at meetings of the Conference of the Parties. Any other body or agency, whether national or international, governmental or non-governmental, qualified in fields relating to hazardous wastes or other wastes which has informed the Secretariat of its wish to be represented as an observer at a meeting of the Conference of the Parties, may be admitted unless at least one third of the Parties present object. The admission and participation of observers shall be subject to the rules of procedure adopted by the conference of the Parties.

7. The Conference of the Parties shall undertake three years after the entry into force of this Convention, and at least every six years thereafter, an evaluation of its effectiveness and, if deemed necessary, to consider the adoption of a complete or partial ban of transboundary movements of hazardous wastes and other wastes in light of the latest scientific, environmental, technical and economic information.

Article 16
Secretariat

1. The functions of the Secretariat shall be:
   (a) To arrange for and service meetings provided for in Articles 15 and 17;
   (b) To prepare and transmit reports based upon information received in accordance with Articles 3, 4, 6, 11 and 13 as well as upon information derived from meetings of subsidiary bodies established under Article 15 as well as upon, as appropriate, information provided by relevant intergovernmental and non-governmental entities;

(c) To prepare reports on its activities carried out in implementation of its functions under this Convention and present them to the Conference of the Parties;

(d) To ensure the necessary coordination with relevant international bodies, and in particular to enter into such administrative and contractual arrangements as may be required for the effective discharge of its functions;

(e) To communicate with focal points and competent authorities established by the Parties in accordance with Article 5 of this Convention;

(f) To compile information concerning authorized national sites and facilities of Parties available for the disposal of their hazardous wastes and other wastes and to circulate this information among Parties;

(g) To receive and convey information from and to Parties on;
   • sources of technical assistance and training;
   • available technical and scientific know-how;
   • sources of advice and expertise; and
   • availability of resources
   with a view to assisting them, upon request, in such areas as:
   • the handling of the notification system of this Convention;
   • the management of hazardous wastes and other wastes;
   • environmentally sound technologies relating to hazardous wastes and
   • other wastes, such as low- and non-waste technology;
   • the assessment of disposal capabilities and sites;
   • the monitoring of hazardous wastes and other wastes; and
   • emergency responses;

(h) To provide Parties, upon request, with information on consultants or consulting firms having the necessary technical competence in the field, which can assist them to examine a notification for a transboundary movement, the concurrence of a shipment of hazardous wastes or other wastes with the relevant notification, and/or the fact that the proposed disposal facilities for hazardous wastes or other wastes are environmentally sound, when they have reason to believe that the wastes in question will not be managed in an environmentally sound manner. Any such examination would not be at the expense of the Secretariat;

(i) To assist Parties upon request in their identification of cases of illegal traffic and to circulate immediately to the Parties concerned any information it has received regarding illegal traffic;

(j) To cooperate with Parties and with relevant and competent international organizations and agencies in the provision of experts and equipment for the purpose of rapid assistance to States in the event of an emergency situation; and

(k) To perform such other functions relevant to the purposes of this Convention as may be determined by the Conference of the Parties.

2. The secretariat functions will be carried out on an interim basis by UNEP until the completion of the first meeting of the Conference of the Parties held pursuant to Article 15.

3. At its first meeting, the Conference of the Parties shall designate the Secretariat from among those existing competent intergovernmental organizations which have signified their willingness to carry out the secretariat functions under this Convention. At this meeting, the Conference of the Parties shall also evaluate the implementation by the interim Secretariat of the functions assigned to it, in particular under paragraph 1 above, and decide upon the structures appropriate for those functions.

Article 17
Amendment of the Convention

1. Any Party may propose amendments to this Convention and any Party to a protocol may propose amendments to that protocol. Such amendments shall take due account, inter alia, of relevant scientific and technical considerations.

2. Amendments to this Convention shall be adopted at a meeting of the Conference of the Parties. Amendments to any protocol shall be adopted at a meeting of the Parties to the protocol in question. The text of any proposed amendment to this Convention or to any protocol, except as may otherwise be provided in such protocol, shall be communicated to the Parties by the Secretariat at least six months before the meeting at which it is proposed for adoption. The Secretariat shall also communicate proposed amendments to the Signatories to this Convention for information.

3. The Parties shall make every effort to reach agreement on any proposed amendment to this Convention by consensus. If all efforts at consensus have been exhausted, and no agreement reached, the amendment shall as a last resort be adopted by a three-fourths majority vote of the Parties present and voting at the meeting, and shall be submitted by the Depositary to all Parties for ratification, approval, formal confirmation or acceptance.

4. The procedure mentioned in paragraph 3 above shall apply to amendments to any protocol, except that a two-thirds majority of the Parties to that protocol present and voting at the meeting shall suffice for their adoption.

5. Instruments of ratification, approval, formal confirmation or acceptance of amendments shall be deposited with the Depositary. Amendments adopted in accordance with paragraphs 3 or 4 above shall enter into force between Parties having accepted them on the ninetieth day after the receipt by the Depositary of their instrument of ratification, approval, formal confirmation or acceptance by at least three-fourths of the Parties who accepted the amendments to the protocol concerned, except as may otherwise be provided in such protocol. The amendments shall enter into force for any other Party on the ninetieth day after that Party deposits its instrument of ratification, approval, formal confirmation or acceptance of the amendments.

6. For the purpose of this Article, "Parties present and voting" means Parties present and casting an affirmative or negative vote.

Article 18
Adoption and Amendment of Annexes

1. The annexes to this Convention or to any protocol shall form an integral part of this Convention or of such protocol, as the case may be and, unless expressly provided otherwise, a reference to this Convention or its protocols constitutes at the same time a reference to any annexes thereto. Such annexes shall be restricted to scientific, technical and administrative matters.

2. Except as may be otherwise provided in any protocol with respect to its annexes, the following procedure shall apply to the proposal, adoption and entry into force of additional annexes to this Convention or of annexes to a protocol:

    (a) Annexes to this Convention and its protocols shall be proposed and adopted according to the procedure laid down in Article 17, paragraphs 2, 3 and 4;

    (b) Any Party that is unable to accept an additional annex to this Convention or an annex to any protocol to which it is party shall so notify the Depositary, in writing, within six months from the date of the communication of the adoption by the Depositary. The Depositary shall without delay notify all Parties of any such notification received. A Party may at any time substitute an acceptance for a previous declaration of objection and the annexes shall thereupon enter into force for that Party;

(c) On the expiry of six months from the date of the circulation of the communication by the Depositary, the annex shall become effective for all Parties to this Convention or to any protocol concerned, which have not submitted a notification in accordance with the provision of subparagraph (b) above.

3. The proposal, adoption and entry into force of amendments to annexes to this Convention or to any protocol shall be subject to the same procedure as for the proposal, adoption and entry into force of annexes to the Convention or annexes to a protocol. Annexes and amendments thereto shall take due account, inter alia, of relevant scientific and technical considerations.

4. If an additional annex or an amendment to an annex involves an amendment to this Convention or to any protocol, the additional annex or amended annex shall not enter into force until such time as the amendment to this Convention or to the protocol enters into force.

Article 19
Verification

Any Party which has reason to believe that another Party is acting or has acted in breach of its obligations under this Convention may inform the Secretariat thereof, and in such an event, shall simultaneously and immediately inform, directly or through the Secretariat, the Party against whom the allegations are made. All relevant information should be submitted by the Secretariat to the Parties.

Article 20
Settlement of Disputes

1. In case of a dispute between Parties as to the interpretation or application of, or compliance with, this Convention or any protocol thereto, they shall seek a settlement of the dispute through negotiation or any other peaceful means of their own choice.

2. If the Parties concerned cannot settle their dispute through the means mentioned in the preceding paragraph, the dispute, if the parties to the dispute agree, shall be submitted to the International Court of Justice or to arbitration under the conditions set out in Annex VI on Arbitration. However, failure to reach common agreement on submission of the dispute to the International Court of Justice or to arbitration shall not absolve the Parties from the responsibility of continuing to seek to resolve it by the means referred to in paragraph 1.

3. When ratifying, accepting, approving, formally confirming or acceding to this Convention, or at any time thereafter, a State or political and/or economic integration organization may declare that it recognizes as compulsory ipso facto and without special agreement, in relation to any Party accepting the same obligation:

(a) submission of the dispute to the International Court of Justice; and/or

(b) arbitration in accordance with the procedures set out in Annex VI. Such declaration shall be notified in writing to the Secretariat which shall communicate it to the Parties.

Article 21
Signature

This Convention shall be open for signature by States, by Namibia represented by the United Nations Council for Namibia and by political and/or economic integration organizations, in Basel on 22 March 1989, at the Federal Department of Foreign Affairs of Switzerland in Berne from 23 March 1989 to 30 June 1989, and at United Nations Headquarters in New York from 1 July 1989 to 22 March 1990.

Article 22
Ratification, Acceptance, Formal Confirmation or Approval

1. This Convention shall be subject to ratification, acceptance or approval by States and by Namibia, represented by the United Nations Council for Namibia and to formal confirmation or approval by political and/or economic integration organizations. Instruments of ratification, acceptance, formal confirmation, or approval shall be deposited with the Depositary.
2. Any organization referred to in paragraph 1 above which becomes a Party to this Convention without any of its member States being a Party shall be bound by all the obligations under the Convention. In the case of such organizations, one or more of whose member States is a Party to the Convention, the organization and its member States shall decide on their respective responsibilities for the performance of their obligations under the Convention. In such cases, the organization and the member States shall not be entitled to exercise rights under the Convention concurrently.
3. In their instruments of formal confirmation or approval, the organizations referred to in paragraph 1 above shall declare the extent of their competence with respect to the matters governed by the Convention. These organizations shall also inform the Depositary, who will inform the Parties of any substantial modification in the extent of their competence.

Article 23
Accession

1. This Convention shall be open for accession by States, by Namibia, represented by the United Nations Council for Namibia, and by political and/or economic integration organizations from the day after the date on which the Convention is closed for signature. The instruments of accession shall be deposited with the Depositary.
2. In their instruments of accession, the organizations referred to in paragraph 1 above shall declare the extent of their competence with respect to the matters governed by the Convention. These organizations shall also inform the Depositary of any substantial modification in the extent of their competence.
3. The provisions of Article 22 paragraph 2, shall apply to political and/or economic integration organizations which accede to this Convention.

Article 24
Right to Vote

1. Except as provided for in paragraph 2 below, each Contracting Party to this Convention shall have one vote.
2. Political and/or economic integration organizations, in matters within their competence, in accordance with Article 22, paragraph 3, and Article 23, paragraph 2, shall exercise their right to vote with a number of votes equal to the number of their member States

which are Parties to the Convention or the relevant protocol. Such organizations shall not exercise their right to vote if their member States exercise theirs, and vice versa.

Article 25
Entry into Force

1. This Convention shall enter into force on the ninetieth day after the date of deposit of the twentieth instrument of ratification, acceptance, formal confirmation, approval or accession.
2. For each State or political and/or economic integration organization which ratifies, accepts, approves or formally confirms this Convention or accedes thereto after the date of the deposit of the twentieth instrument of ratification, acceptance, approval, formal confirmation or accession, it shall enter into force on the ninetieth day after the date of deposit by such State or political and/or economic integration organization of its instrument of ratification, acceptance, approval, formal confirmation or accession.
3. For the purposes of paragraphs 1 and 2 above, any instrument deposited by a political and/or economic integration organization shall not be counted as additional to those deposited by member States of such organization.

Article 26
Reservations and Declarations

1. No reservation or exception may be made to this Convention.
2. Paragraph 1 of this Article does not preclude a State or political and/or economic integration organizations, when signing, ratifying, accepting, approving, formally confirming or acceding to this Convention, from making declarations or statements, however phrased or named, with a view, inter alia, to the harmonization of its laws and regulations with the provisions of this Convention, provided that such declarations or statements do not purport to exclude or to modify the legal effects of the provisions of the Convention in their application to that State.

Article 27
Withdrawal

1. At any time after three years from the date on which this Convention has entered into force for a Party, that Party may withdraw from the Convention by giving written notification to the Depositary.
2. Withdrawal shall be effective one year from receipt of notification by the Depositary, or on such later date as may be specified in the notification.

Article 28
Depository

The Secretary-General of the United Nations shall be the Depository of this Convention and of any protocol thereto.

Article 29
Authentic texts

The original Arabic, Chinese, English, French, Russian and Spanish texts of this Convention are equally authentic.

IN WITNESS WHEREOF the undersigned, being duly authorized to that effect, have signed this Convention.

Done at on the day of 1989

Annex I
Categories of Wastes Controlled

Waste Streams

Y1   Clinical wastes from medical care in hospitals, medical centers and clinics
Y2   Wastes from the production and preparation of pharmaceutical products
Y3   Waste pharmaceuticals, drugs and medicines
Y4   Wastes from the production, formulation and use of biocides and phytopharmaceuticals
Y5   Wastes from the manufacture, formulation and use of wood preserving chemicals
Y6   Wastes from the production, formulation and use of organic solvents
Y7   Wastes from heat treatment and tempering operations containing cyanides
Y8   Waste mineral oils unfit for their originally intended use
Y9   Waste oils/water, hydrocarbons/water mixtures, emulsions
Y10   Waste substances and articles containing or contaminated with polychlorinated biphenyls (PCBs) and/or polychlorinated terphenyls (PCTs) and/or polybrominated biphenyls (PBBs)
Y11   Waste tarry residues arising from refining, distillation and any pyrolytic treatment
Y12   Wastes from production, formulation and use of inks, dyes, pigments, paints, lacquers, varnish
Y13   Wastes from production, formulation and use of resins, latex, plasticizers, glues/adhesives
Y14   Waste chemical substances arising from research and development or teaching activities which are not identified and/or are new and whose effects on man and/or the environment are not known
Y15   Wastes of an explosive nature not subject to other legislation
Y16   Wastes from production, formulation and use of photographic chemicals and processing materials
Y17   Wastes resulting from surface treatment of metals and plastics
Y18   Residues arising from industrial waste disposal operations

Wastes having as constituents

Y19   Metal carbonyls
Y20   Beryllium; beryllium compounds
Y21   Hexavalent chromium compounds
Y22   Copper compounds
Y23   Zinc compounds
Y24   Arsenic; arsenic compounds
Y25   Selenium, selenium compounds
Y26   Cadmium; cadmium compounds
Y27   Antimony; antimony compounds
Y28   Tellurium; tellurium compounds
Y29   Mercury; mercury compounds
Y30   Thallium; thallium compounds
Y31   Lead, lead compounds
Y32   Inorganic fluorine compounds excluding calcium fluoride
Y33   Inorganic cyanides

Y34   Acidic solutions or acids in solid form
Y35   Basic solutions or bases in solid form
Y36   Asbestos (dust and fibres)
Y37   Organic phosphorous compounds
Y38   Organic cyanides
Y39   Phenols; phenol compounds including chlorophenols
Y40   Ethers
Y41   Halogenated organic solvents
Y42   Organic solvents excluding halogenated solvents
Y43   Any congenor of polychlorinated dibenzo-furan
Y44   Any congenor of polychlorinated dibenzo-p-dioxin
Y45   Organohalogen compounds other than substances referred to in this Annex (e.g., Y39, Y41, Y42, Y43, Y44).

Annex II
Categories of Wastes Requiring Special Consideration

Y46   Wastes collected from households
Y47   Residues arising from the incineration of household

Annex III
List of Hazardous Characteristics

UN Class*	Code	Characteristics
1	H1	Explosive
		An explosive substance or waste is a solid or liquid substance or waste (or mixture of substances or wastes) which is in itself capable by chemical reaction of producing gas at such a temperature and pressure and at such a speed as to cause damage to the surroundings.
3	H3	Flammable liquids
		The word "flammable" has the same meaning as "inflammable". Flammable liquids are liquids, or mixtures of liquids, or liquids containing solids in solution or suspension (for example, paints, varnishes, lacquers, etc., but not including substances or wastes otherwise classified on account of their dangerous characteristics) which give off a flammable vapour at temperatures of not more than 60.5 deg. C, closed-cup test, or not more than 65.6 deg C, open-cup test. (Since the results of open-cup tests and of closed-cup tests are not strictly comparable and even individual results by the same test are often variable, regulations varying from the above figures to make allowance for such differences would be within the spirit of this definition.)

*Corresponds to the hazard classification system included in the United Nations Recommendations on the Transport of Dangerous Goods (ST/SG/ AC.10/1/Rev.5, United Nations, New York, 1988).

UN Class*	Code	Characteristics
4.1	H4.1	Flammable solids Solids, or waste solids, other than those classed as explosives, which under conditions encountered in transport are readily combustible, or may cause or contribute to fire through friction.
4.2	H4.2	Substances or wastes liable to spontaneous combustion. Substances or wastes which are liable to spontaneous heating under normal conditions encountered in transport, or to heating up on contact with air, and then being liable to catch fire.
1.3	H4.2	Substances or wastes which, in contact with water emit flammable gases. Substances or wastes which, by interaction with water, are liable to become spontaneously flammable or to give off flammable gases in dangerous quantities.
5.1	H5.1	Oxidizing Substances or wastes which, while in themselves not necessarily combustible, may, generally by yielding oxygen cause, or contribute to, the combustion of other materials.
5.2	H5.2	Organic Peroxides Organic substances or wastes which contain the bivalent-o-o-structure are thermally unstable substances which may undergo exothermic self-accelerating decomposition.
6.1	H6.1	Poisonous (acute) Substances or wastes liable either to cause death or serious injury or to harm human health if swallowed or inhaled or by skin contact.
6.2	H6.2	Infectious substances Substances or wastes containing viable micro organisms or their toxins which are known or suspected to cause disease in animals or humans.
8	H8	Corrosives Substances or wastes which, by chemical action, will cause severe damage when in contact with living tissue, or, in the case of leakage, will materially damage, or even destroy, other goods or the means of transport; they may also cause other hazards.
9	H10	Liberation of toxic gases in contact with air or water Substances or wastes which, by interaction with air or water, are liable to give off toxic gases in dangerous quantities.
9	H11	Toxic (delayed or chronic) Substances or wastes which, if they are inhaled or ingested or if they penetrate the skin, may involve delayed or chronic effects, including carcinogenicity.
9	H12	Ecotoxic Substances or wastes which if released present or may present immediate or delayed adverse impacts to the environment by means of bioaccumulation and/or toxic effects upon biotic systems.
9	H13	Capable, by any means, after disposal, of yielding another material, e.g., leachate, which possesses any of the characteristics listed above.

Tests

The potential hazards posed by certain types of wastes are not yet fully documented; tests to quantitatively define these hazards do not exist. Further research is necessary in order to develop means to characterize potential hazards posed to man and/or the environment by these wastes. Standardized tests have been derived with respect to pure substances and materials. Many countries have developed national tests which can be applied to materials listed in Annex 1, in order to decide if these materials exhibit any of the characteristics listed in this Annex.

Annex IV
Disposal Operations

A. Operations which do not lead to the possibility of resource recovery, recycling, reclamation, direct reuse or alternative uses.

Section A encompasses all such disposal operations which occur in practice.

D1   Deposit into or onto land, (e.g., landfill, etc.)

D2   Land treatment, (e.g., biodegradation of liquid or sludgy discards in soils, etc.)

D3   Deep injection, (e.g., injection of pumpable discards into wells, salt domes or naturally occurring repositories, etc.)

D4   Surface impoundment, (e.g., placement of liquid or sludge discards into pits, ponds or lagoons, etc.)

D5   Specially engineered landfill, (e.g., placement into lined discrete cells which are capped and isolated from one another and the environment, etc.)

D6   Release into a water body except seas/oceans

D7   Release into seas/oceans including sea-bed insertion

D8   Biological treatment not specified elsewhere in this Annex which results in final compounds or mixtures which are discarded by means of any of the operations in Section A

D9   Physico chemical treatment not specified elsewhere in this Annex which results in final compounds or mixtures which are discarded by means of any of the operations in Section A, (e.g., evaporation, drying, calcination, neutralisation, precipitation, etc.)

D10   Incineration on land

D11   Incineration at sea

D12   Permanent storage (e.g., emplacement of containers in a mine, etc.)

D13   Blending or mixing prior to submission to any of the operations in Section A

D14   Repackaging prior to submission to any of the operations in Section A

D15   Storage pending any of the operations in Section A

B. Operations which may lead to resource recovery, recycling, reclamation, direct reuse or alternative uses.

Section B encompasses all such operations with respect to materials legally defined as or considered to be hazardous wastes and which otherwise would have been destined for operations included in Section A.

R1   Use as a fuel (other than in direct incineration) or other means to generate energy

R2   Solvent reclamation/regeneration

R3   Recycling/reclamation of organic substances which are not used as solvents

R4   Recycling/reclamation of metals and metal compounds

R5   Recycling/reclamation of other inorganic materials

R6   Regeneration of acids or bases

R7   Recovery of components used for pollution abatement

R8   Recovery of components from catalysts

R9   Used oil re-refining or other reuses of previously used oil

R10   Land treatment resulting in benefit to agriculture or ecological improvement

R11   Uses of residual materials obtained from any of the operations numbered R1–R10

R12   Exchange of wastes for submission to any of the operations numbered R1–R11

R13   Accumulation of material intended for any operation in Section B

Annex V A
Information to be Provided on Notification

1. Reason for waste export
2. Exporter of the waste/1
3. Generator(s) of the waste and site of generation/1
4. Disposer of the waste and actual site of disposal/1
5. Intended carrier(s) of the waste or their agents, if known/1
6. Country of export of the waste Competent authority/2
7. Expected countries of transit Competent authority/2
8. Country of import of the waste Competent authority/2
9. General or single notification
10. Projected date(s) of shipment(s) and period of time over which waste is to be exported and proposed itinerary (including point of entry and exit)/3
11. Means of transport envisaged (road, rail, sea, air, inland waters)
12. Information relating to insurance/4
13. Designation and physical description of the waste including Y number and UN number and its compositions/5 and information on any special handling requirements including emergency provisions in case of accidents
14. Type of packaging envisaged (e.g., bulk, drummed, tanker)
15. Estimated quantity in weight/volume/6
16. Process by which the waste is generated/7
17. For wastes listed in Annex III, classifications from Annex II: hazardous characteristic, H number, and UN class.
18. Method of disposal as per Annex IV
19. Declaration by the generator and exporter that the information is correct
20. Information transmitted (including technical description of the plant) to the exporter or generator from the disposer of the waste upon which the latter has based his assessment that there was no reason to believe that the wastes will not be managed in an environmentally sound manner in accordance with the laws and regulations of the country of import.
21. Information concerning the contract between the exporter and disposer.

Notes

1/ Full name and address, telephone, telex or telefax number and the name, address, telephone, telex or telefax number of the person to be contacted.

2/ Full name and address, telephone, telex or telefax number.

3/ In the case of a general notification covering several shipments, either the expected dates of each shipment or, if this is not known, the expected frequency of the shipments will be required.

4/ Information is to be provided on relevant insurance requirements and how they are met by exporter, carrier and disposer.

5/ The nature and the concentration of the most hazardous components, in terms of toxicity and other dangers presented by the waste both in handling and in relation to the proposed disposal method.

6/ In the case of a general notification covering several shipments, both the estimated total quantity and the estimated quantities for each individual shipment will be required.

7/ Insofar as this is necessary to assess the hazard and determine the appropriateness of the proposed disposal operation.

Annex V B
Information to be Provided on the Movement Document

1. Exporter of the waste/1
2. Generator(s) of the waste and site of generation/1
3. Disposer of the waste and actual site of disposal/1
4. Carrier(s) of the waste/1 or his agent(s)
5. Subject of general or single notification
6. The date the transboundary movement started and date(s) and signature on receipt by each person who takes charge of the waste
7. Means of transport (road, rail, inland waterway, sea, air) including countries of export, transit and import, also point of entry and exit where these have been designated
8. General description of the waste (physical state, proper UN shipping name and class, UN number, Y number and H number as applicable)
9. Information on special handling requirements including emergency provision in case of accidents
10. Type and number of packages
11. Quantity in weight/volume
12. Declaration by the generator or exporter that the information is correct
13. Declaration by the generator or exporter indicating no objection from the competent authorities of all States concerned which are Parties.
14. Certification by disposer of receipt at designated disposal facility and indication of method of disposal and of the approximate date of disposal.

Notes

The information required on the movement document shall, where possible, be integrated in one document with that required under transport rules. Where this is not possible, the information should complement rather than duplicate that required under the transport rules. The movement document shall carry instructions as to who is to provide information and fill out any form.

1/ Full name and address, telephone, telex or telefax number and the name, address, telephone, telex or telefax number of the person to be contacted in case of emergency.

Annex VI
Arbitration

Article 1

Unless the agreement referred to in Article 20 of the Convention provides otherwise, the arbitration procedure shall be conducted in accordance with Articles 2 to 10 below.

Article 2

The claimant party shall notify the Secretariat that the parties have agreed to submit the dispute to arbitration pursuant to paragraph 2 or paragraph 3 of Article 20 and include, in particular, the Articles of the Convention the interpretation or application of which are at issue. The Secretariat shall forward the information thus received to all Parties to the Convention.

Article 3

The arbitral tribunal shall consist of three members. Each of the Parties to the dispute shall appoint an arbitrator, and the two arbitrators so appointed shall designate by common agreement the third arbitrator, who shall be the chairman of the tribunal. The latter shall not be a national of one of the parties to the dispute, nor have his usual place of residence in the territory of one of these parties nor be employed by any of them, nor have dealt with the case in any other capacity.

Article 4

1. If the chairman of the arbitral tribunal has not been designated within two months of the appointment of the second arbitrator, the Secretary-General of the United Nations shall, at the request of either party, designate him within a further two months' period.
2. If one of the parties to the dispute does not appoint an arbitrator within two months of the receipt of the request, the other party may inform the Secretary-General of the United Nations who shall designate the chairman of the arbitral tribunal within a further two months' period. Upon designation, the chairman of the arbitral tribunal shall request the party which has not appointed an arbitrator to do so within two months. After such period, he shall inform the Secretary-General of the United Nations, who shall make this appointment within a further two months' period.

Article 5

1. The arbitral tribunal shall render its decision in accordance with international law and in accordance with the provisions of this Convention.
2. Any arbitral tribunal constituted under the provisions of this Annex shall draw up its own rules of procedure.

Article 6

1. The decisions of the arbitral tribunal both on procedure and on substance, shall be taken by majority vote of its members.
2. The tribunal may take all appropriate measures in order to establish the facts. It may, at the request of one or the parties, recommend essential interim measures of protection.
3. The parties to the dispute shall provide all facilities necessary for the effective conduct of the proceedings.
4. The absence or default of a party in the dispute shall not constitute an impediment to the proceedings.

Article 7

The tribunal may hear and determine counter claims arising directly out of the subject matter of the dispute.

Article 8

Unless the arbitral tribunal determines otherwise because of the particular circumstances of the case, the expenses of the tribunal, including the remuneration of its members, shall be borne by the parties to the dispute in equal shares. The tribunal shall keep a record of all its expenses, and shall furnish a final statement thereof to the parties.

Article 9

Any Party that has an interest of a legal nature in the subject matter of the dispute which may be affected by the decision in the case, may intervene in the proceedings with the consent of the tribunal.

Article 10

1. The tribunal shall render its award within five months of the date on which it is established unless it finds it necessary to extend the time-limit for a period which should not exceed five months.
2. The award of the arbitral tribunal shall be accompanied by a statement of reasons. It shall be final and binding upon the parties to the dispute.
3. Any dispute which may arise between the parties concerning the interpretation or execution of the award may be submitted by either party to the arbitral tribunal which made the award or, if the latter cannot be seized thereof, to another tribunal constituted for this purpose in the same manner as the first.

OCEANS AND THE MARINE ENVIRONMENT

Convention on the Prevention of Marine Pollution by Dumping of Wastes and Other Matter (London, Mexico City, Moscow and Washington, 29 December 1972)

Preamble

The Contracting Parties to this Convention

Recognizing that the marine environment and the living organisms which it supports are of vital importance to humanity, and all people have an interest in assuring that it is so managed that its quality and resources are not impaired;

Recognizing that the capacity of the sea to assimilate wastes and render them harmless, and its ability to regenerate natural resources, is not unlimited;

Recognizing that States have, in accordance with the Charter of the United Nations and the principles of international law, the sovereign right to exploit their own resources pursuant to their own environmental policies, and the responsibility to ensure that activities within their jurisdiction or control do not cause damage to the environment of other States or of areas beyond the limits of national jurisdiction;

Recalling Resolution 2749 (XXV) of the General Assembly of the United Nations on the principles governing the sea-bed and the ocean floor and the subsoil thereof, beyond the limits of national jurisdiction;

Noting that marine pollution originates in many sources, such as dumping and discharges through the atmosphere, rivers, estuaries, outfalls and pipelines, and that it is important that States use the best practicable means to prevent such pollution and develop products and processes which will reduce the amount of harmful wastes to be disposed of;

Being convinced that international action to control the pollution of the sea by dumping can and must be taken without delay but that this action should not preclude discussion of measures to control other sources of marine pollution as soon as possible and;

Wishing to improve protection of the marine environment by encouraging States with a common interest in particular geographical areas to enter into appropriate agreements supplementary to this Convention;

Have agreed as follows:

Article 1

Contracting Parties shall individually and collectively promote the effective control of all sources of pollution of the marine environment, and pledge themselves especially to take all practicable steps to prevent the pollution of the sea by the dumping of waste and other matter that is liable to create hazards to human health, to harm living resources and marine life, to damage amenities or to interfere with other legitimate uses of the sea.

Article 2

Contracting Parties shall, as provided for in the following Articles, take effective measures individually, according to their scientific, technical and economic capabilities, and collectively, to prevent marine pollution caused by dumping and shall harmonize their policies in this regard.

Article 3

For the purposes of this Convention:

1. a. "Dumping" means:
    - (i) any deliberate disposal at sea of wastes or other matter from vessels, aircraft, platforms or other man-made structures at sea;
    - (ii) any deliberate disposal at sea of vessels, aircraft, platforms or other man-made structures at sea;
    b. "Dumping" does not include:
    - (i) the disposal at sea of wastes or other matter incidental to, or derived from the normal operations of vessels, aircraft, platforms or other man-made structures at sea and their equipment, other than wastes or other matter transported by or to vessels, aircraft, platforms or other man-made structures at sea, operating for the purpose of disposal of such matter or derived from the treatment of such wastes or other matter on such vessels, aircraft, platforms or structures;
    - (ii) placement of matter for a purpose other than the mere disposal thereof, provided that such placement is not contrary to the aims of this convention;
    c. The disposal of wastes or other matter directly arising from, or related to the exploration, exploitation and associated off-shore processing of seabed mineral resources will not be covered by the provisions of this Convention.
2. "Vessels and aircraft" means waterborne or airborne craft of any type whatsoever. This expression includes air cushioned craft and floating craft, whether self-propelled or not.
3. "Sea" means all marine waters other than the internal waters of States.
4. "Wastes or other matter" means material and substance of any kind, form or description.
5. "Special permit" means permission granted specifically on application in advance and in accordance with Annex II and Annex III.
6. "General permit" means permission granted in advance and in accordance with Annex III.
7. "The Organisation" means the organisation designated by the Contracting Parties in accordance with Article 14.2.

Article 4

1. In accordance with the provisions of this Convention, Contracting Parties shall prohibit the dumping of any wastes or other matter in whatever form or condition except as otherwise specified below:
   a. the dumping of wastes or other matter listed in Annex I is prohibited;
   b. the dumping of wastes or other matter listed in Annex II requires a prior special permit;
   c. the dumping of all other wastes or matter requires a prior general permit.
2. Any permit shall be issued only after careful consideration of all the factors set forth in Annex III, including prior studies of the characteristics of the dumping site, as set forth in Sections B and C of that Annex.
3. No provision of this Convention is to be interpreted as preventing a Contracting Party from prohibiting, insofar as that Party is concerned, the dumping of wastes or other matter not mentioned in Annex I. That Party shall notify such measures to the Organisation.

Article 5

1. The provisions of Article 4 shall not apply when it is necessary to secure the safety of human life or of vessels, aircraft, platforms or other man-made structures at sea in cases of force majeure caused by stress of weather, or in any case which constitutes a danger to human life or a real threat to vessels, aircraft, platforms or other man-made structures at sea, if dumping appears to be the only way of averting the threat and if there is every probability that the damage consequent upon such dumping will be less than would otherwise occur. Such dumping shall be so conducted as to minimize the likelihood of damage to human or marine life and shall be reported forthwith to the Organisation.
2. A Contracting Party may issue a special permit as an exception to Article 4.1.a, in emergencies, posing unacceptable risk relating to human health and admitting no other feasible solution. Before doing so the Party shall consult any other country or countries that are likely to be affected and the Organisation which, after consulting other Parties, and international organizations as appropriate, shall, in accordance with Article 14 promptly recommend to the Party the most appropriate procedures to adopt. The Party shall follow these recommendations to the maximum extent feasible consistent with the time within which action must be taken and with the general obligation to avoid damage to the marine environment and shall inform the Organisation of the action it takes. The Parties pledge themselves to assist one another in such situations.
3. Any Contracting Party may waive its rights under paragraph 2 at the time of, or subsequent to ratification of, or accession to this Convention.

Article 6

1. Each Contracting Party shall designate an appropriate authority or authorities to:
   a. issue special permits which shall be required prior to, and for, the dumping of matter listed in Annex II and in the circumstances provided for in Article 5.2;
   b. issue general permits which shall be required prior to, and for, the dumping of all other matter;
   c. keep records of the nature and quantities of all matter permitted to be dumped and the location, time and method of dumping;
   d. monitor individually, or in collaboration with other Parties and competent international organisations, the condition of the seas for the purposes of this Convention.

2. The appropriate authority or authorities of a Contracting Party shall issue prior special or general permits in accordance with paragraph 1 with respect to the matter intended for dumping:
   a. loaded in its territory;
   b. loaded by a vessel or aircraft registered in its territory or flying its flag, when the loading occurs in the territory of a State not party to this Convention.
3. In issuing permits under subparagraphs 1.a and b above, the appropriate authority or authorities shall comply with Annex III, together with such additional criteria, measures and requirements as they may consider relevant.
4. Each Contracting Party, directly or through a Secretariat established under a regional agreement, shall report to the Organisation, and where appropriate to other Parties, the information specified in subparagraphs c and d of paragraph 1 above, and the criteria, measures and requirements it adopts in accordance with paragraph 3 above. The procedure to be followed and the nature of such reports shall be agreed by the Parties in consultation.

Article 7

1. Each Contracting Party shall apply the measures required to implement the present Convention to all:
   a. vessels and aircraft registered in its territory or flying its flag;
   b. vessels and aircraft loading in its territory or territorial seas matter which is to be dumped;
   c. vessels and aircraft and fixed or floating platforms under its jurisdiction believed to be engaged in dumping.
2. Each Party shall take in its territory appropriate measures to prevent and punish conduct in contravention of the provisions of this Convention.
3. The Parties agree to cooperate in the development of procedures for the effective application of this Convention particularly on the high seas, including procedures for the reporting of vessels and aircraft observed dumping in contravention of the Convention.
4. This Convention shall not apply to those vessels and aircraft entitled to sovereign immunity under international law. However each Party shall ensure by the adoption of appropriate measures that such vessels and aircraft owned or operated by it act in a manner consistent with the object and purpose of this Convention, and shall inform the Organisation accordingly.
5. Nothing in this Convention shall affect the right of each Party to adopt other measures, in accordance with the principles of international law, to prevent dumping at sea.

Article 8

In order to further the objectives of this Convention, the Contracting Parties with common interests to protect the marine environment in a given geographical area shall endeavour, taking into account characteristic regional features, to enter into regional agreements consistent with this Convention for the prevention of pollution, especially by dumping. The Contracting Parties to the present Convention shall endeavour to act consistently with the objectives and provisions of such regional agreements, which shall be notified to them by the Organisation. Contracting Parties shall seek to cooperate with the Parties to regional agreements in order to develop harmonized procedures to be followed by Contracting Parties to the different conventions concerned. Special attention shall be given to co-operation in the field of monitoring and scientific research.

Article 9

The Contracting Parties shall promote, through collaboration within the Organisation and other international bodies, support for those Parties which request it for:

   a. the training of scientific and technical personnel;
   b. the supply of necessary equipment and facilities for research and monitoring;
   c. the disposal and treatment of waste and other measures to prevent or mitigate pollution caused by dumping; preferably within the countries concerned, so furthering the aims and purposes of this Convention.

Article 10

In accordance with the principles of international law regarding State responsibility for damage to the environment of other States or to any other area of the environment, caused by dumping of wastes and other matter of all kinds, the Contracting Parties undertake to develop procedures for the assessment of liability and the settlement of disputes regarding dumping.

Article 11

The Contracting Parties shall at their first consultative meeting consider procedures for the settlement of disputes concerning the interpretation and application of this Convention.

Article 12

The Contracting Parties pledge themselves to promote, within the competent specialised agencies and other international bodies, measures to protect the marine environment against pollution caused by:

   a. hydrocarbons, including oil, and their wastes;
   b. other noxious or hazardous matter transported by vessels for purposes other than dumping;
   c. wastes generated in the course of operation of vessels, aircraft, platforms and other man-made structures at sea;
   d. radio-active pollutants from all sources, including vessels;
   e. agents of chemical and biological warfare;
   f. wastes or other matter directly arising from, or related to the exploration, exploitation and associated off-shore processing of sea-bed mineral resources.

The Parties will also promote, within the appropriate international organisation, the codification of signals to be used by vessels engaged in dumping.

Article 13

Nothing in this Convention shall prejudice the codification and development of the law of the sea by the United Nations Conference on the Law of the Sea convened pursuant to Resolution 2750 C (XXV) of the General Assembly of the United Nations nor the present or future claims and legal views of any State concerning the law of the sea and the nature and extent of coastal and flag State jurisdiction. The Contracting Parties agree to consult at a meeting to be convened by the Organisation after the Law of the Sea Conference, and in any case not later than 1976, with a view to defining the nature and extent of the right and the responsibility of a coastal State to apply the Convention in a zone adjacent to its coast.

Article 14

1. The Government of the United Kingdom of Great Britain and Northern Ireland as a depositary shall call a meeting of the Contracting Parties not later than three months after the entry into force of this Convention to decide on organisational matters.
2. The Contracting Parties shall designate a competent Organisation existing at the time of that meeting to be responsible for Secretariat duties in relation to this Convention. Any Party to this Convention not being a member of this Organisation shall make an appropriate contribution to the expenses incurred by the Organisation in performing these duties.
3. The Secretariat duties of the Organisation shall include:
   a. the convening of consultative meetings of the Contracting Parties not less frequently than once every two years and of special meetings of the Parties at any time on the request of two-thirds of the Parties;
   b. preparing and assisting, in consultation with the Contracting Parties and appropriate International Organisations, in the development and implementation of procedures referred to in subparagraph 4.e of this Article;
   c. considering inquiries by, and information from the Contracting Parties, consulting with them and with the appropriate International Organisations, and providing recommendations to the Parties on questions related to, but not specifically covered by the Convention;
   d. conveying to the Parties concerned all notifications received by the Organisations in accordance with Articles 4.3, 5.1 and 2, 6.4, 15, 20 and 21.

Prior to the designation of the Organisation these functions shall, as necessary, be performed by the depositary, who for this purpose shall be the Government of the United Kingdom of Great Britain and Northern Ireland.

4. Consultative or special meetings of the Contracting Parties shall keep under continuing review the implementation of this Convention and may, inter alia:
   a. review and adopt amendments to this Convention and its Annexes in accordance with Article 15;
   b. invite the appropriate scientific body or bodies to collaborate with and to advise the Parties or the Organisation on any scientific or technical aspect relevant to this Convention, including particularly the content of the Annexes;
   c. receive and consider reports made pursuant to Article 6.4;
   d. promote co-operation with and between regional organisations concerned with the prevention of marine pollution;
   e. develop or adopt, in consultation with appropriate International Organisations, procedures referred to in Article 5.2, including basic criteria for determining exceptional and emergency situations, and procedures for consultative advice and the safe disposal of matter in such circumstances, including the designation of appropriate dumping areas, and recommend accordingly;
   f. consider any additional action that may be required.
5. The Contracting Parties at their first consultative meeting shall establish rules of procedure as necessary.

Article 15

1. a. At meetings of the Contracting Parties called in accordance with Article 14 amendments to this Convention may be adopted by a two-thirds majority of those present. An amendment shall enter into force for the Parties which have accepted it on the sixtieth day after two-thirds of the Parties shall have deposited an instrument of acceptance of the amendment with the Organisation. Thereafter the amendment shall enter into force for any other Party 30 days after that Party deposits its instrument of acceptance of the amendment.
   b. The Organisation shall inform all Contracting Parties of any requests made for a special meeting under Article 14 and of any amendments adopted at meetings of the Parties and of the date on which each such amendment enters into force for each Party.
2. Amendments to the Annexes will be based on scientific or technical considerations. Amendments to the Annexes approved by a two-thirds majority of those present at a meeting called in accordance with Article 14 shall enter into force for each Contracting Party immediately on notification of its acceptance to the Organisation and 100 days after approval by the meeting for all other Parties except for those which before the end of the 100 days make a declaration that they are not able to accept the amendment at that time. Parties should endeavour to signify their acceptance of an amendment to the Organisation as soon as possible after approval at a meeting. A Party may at any time substitute an acceptance for a previous declaration of objection and the amendment previously objected to shall thereupon enter into force for that Party.
3. An acceptance or declaration of objection under this Article shall be made by the deposit of an instrument with the Organisation. The Organisation shall notify all Contracting Parties of the receipt of such instruments.
4. Prior to the designation of the Organisation, the Secretariat functions herein attributed to it, shall be performed temporarily by the Government of the United Kingdom of Great Britain and Northern Ireland, as one of the depositaries of this Convention.

Article 16

This Convention shall be open for signature by any State at London, Mexico City, Moscow and Washington from 29 December 1972 until 31 December 1973.

Article 17

This Convention shall be subject to ratification. The instruments of ratification shall be deposited with the Governments of Mexico, Union of Soviet Socialist Republics, United Kingdom of Great Britain and Northern Ireland, and the United States of America.

Article 18

After 31 December 1973, this Convention shall be open for accession by any State. The instruments of accession shall be deposited with the Governments of Mexico, the Union of Soviet Socialist Republics, the United Kingdom of Great Britain and Northern Ireland, and the United States of America.

Article 19

1. This Convention shall enter into force on the thirtieth day following the date of deposit of the fifteenth instrument of ratification or accession.
2. For each Contracting Party ratifying or acceding to the Convention after the deposit of the fifteenth instrument of ratification or accession, the Convention shall enter into force on the thirtieth day after deposit by such Party of its instrument of ratification or accession.

Article 20

The depositaries shall inform Contracting Parties:

a. of signatures to this Convention and of the deposit of instruments of ratification, accession or withdrawal, in accordance with Articles 16, 17, 18 and 21; and
b. of the date on which this Convention will enter into force, in accordance with Article 19.

Article 21

Any Contracting Party may withdraw from this Convention by giving six months' notice in writing to a depositary, which shall promptly inform all Parties of such notice.

Article 22

The original of this Convention, of which the English, French, Russian and Spanish texts are equally authentic, shall be deposited with the Governments of Mexico, the Union of Soviet Socialist Republics, the United Kingdom of Great Britain and Northern Ireland, and the United States of America who shall send certified copies thereof to all States.

Annex I
omissis

Annex II
omissis

Annex III
omissis

Amendments to Annexes to the Convention Concerning Incineration at Sea
(London, 12 October 1978)
omissis.

Convention on Wetlands of International Importance Especially as Waterfowl Habitat (Ramsar), 1971

The Contracting Parties,
RECOGNIZING the interdependence of man and his environment;
CONSIDERING the fundamental ecological functions of wetlands as regulators of water regimes and as habitats supporting a characteristic flora and fauna, especially waterfowl;
BEING convinced that wetlands constitute a resource of great economic, cultural, scientific and recreational value, the loss of which would be irreparable;
DESIRING to stem the progressive encroachment on and loss of wetlands now and in the future;

RECOGNIZING that waterfowl in their seasonal migrations may transcend frontiers and so should be regarded as an international resource;

BEING confident that the conservation of wetlands and their flora and fauna can be ensured by combining far-sighted national policies with coordinated international action;

HAVE AGREED as follows:

Article 1

1. For the purpose of this Convention wetlands are areas of marsh, fen, peatland or water, whether natural or artificial, permanent or temporary, with water that is static or flowing, fresh, brackish or salt, including areas of marine water the depth of which at low tide does not exceed six metres.
2. For the purpose of this Convention waterfowl are birds ecologically dependent on wetlands.

Article 2

1. Each Contracting Party shall designate suitable wetlands within its territory for inclusion in a List of Wetlands of International Importance, hereinafter referred to as "the List" which is maintained by the bureau established under Article 8. The boundaries of each wetland shall be precisely described and also delimited on a map and they may incorporate riparian and coastal zones adjacent to the wetlands, and islands or bodies of marine water deeper than six metres at low tide lying within the wetlands, especially where these have importance as waterfowl habitat.
2. Wetlands should be selected for the List on account of their international significance in terms of ecology, botany, zoology, limnology or hydrology. In the first instance, wetlands of international importance to waterfowl at any season should be included.
3. The inclusion of a wetland in the List does not prejudice the exclusive sovereign rights of the Contracting Party in whose territory the wetland is situated.
4. Each Contracting Party shall designate at least one wetland to be included in the List when signing this Convention or when depositing its instrument of ratification or accession, as provided in Article 9.
5. Any Contracting Party shall have the right to add to the List further wetlands situated within its territory, to extend the boundaries of those wetlands already included by it in the List, or, because of its urgent national interests, to delete or restrict the boundaries of wetlands already included by it in the List and shall, at the earliest possible time, inform the organization or government responsible for the continuing bureau duties specified in Article 8 of any such changes.
6. Each Contracting Party shall consider its international responsibilities for the conservation, management and wise use of migratory stocks of waterfowl, both when designating entries for the List and when exercising its right to change entries in the List relating to wetlands within its territory.

Article 3

1. The Contracting Parties shall formulate and implement their planning so as to promote the conservation of the wetlands included in the List, and as far as possible the wise use of wetlands in their territory.
2. Each Contracting Party shall arrange to be informed at the earliest possible time if the ecological character of any wetland in its territory and included in the List has changed, is changing or is likely to change as the result of technological developments, pollution

or other human interference. Information on such changes shall be passed without delay to the organization or government responsible for the continuing bureau duties specified in Article 8.

Article 4

1. Each Contracting Party shall promote the conservation of wetlands and waterfowl by establishing nature reserves on wetlands, whether they are included in the List or not, and provide adequately for their wardening.
2. Where a Contracting Party in its urgent national interest, deletes or restricts the boundaries of a wetland included in the List, it should as far as possible compensate for any loss of wetland resources, and in particular it should create additional nature reserves for waterfowl and for the protection, either in the same area or elsewhere, of an adequate portion of the original habitat.
3. The Contracting Parties shall encourage research and the exchange of data and publications regarding wetlands and their flora and fauna.
4. The Contracting Parties shall endeavour through management to increase waterfowl populations on appropriate wetlands.
5. The Contracting Parties shall promote the training of personnel competent in the fields of wetland research, management and wardening.

Article 5

The Contracting Parties shall consult with each other about implementing obligations arising from the Convention especially in the case of a wetland extending over the territories of more than one Contracting Party or where a water system is shared by Contracting Parties.

They shall at the same time endeavour to coordinate and support present and future policies and regulations concerning the conservation of wetlands and their flora and fauna.

Article 6

1. The Contracting Parties shall, as the necessity arises, convene Conferences on the Conservation of Wetlands and Waterfowl.
2. These Conferences shall have an advisory character and shall be competent, inter alia:
   a) to discuss the implementation of this Convention;
   b) to discuss additions to and changes in the List;
   c) to consider information regarding changes in the ecological character of wetlands included in the List provided in accordance with paragraph 2 of Article 3;
   d) to make general or specific recommendations to the Contracting Parties regarding the conservation, management and wise use of wetlands and their flora and fauna;
   e) to request relevant international bodies to prepare reports and statistics on matters which are essentially international in character affecting wetlands.
3. The Contracting Parties shall ensure that those responsible at all levels for wetlands management shall be informed of, and take into consideration, recommendations of such Conferences concerning the conservation, management and wise use of wetlands and their flora and fauna.

Article 7

1. The representatives of the Contracting Parties at such Conferences should include persons who are experts on wetlands or waterfowl by reason of knowledge and experience gained in scientific, administrative or other appropriate capacities.
2. Each of the Contracting Parties represented at a Conference shall have one vote, recommendations being adopted by a simple majority of the votes cast, provided that not less than half the Contracting Parties cast votes.

Article 8

1. The International Union for the Conservation of Nature and Natural Resources shall perform the continuing bureau duties under this Convention until such time as another organization or government is appointed by a majority of two-thirds of all Contracting Parties.
2. The continuing bureau duties shall be, inter alia:
   a) to assist in the convening and organizing of Conferences specified in Article 6;
   b) to maintain the List of Wetlands of International Importance and to be informed by the Contracting Parties of any additions, extensions, deletions or restrictions concerning wetlands included in the List provided in accordance with paragraph 5 of Article 2;
   c) to be informed by the Contracting Parties of any changes in the ecological character of wetlands included in the List provided in accordance with paragraph 2 of Article 3;
   d) to forward notification of any alterations to the List, or changes in character of wetlands included therein, to all Contracting Parties and to arrange for these matters to be discussed at the next Conference;
   e) to make known to the Contracting Party concerned, the recommendations of the Conferences in respect of such alterations to the List or of changes in the character of wetlands included therein.

Article 9

1. This Convention shall remain open for signature indefinitely.
2. Any member of the United Nations or of one of the Specialized Agencies or of the International Atomic Energy Agency or Party to the Statute of the International Court of Justice may become a Party to this Convention by:
   a) signature without reservation as to ratification;
   b) signature subject to ratification followed by ratification;
   c) accession.
3. Ratification or accession shall be effected by the deposit of an instrument of ratification or accession with the Director-General of the United Nations Educational, Scientific and Cultural Organization, (hereinafter referred to as "the Depository").

Article 10

1. This Convention shall enter into force four months after seven States have become Parties to this Convention in accordance with paragraph 2 of Article 9.
2. Thereafter this Convention shall enter into force for each Contracting Party four months after the day of its signature without reservation as to ratification, or its deposit of an instrument of ratification or accession.

Article 11

1. This Convention shall continue in force for an indefinite period.
2. Any Contracting Party may denounce this Convention after a period of five years from the date on which it entered into force for that Party by giving written notice thereof to the Depository. Denunciation shall take effect four months after the day on which notice thereof is received by the Depository.

Article 12

1. The Depository shall inform all States that have signed and acceded to this Convention as soon as possible of
   a) signatures to the Convention;
   b) deposits of instruments of ratification of this Convention;
   c) deposits of instruments of accession to this Convention;
   d) the date of entry into force of this Convention;
   e) notifications of denunciation of this Convention.
2. When this Convention has entered into force, the Depository shall have it registered with the Secretariat of the United Nations in accordance with Article 102 of the Charter.

IN WITNESS WHEREOF, the undersigned, being duly authorized to that effect, have signed this Convention.

DONE at Ramsar this 2nd day of February 1971, in a single original in the English, French, German and Russian languages, in any case of divergency the English text prevailing, which shall be deposited with the Depository which shall send true copies thereof to all Contracting Parties.

Protocol of Amendment

A Protocol to amend the Convention on Wetlands of International Importance Especially as Waterfowl Habitat was adopted in Paris on 3 December 1982. The Protocol provides as follows:

The Contracting Parties,

CONSIDERING that for the effectiveness of the Convention on Wetlands of International Importance especially as Waterfowl Habitat, done at Ramsar on 2 February 1971 (hereinafter referred to as "the Convention"), it is indispensable to increase the number of Contracting Parties;

AWARE that the addition of authentic language versions would facilitate wider participation in the Convention;

CONSIDERING furthermore that the text of the Convention does not provide for an amendment procedure, which makes it difficult to amend the text as may be considered necessary;

HAVE AGREED as follows:

Article 1

The following Article shall be added between Article 10 and Article 11 of the Convention:
"Article 10 Bis"

1. This Convention may be amended at a meeting of the Contracting Parties convened for that purpose in accordance with this Article.
2. Proposals for amendment may be made by any Contracting Party.
3. The text of any proposed amendment and the reasons for it shall be communicated to the organization or government performing the continuing bureau duties under the Convention (hereinafter referred to as "the Bureau") and shall promptly be communicated

by the Bureau to all Contracting Parties. Any comments on the text by the Contracting Parties shall be communicated to the Bureau within three months of the date on which amendments were communicated to the Contracting Parties by the Bureau. The Bureau shall, immediately after the last day for submission of comments, communicate to the Contracting Parties all comments submitted by that day.

4. A meeting of Contacting Parties to consider an amendment communicated in accordance with paragraph 3 shall be convened by the Bureau upon the written request of one third of the Contracting Parties. The Bureau shall consult the Parties concerning the time and venue of the meeting.

5. Amendments shall be adopted by a two-thirds majority of the Contracting Parties present and voting.

6. An amendment adopted shall enter into force for the Contracting Parties which have accepted it on the first day of the fourth month following the date on which two-thirds of the Contracting Parties have deposited an instrument of acceptance with the Depository. For each Contracting Party which deposits an instrument of acceptance after the date on which two-thirds of the Contracting Parties have deposited an instrument of acceptance, the amendment shall enter into force on the first day of the fourth month following the date of the deposit of its instrument of acceptance.

Article 2

In the testimonium following Article 12 of the Convention, the words "in any case of divergency the English text prevailing" shall be deleted and replaced by the words "all texts being equally authentic".

Article 3

The revised text of the original French version of the Convention is reproduced in the Annex to this Protocol.

Article 4

This Protocol shall be open for signature at UNESCO headquarters in Paris from 3 December 1982.

Article 5

1. Any State referred to in Article 9, paragraph 2, of the Convention may become a Contracting Party to this Protocol by:
   a) signature without reservation as to ratification, acceptance or approval;
   b) signature subject to ratification, acceptance or approval, followed by ratification, acceptance or approval;
   c) accession.

2. Ratification, acceptance, approval or accession shall be effected by the deposit of an instrument of ratification, acceptance, approval or accession with the Director-General of the United Nations Educational, Scientific and Cultural Organization (hereinafter referred to as "the Depository").

3. Any State which becomes a Contracting Party to the Convention after the entry into force of this Protocol shall, failing an expression of a different intention at the time of

signature or of the deposit of the instrument referred to in Article 9 of the Convention, be considered as a Party to the Convention as amended by this Protocol.

4. Any State which becomes a Contracting Party to this Protocol without being a Contracting Party to the Convention, shall be considered as a Party to the Convention as amended by this Protocol as of the date of entry into force of this Protocol for that State.

Article 6

1. This Protocol shall enter into force the first day of the fourth month following the date on which two-thirds of the States which are Contracting Parties to the Convention on the date on which this Protocol is opened for signature have signed it without reservation as to ratification, acceptance or approval, or have ratified, accepted, approved or acceded to it.

2. With regard to any State which becomes a Contracting Party to this Protocol in the manner described in paragraph 1 and 2 of Article 5 above, after the date of its entry into force, this Protocol shall enter into force on the date of its signature without reservation as to ratification, acceptance or approval, or of its ratification, acceptance, approval or accession.

3. With regard to any State which becomes a Contracting Party to this Protocol in the manner described in paragraph 1 and 2 of Article 5 above, during the period between the date on which this Protocol is opened for signature and its entry into force, this Protocol shall enter into force on the date determined in paragraph 1 above.

Article 7

1. The original of this Protocol, in the English and French languages, each version being equally authentic, shall be deposited with the Depository. The Depository shall transmit certified copies of each of these versions to all States that have signed this Protocol or deposited instruments of accession to it.

2. The Depository shall inform all Contracting Parties of the Convention and all States that have signed and acceded to this Protocol as soon as possible of:
   a) signatures to this Protocol;
   b) deposits of instruments of ratification, acceptance or approval of this Protocol;
   c) deposits of instruments of accession to this Protocol;
   d) the date of entry into force of this Protocol.

3. When this Protocol has entered into force, the Depository shall have it registered with the Secretariat of the United Nations in accordance with Article 102 of the Charter.

IN WITNESS WHEREOF, the undersigned, being duly authorized to that effect, have signed this Protocol.
DONE at Paris on 3 December 1982.

LISTING OF INTERNATIONAL TREATIES RELATING TO THE ENVIRONMENT

Organization by Topic

- Air Pollution
- Biological Diversity
- Culture and Natural Heritage
- Deforestation
- Desertification

- Economics
- Environmental Impact Assessment
- Global Climate Change
- Nature, Wildlife and Habitat
- Oceans and the Marine Environment
- Prevention and Preparedness
- Stratospheric Ozone Depletion
- Trade, Commerce and Industry
- Waste

Air Pollution

Convention on the Transboundary Effects of Industrial Accidents, 17 March 1992

Convention on Environmental Impact Assessment in a Transboundary Context, 25 February 1991

Basel Convention on the Control of Transboundary Movements of Hazardous Wastes and Their Disposal, 22 March 1989

Protocol to the 1979 Convention on Long-Range Transboundary Air Pollution on Long-Term Financing of Cooperative Programme for Monitoring and Evaluation of the Long-Range Transmissions of Air Pollutants in Europe (EMEP), 28 September 1984

Protocol Concerning Cooperation in Combating Pollution in Cases of Emergency, 21 March 1981

Convention on Long-Range Transboundary Air Pollution, 13 November 1979

(OECD) Control of Air Pollution from Fossil Fuel Combustion, 20 June 1985

(OECD) Measures to Reduce All Man-Made Emissions of Mercury to the Environment, 18 September 1973

Biological Diversity

Convention on Biological Diversity, 5 June 1992

Culture and Natural Heritage

Convention for the Protection of the World Cultural and Natural Heritage, 23 November 1972

Deforestation

International Tropical Timber Agreement 26 January 1994 International Tropical Timber Agreement, 18 November 1983

Conservation of Biological Diversity Agreement Establishing the Inter-American Institute for Global Change Research, 13 May 1992

The Convention on Biological Diversity, 5 June 1992

The Rio Declaration, 3-14 June 1992

Agenda 21, 3-14 June 1992

Protocol on Environmental Protection to the Antarctic Treaty, 4 October 1991

European Convention for the Protection of Pet Animals, 13 November 1987

ASEAN Agreement on the Conservation of Nature and Natural Resources, 9 July 1985

International Tropical Timber Agreement, 18 November 1983

Protocol to Amend the Convention on Wetlands of International Importance Especially as Waterfowl Habitat, 3 December 1982

Benelux Convention on Nature Conservation and Landscape Protection, 8 June 1982

World Charter for Nature, 1982
Protocol Concerning Mediterranean Specially Protected Areas, 3 April 1982
Convention on the Conservation of Antarctic Marine Living Resources, 20 May 1980
Convention for the Conservation and Management of the Vicuna, 20 December 1979
Convention on the Conservation of European Wildlife and Natural Habitats, 19 September 1979
Convention on the Conservation of Migratory Species of Wild Animals, 23 June 1979
European Convention for the Protection of Animals for Slaughter, 10 May 1979
Treaty for Amazonian Cooperation, 3 July 1978
Convention on Conservation of Nature in the South Pacific, 12 June 1976
European Convention for the Protection of Animals Kept for Farming Purposes, 10 March 1976
Agreement on Conservation of Polar Bears, 15 November 1973
Convention on International Trade in Endangered Species of Wild Fauna and Flora, 3 March 1973
Convention for the Protection of the World Cultural and Natural Heritage, 23 November 1972
Convention for the Conservation of Antarctic Seals, 1 June 1972
Convention on Wetlands of International Importance Especially as Waterfowl Habitat, 2 February 1971
Benelux Convention on the Hunting and Protection of Birds (as amended), 10 June 1970
Agreed Measures for the Conservation of Antarctic Fauna and Flora, 2 June 1964
Plant Protection Agreement for the South-East Asia and Pacific Region (as amended), 27 February 1956
The Antarctic Treaty, 1 December 1959
International Plant Protection Convention, 6 December 1951
Convention for the Establishment of the European and Mediterranean Plant Protection Organization (as amended 18 April 1951 PARIS), 18 April 1951
International Convention for the Protection of Birds, 18 October 1950
Convention on Nature Protection and Wildlife Preservation in the Western Hemisphere, 12 October 1940

Desertification

United Nations Convention to Combat Desertification in those Countries Experiencing Serious Drought and/or Desertification, Particularly in Africa, 12 September 1994
United Nations Conference on Desertification (UNCOD) Plan of Action to Combat Desertification and General Assembly Resolutions, 29-9 August 1977 (200k file)
Agreement for the Establishment of a Commission for Controlling the Desert Locust in North-West Africa (as amended), 1 November 1970
Agreement for the Establishment of a Commission for Controlling the Desert Locust in the Near East (as amended), 2 July 1965
Agreement for the Establishment of a Commission for Controlling the Desert Locust in the Eastern Region of its Distribution Area in South-West Asia (as amended), 3 December 1963

Economics

(OECD) Environment and Economics Guiding Principles Concerning International Economic Aspects of Environmental Policies, 26 May 1972
(OECD) Implementation of the Polluter-Pays Principle, 14 November 1974
(OECD) International Conference on Environment and Economics: Conclusions, 21 July 1984
Protocol to Amend the International Convention on the Establishment of an International Fund for Compensation for Oil Pollution Damage, 25 May 1984

Protocol to the International Convention on the Establishment of an International Fund for Compensation for Oil Pollution Damage, 19 November 1976

Environmental Impact Assessment

Convention on Environmental Impact Assessment in a Transboundary Context, 25 February 1991

(OECD) Assessment of Projects with Significant Impact on the Environment, 8 May 1979

Global Climate Change

United Nations Framework Convention on Climate Change, 9 May 1992
Agreement Establishing the Inter-American Institute for Global Change Research, 13 May 1992
The Rio Declaration, 3-14 June 1992
Agenda 21, 3-14 June 1992
Kyoto Protocol to the United Nations Framework Convention on Climate Change, 1997

Nature, Wildlife and Habitat

The Rio Declaration, 3-14 June 1992
Agenda 21, 3-14 June 1992
Agreed Measures for the Conservation of Antarctic Fauna and Flora, 2 June 1964
Agreement Concerning Cooperation in Marine Fishing, 28 July 1962
Agreement Concerning Cooperation in the Quarantine of Plants and Their Protection Against Pests and Diseases, 14 December 1959
Agreement Concerning Interim Arrangements Relating to Polymetallic Nodules of the Deep Sea, 2 September 1982
Agreement Concerning Measures for Protection of the Stocks of Deep-Sea Prawns (Pandalus borealis), European Lobsters (Homarus vulaaris), Norway Lobsters (Nephropsnorveaicus) and Crabs (Cancer Paqurus) (as amended), 7 March 52
Agreement on Conservation of Polar Bears, 15 November 1973
Agreement to Promote Compliance with International Conservation and Management Measures by Fishing Vessels on the High Seas, 29 November 1993
ASEAN Agreement on the Conservation of Nature and Natural Resources, 9 July 1985
Benelux Convention on Nature Conservation and Landscape Protection, 8 June 1982
Benelux Convention on the Hunting and Protection of Birds (as amended), 10 June 1970
Convention Concerning Fishing in the Black Sea (as amended), 7 July 1959
Convention for the Conservation and Management of the Vicuna, 20 December 1979
Convention for the Conservation of Antarctic Seals, 1 June 1972
Convention for the Conservation of Salmon in the North Atlantic Ocean, 2 March 1982
Convention for the Conservation of Southern Bluefin Tuna, 10 May 1993
Convention for the Establishment of an Inter-American Tropical Tuna Commission, 31 May 1949
Convention for the Establishment of the European and Mediterranean Plant Protection Organization (as amended 18 April 1951 PARIS), 18 April 1951
Convention for the Protection of the Natural Resources and Environment of the South Pacific Region, 24 November 1986
Convention on Conservation of Nature in the South Pacific, 12 June 1976
Convention on Fishing and Conservation of the Living Resources in the Baltic Sea and Belts, 13 September 1973

Convention on Fishing and Conservation of the Living Resources of the High Seas, 29 April 1958

Convention on International Trade in Endangered Species of Wild Fauna and Flora, 3 March 1973

Convention on the Conservation of Antarctic Marine Living Resources, 20 May 1980

Convention on the Conservation of European Wildlife and Natural Habitats, 19 September 1979

Convention on the Conservation of Migratory Species of Wild Animals, 23 June 1979

Convention on the Continental Shelf, 29 April 1958

Draft Agreement for the Implementation of the Provisions of the United Nations Convention on the Law of the Sea of 10 December 1982 Relating to the Conservation and Management of Straddling Fish Stocks and Highly Migratory Fish Stocks, 4 August 1995

European Convention for the Protection of Animals During International Transport, 13 December 1968

European Convention for the Protection of Animals for Slaughter, 10 May 1979

European Convention for the Protection of Animals Kept for Farming Purposes, 10 March 1976

European Convention for the Protection of Pet Animals, 13 November 1987

Fisheries Convention, 9 March 1964

Interim Convention on Conservation of North Pacific Fur Seals, 9 February 1957

1980 Protocol Amending the Interim Convention on Conservation of North Pacific Fur Seals

1984 Protocol Amending the Interim Convention on Conservation of North Pacific Fur Seals, 12 October 1984

International Convention for the Protection of Birds, 18 October 1950

International Convention for the Regulation of Whaling, 2 December 1946

International Plant Protection Convention, 6 December 1951

International Tropical Timber Agreement, 18 November 1983

International Tropical Timber Agreement, 26 January 1994

Kuwait Regional Convention for Cooperation on the Protection of the Marine Environment from Pollution, 24 April 1978

Oceans and the Marine Environment

Draft Agreement for the Implementation of the Provisions of the United Nations Convention on the Law of the Sea of 10 December 1982 Relating to the Conservation and Management of Straddling Fish Stocks and Highly Migratory Fish Stocks, 4 August 1995

Agreement Relating to the Implementation of Part XI of the 1982 United Nations Convention on the Law of the Sea, 28 July 1994

Agreement to Promote Compliance with International Conservation and Management Measures by Fishing Vessels on the High Seas, 29 November 1993

Agreement establishing the South Pacific Regional Environment Programme (SPREP), 16 June 1993

Convention for the Conservation of Southern Bluefin Tuna, 10 May 1993

Annex III to the Protocol of 17 February 1978 relating to the International Convention for the Prevention of Pollution from Ships of 2 November 1973 (MARPOL 73/78), as amended on 30 October 1992

Convention for the Protection of the Marine Environment of the North East Atlantic, 22 September 1992

Niue Treaty on Cooperation in Fisheries Surveillance and Law Enforcement in the South Pacific Region, 9 July 1992

Agreement Establishing the Inter-American Institute for Global Change Research, 13 May 1992

Protocol on Protection of the Black Sea Marine Environment Against Pollution from Land Based Sources, 21 April 1992

Convention for a North Pacific Marine Science Organisation (PICES), 12 December 1990

International Convention on Oil Pollution Preparedness, Response and Cooperation, 29 November 1990

Agreement on the Organization for Indian Ocean Marine Affairs Cooperation, 17 September 1990

Protocol to the Kuwait Regional Convention for the Protection of the Marine Environment Against Pollution from Land-Based Sources, 21 February 1990

Amendment to the Annex to the Convention for the Prevention of Marine Pollution by Dumping of Wastes and Other Matter, 3 November 1989

Protocol Concerning Marine Pollution Resulting from Exploration and Exploitation of the Continental Shelf, 29 March 1989

Convention for the Suppression of Unlawful Acts Against the Safety of Maritime Navigation, 10 March 1988

Protocol for the Suppression of Unlawful Acts against the Safety of Fixed Platforms Located on the Continental Shelf, 10 March 1988

Agreement on the Network of Aquaculture Centres in Asia and the Pacific, 8 January 1988

Protocol for the Prevention of Pollution of the South Pacific Region by Dumping, 25 November 1986

Protocol Concerning Cooperation in Combating Pollution Emergencies in the South Pacific Region, 25 November 1986

Convention for the Protection of the Natural Resources and Environment of the South Pacific Region, 24 November 1986

International Agreement on the Use of INMARSAT Ship Earth Stations within the Territorial Sea and Ports, 16 October 1985

Convention for the Protection, Management and Development of the Marine and Coastal Environment of the Eastern African Region, 21 June 1985

1984 Protocol Amending the Interim Convention on Conservation of North Pacific Fur Seals

Protocol to Amend the International Convention on Civil Liability for Oil Pollution Damage, 25 May 1984

Protocol to Amend the International Convention on the Establishment of an International Fund for Compensation for Oil Pollution Damage, 25 May 1984

Agreement for Cooperation in Dealing with Pollution of the North Sea by Oil and Other Harmful Substances, 13 September 1983

Supplementary Protocol to the Agreement on Regional Co-Operation in Combating Pollution of the South-East Pacific by Hydrocarbons or Other Harmful Substances, 22 July 1983

Protocol Concerning Cooperation in Combating Oil Spills in the Wider Caribbean Region, 24 March 1983

Convention for the Protection and Development of the Marine Environment of the Wider Caribbean Region, 24 March 1983

United Nations Convention on the Law of the Sea, 10 December 1982

Agreement Concerning Interim Arrangements Relating to Polymetallic Nodules of the Deep Sea, 2 September 1982

Convention for the Conservation of Salmon in the North Atlantic Ocean, 2 March 1982

Regional Convention for the Conservation of the Red Sea and Gulf of Aden Environment, 14 February 1982

Protocol Concerning Regional Cooperation in Combating Pollution by Oil and Other Harmful Substances in Cases of Emergency, 14 February 1982

Protocol Concerning Mediterranean Specially Protected Areas, 3 April 1982

Agreement on Regional Cooperation in Combating Pollution of the South-East Pacific by Oil and Other Harmful Substances in Cases of Emergency, 12 November 1981

Convention for the Protection of the Marine Environment and Coastal Area of the South-East Pacific, 12 November 1981

Convention for Cooperation in the Protection and Development of the Marine and Coastal Environment of the West and Central African Region, 23 March 1981

Convention on Future Multilateral Cooperation in North-East Atlantic Fisheries, 18 November 1980

Amendment to the Annex to the Convention on the Prevention of Marine Pollution by Dumping of Wastes and Other Matter, 24 September 1980

Protocol for the Protection of the Mediterranean Sea Against Pollution from Land-Based Sources, 17 May 1980

1980 Protocol Amending the Interim Convention on Conservation of North Pacific Fur Seals

Protocol Amending the International Convention Relating to the Limitation of the Liability of Owners of Sea-Going Ships, 21 December 1979

South Pacific Forum Fisheries Agency Convention, 10 July 1979

Amendments to the Convention on the Prevention of Marine Pollution by Dumping of Wastes and Other Matter Concerning Settlement of Disputes, 12 October 1978

Kuwait Regional Convention for Cooperation on the Protection of the Marine Environment from Pollution, 24 April 1978

Protocol of 1978 Relating to the International Convention for the Prevention of Pollution from Ships, 17 February 1978

Protocol to the International Convention on Civil Liability for Oil Pollution Damage, 19 November 1976

Protocol to the International Convention on the Establishment of an International Fund for Compensation for Oil Pollution Damage, 19 November 1976

Convention on Conservation of Nature in the South Pacific, 12 June 1976

Agreement Concerning the Protection of the Waters of the Mediterranean Shores, 10 May 1976

Interim Convention on Conservation of North Pacific Fur Seals 9 February 1957 (as amended May 7, 1976)

Protocol for the Prevention of Pollution of the Mediterranean Sea by Dumping from Ships and Aircraft, 16 February 1976

Convention for the Protection of the Mediterranean Sea Against Pollution, 16 February 1976

Protocol Concerning Co-operation in Combating Pollution of the Mediterranean Sea by Oil and Other Harmful Substances in Cases of Emergency, 2 February 1976

Convention on the Prevention of Marine Pollution from Land-based Sources, 4 June 1974

The Nordic Environmental Protection Convention, 19 February 1974

Protocol Relating to Intervention on the High Seas in Cases of Pollution by Substances Other than Oil, 2 November 1973

International Convention for the Prevention of Pollution from Ships, 2 November 1973

Convention on Fishing and Conservation of the Living Resources in the Baltic Sea and Belts, 13 September 1973

Convention on the Prevention of Marine Pollution by Dumping of Wastes and Other Matter, 29 December 1972

Convention for the Prevention of Marine Pollution by Dumping from Ships and Aircraft (as amended), 15 February 1972

International Convention on the Establishment of an International Fund for Compensation for Oil Pollution Damage, 18 December 1971

Amendments to the International Convention for the Prevention of Pollution of the Sea by Oil Concerning the Protection of the Great Barrier Reef, 12 October 1971

Agreement Concerning Cooperation in Taking Measures Against Pollution of the Sea by Oil, 16 September 1971

International Convention Relating to Intervention on the High Seas in Cases of Oil Pollution Casualties, 29 November 1969

International Convention on Civil Liability for Oil Pollution Damage, 29 November 1969

Agreement for Cooperation in Dealing with Pollution of the North Sea by Oil, 9 June 1969

International Convention for the Conservation of Atlantic Tunas, 14 May 1966

Convention for the International Council for the Exploration of the Sea (as amended), 12 September 1964

Agreed Measures for the Conservation of Antarctic Fauna and Flora, 2 June 1964

Fisheries Convention, 9 March 1964

Treaty Banning Nuclear Weapon Tests in the Atmosphere, in Outer Space and Under Water, 10 October 1963

Agreement Concerning Cooperation in Marine Fishing, 28 July 1962

International Convention for the Safety of Life at Sea, 17 June 1960

The Antarctic Treaty, 1 December 1959

Convention Concerning Fishing in the Black Sea (as amended), 7 July 1959

North-East Atlantic Fisheries Convention, 24 January 1959

Convention on Fishing and Conservation of the Living Resources of the High Seas, 29 April 1958

Optional Protocol of Signature Concerning the Compulsory Settlement of Disputes Arising out of the United Nations Conference on the Law of the Sea, 29 April 1958

Convention on the Territorial Sea & the Contiguous Zone, 29 April 1958

Convention on the Continental Shelf, 29 April 1958

Convention on the High Seas, 29 April 1958

International Convention Relating to the Limitation of the Liability of Owners of Sea-Going Ships, 10 October 1957

Interim Convention on Conservation of North Pacific Fur Seals, 9 February 1957

International Convention for the Prevention of Pollution of the Sea by Oil (as amended on 11 April 1962 and 21 October 1969), 12 May 1954

International Convention for the High Seas Fisheries of the North Pacific Ocean (as amended), 9 May 1952

Agreement Concerning Measures for Protection of the Stocks of Deep-Sea Prawns (Pandalus borealis), European Lobsters (Homarus vulaaris), Norway Lobsters (Nephropsnorveaicus) and Crabs (Cancer Paqurus) (as amended), 7 March 52

Agreement for the Establishment of a General Fisheries Council for the Mediterranean (as amended), 24 September 1949

Convention for the Establishment of an Inter-American Tropical Tuna Commission, 31 May 1949

International Convention for the Regulation of Whaling, 2 December 1946

MARPOL Optional Annex Annex IV: Regulations for the Prevention of Pollution by Sewage from Ships

Prevention and Planning

(OECD) Declaration of Anticipatory Environmental Policies, 8 May 1979

International Convention on Oil Pollution Preparedness, Response and Cooperation, 29 November 1990

International Convention for the Prevention of Pollution from Ships, 2 November 1973

Stratospheric Ozone Protection

Amendment to the Montreal Protocol on Substances that Deplete the Ozone Layer Copenhagen, 1992
Adjustments and Amendment to the Montreal Protocol on Substances that Deplete the Ozone Layer, London, 23 June 1990
Kyoto Protocol to the United Nations Framework Convention on Climate Change, 1997
Montreal Protocol on Substances that Deplete the Ozone Layer, 16 September 1987
Vienna Convention for the Protection of the Ozone Layer, 22 March 1985

Sustainable Development

ASEAN Resolution on Sustainable Development, 30 October 1987
The Rio Declaration, 3-14 June 1992
Agenda 21, 3-14 June 1992
Declaration of the United Nations Conference on the Human Environment, 5 to 16 June 1972

Trade, Commerce and Industry

Convention on the Transboundary Effects of Industrial Accidents, 17 March 1992
The Rio Declaration, 3-14 June 1992
Agenda 21, 3-14 June 1992
Convention on the Ban of the Import into Africa and the Control of Transboundary Movement and Management of Hazardous Wastes Within Africa, 30 January 1991
Convention on the Regulation of Antarctic Mineral Resource Activities, 2 June 1988
Convention on the Transboundary Effects of Industrial Accidents, 17 March 1992
International Tropical Timber Agreement, 18 November 1983
European Convention for the Protection of Animals Kept for Farming Purposes, 10 March 1976
Convention on International Trade in Endangered Species of Wild Fauna and Flora, 3 March 1973
European Agreement on the Restriction of the Use of Certain Detergents in Washing and Cleaning Products, 16 September 1968

Waste

Agreement Between the Government of the United States of America and the Government of Canada Concerning the Transboundary Movement of Hazardous Wastes, 1986
Basel Convention on the Control of Transboundary Movements of Hazardous Wastes and Their Disposal, 22 March 1989
Convention for the Prevention of Marine Pollution by Dumping from Ships and Aircraft (as amended), 15 February 1972
Convention on the Ban of the Import into Africa and the Control of Transboundary Movement and Management of Hazardous Wastes Within Africa, 30 January 1991
(OECD) Comprehensive Waste Management Policy, 28 September 1976
(OECD) Comprehensive Waste Management Policy, 28 September 1976
(OECD) Control of Transfrontier Movements of Wastes Destined for Recovery Operations, 30 March 1992
(OECD) Control of Transfrontier Movements of Wastes Destined for Recovery Operations, 30 March 1992

(OECD) Recommendation of the Council on the Reduction of Transfrontier Movements of Wastes, 31 January 1991

(OECD) Waste Paper Recovery, 30 January 1980

INTERNATIONAL TREATIES RELATING TO THE ENVIRONMENT

Alphabetical Listing

Adjustments and Amendment to the Montreal Protocol on Substances that Deplete the Ozone Layer, 29 June 1990

Adjustments and Amendment to the Montreal Protocol on Substances that Deplete the Ozone Layer, 23-25 November 1992

Agenda 21, 3-14 June 1992

Agreed Measures for the Conservation of Antarctic Fauna and Flora, 2 June 1964

Agreement Between the Government of Canada and the Government of the United States of America on Air Quality, 1991

Agreement Between the Government of the United States of America and the Goverment of Canada Concerning the Transboundary Movement of Hazardous Wastes, 1986

Agreement Between The United States and Canada on Great Lakes Water Quality, 1978

Agreement Concerning Cooperation in Marine Fishing, 28 July 1962

Agreement Concerning Cooperation in Taking Measures Against Pollution of the Sea by Oil, 16 September 1971

Agreement Concerning Cooperation in the Quarantine of Plants and Their Protection Against Pests and Diseases, 14 December 1959

Agreement Concerning Interim Arrangements Relating to Polymetallic Nodules of the Deep Sea, 2 September 1982

Agreement Concerning Measures for Protection of the Stocks of Deep-Sea Prawns (Pandalus borealis), European Lobsters (Homarus vulaaris), Norway Lobsters (Nephropsnorveaicus) and Crabs (Cancer Paqurus) (as amended), 7 March 52

Agreement Concerning the Protection of the Waters of the Mediterranean Shores, 10 May 1976

Agreement Establishing the Inter-American Institute for Global Change Research, 13 May 1992

Agreement establishing the South Pacific Regional Environment Programme (SPREP), 16 June 1993

Agreement for Cooperation in Dealing with Pollution of the North Sea by Oil, 9 June 1969

Agreement for Cooperation in Dealing with Pollution of the North Sea by Oil and Other Harmful Substances, 13 September 1983

Agreement for the Establishment of a Commission for Controlling the Desert Locust in North-West Africa (as amended), 1 November 1970

Agreement for the Establishment of a Commission for Controlling the Desert Locust in the Eastern Region of its Distribution Area in South-West Asia (as amended), 3 December 1963

Agreement for the Establishment of a Commission for Controlling the Desert Locust in the Near East (as amended), 2 July 1965

Agreement for the Establishment of a General Fisheries Council for the Mediterranean (as amended), 24 September 1949

Agreement on Conservation of Polar Bears, 15 November 1973

Agreement on Regional Cooperation in Combating Pollution of the South-East Pacific by Hydrocarbons and Other Harmful Substances in Cases of Emergency, 12 November 1981

Agreement on the Network of Aquaculture Centres in Asia and the Pacific, 8 January 1988

Agreement on the Organization for Indian Ocean Marine Affairs Cooperation, 17 September 1990

Agreement Relating to the Implementation of Part XI of the United Nations Convention on the Law of the Sea of 10 December 1982—done 28 July 1994

Agreement to Promote Compliance with International Conservation and Management Measures by Fishing Vessels on the High Seas, 29 November 1993

Amendment to the Annex to the Convention for the Prevention of Marine Pollution by Dumping of Wastes and Other Matter, 3 November 1989

Amendment to the Annex to the Convention on the Prevention of Marine Pollution by Dumping of Wastes and Other Matter, 24 September 1980

Amendments to the Convention on the Prevention of Marine Pollution by Dumping of Wastes and Other Matter Concerning Settlement of Disputes, 12 October 1978

Amendments to the International Convention for the Prevention of Pollution of the Sea by Oil Concerning the Protection of the Great Barrier Reef, 12 October 1971

Annex III to the Protocol of 17 February 1978 relating to the International Convention for the Prevention of Pollution from Ships of 2 November 1973 (MARPOL 73/78), as amended on 30 October 1992

Antarctic Treaty, 1 Dec 1959

ASEAN Agreement on the Conservation of Nature and Natural Resources, 9 July 1985

ASEAN Resolution on Sustainable Development, 30 October 1987

Basel Convention on the Control of Transboundary Movements of Hazardous Wastes and Their Disposal, 22 March 1989

Benelux Convention on Nature Conservation and Landscape Protection, 8 June 1982

Benelux Convention on the Hunting and Protection of Birds (as amended), 10 June 1970

Convention Concerning Fishing in the Black Sea (as amended), 7 July 1959

Convention for a North Pacific Marine Science Organisation (PICES), 12 December 1990

Convention for Cooperation in the Protection and Development of the Marine and Coastal Environment of the West and Central African Region, 23 March 1981

Convention for the Conservation and Management of the Vicuna, 20 December 1979

Convention for the Conservation of Antarctic Seals, 1 June 1972

Convention for the Conservation of Salmon in the North Atlantic Ocean, 2 March 1982

Convention for the Conservation of Southern Bluefin Tuna, 10 May 1993

Convention for the Establishment of an Inter-American Tropical Tuna Commission, 31 May 1949

Convention for the Establishment of the European and Mediterranean Plant Protection Organization (as amended 18 April 1951 PARIS), 18 April 1951

Convention for the International Council for the Exploration of the Sea (as amended), 12 September 1964

Convention for the Prevention of Marine Pollution by Dumping from Ships and Aircraft (as amended), 15 February 1972

Convention for the Protection and Development of the Marine Environment of the Wider Caribbean Region, 24 March 1983

Convention for the Protection of the Marine Environment and Coastal Area of the South-East Pacific, 12 November 1981

Convention for the Protection of the Marine Environment of the North East Atlantic, 22 September 1992

Convention for the Protection of the Mediterranean Sea Against Pollution, 16 February 1976

Convention for the Protection of the Natural Resources and Environment of the South Pacific Region, 24 November 1986

Convention for the Protection of the World Cultural and Natural Heritage, 23 November 1972

Convention for the Protection, Management and Development of the Marine and Coastal Environment of the Eastern African Region, 21 June 1985

Convention for the Suppression of Unlawful Acts Against the Safety of Maritime Navigation, 10 March 1988

Convention of the African Migratory Locust Organization, 25 May 1962

Convention on Biological Diversity, 5 June 1992

Convention on Conservation of Nature in the South Pacific, 12 June 1976

Convention on Environmental Impact Assessment in a Transboundary Context, 25 February 1991

Convention on Fishing and Conservation of the Living Resources in the Baltic Sea and Belts, 13 September 1973

Convention on Fishing and Conservation of the Living Resources of the High Seas, 29 April 1958

Convention on Future Multilateral Cooperation in North-East Atlantic Fisheries, 18 November 1980

Convention on International Trade in Endangered Species of Wild Fauna and Flora, 3 March 1973

Convention on Long-Range Transboundary Air Pollution, 13 November 1979

Convention on Nature Protection and Wildlife Preservation in the Western Hemisphere, 12 October 1940

Convention on the Ban of the Import into Africa and the Control of Transboundary Movement and Management of Hazardous Wastes Within Africa, 30 January 1991

Convention on the Conservation of Antarctic Marine Living Resources, 20 May 1980

Convention on the Conservation of European Wildlife and Natural Habitats, 19 September 1979

Convention on the Conservation of Migratory Species of Wild Animals, 23 June 1979

Convention on the Continental Shelf, 29 April 1958

Convention on the High Seas, 29 April 1958

Convention on the Prevention of Marine Pollution by Dumping of Wastes and Other Matter, 29 December 1972

Convention on the Prevention of Marine Pollution from Land-based Sources, 4 June 1974

Convention on the Regulation of Antarctic Mineral Resource Activities, 2 June 1988

Convention on the Territorial Sea & the Contiguous Zone, 29 April 58

Convention on the Transboundary Effects of Industrial Accidents, 17 March 1992

Convention on Wetlands of International Importance Especially as Waterfowl Habitat, 2 February 1971

Draft Agreement for the Implementation of the Provisions of the United Nations Convention on the Law of the Sea of 10 December 1982 Relating to the Conservation and Management of Straddling Fish Stocks and Highly Migratory Fish Stocks, 4 August 1995

European Agreement on the Restriction of the Use of Certain Detergents in Washing and Cleaning Products, 16 September 1968

European Convention for the Protection of Animals During International Transport, 13 December 1968

European Convention for the Protection of Animals for Slaughter, 10 May 1979

European Convention for the Protection of Animals Kept for Farming Purposes, 10 March 1976

European Convention for the Protection of Pet Animals, 13 November 1987

Fisheries Convention, 9 March 1964

Interim Convention on Conservation of North Pacific Fur Seals, 9 February 1957

1980 Protocol Amending the Interim Convention on Conservation of North Pacific Fur Seals

1984 Protocol Amending the Interim Convention on Conservation of North Pacific Fur Seals, 12 October 1984

International Agreement on the Use of INMARSAT Ship Earth Stations within the Territorial Sea and Ports, 16 October 1985

International Convention for the Conservation of Atlantic Tunas, 14 May 1966

International Convention for the High Seas Fisheries of the North Pacific Ocean (as amended), 9 May 1952

International Convention for the Prevention of Pollution from Ships, 2 November 1973

International Convention for the Prevention of Pollution of the Sea by Oil (as amended on 11 April 1962 and 21 October 1969), 12 May 1954

International Convention for the Protection of Birds, 18 October 1950

International Convention for the Regulation of Whaling, 2 December 1946

International Convention for the Safety of Life at Sea, 17 June 1960

International Convention on Civil Liability for Oil Pollution Damage, 29 November 1969

International Convention on Oil Pollution Preparedness, Response and Cooperation, 29 November 1990

International Convention on the Establishment of an International Fund for Compensation for Oil Pollution Damage, 18 December 1971

International Convention Relating to Intervention on the High Seas in Cases of Oil Pollution Casualties, 29 November 1969

International Convention Relating to the Limitation of the Liability of Owners of Sea-Going Ships, 10 October 1957

International Plant Protection Convention, 6 December 1951

International Tropical Timber Agreement, 18 November 1983

International Tropical Timber Agreement 26 January 1994 Kuwait Regional Convention for Cooperation on the Protection of the Marine Environment from Pollution, 24 April 1978

MARPOL Optional Annex Annex IV: Regulations for the Prevention of Pollution by Sewage from Ships, 1973, 1978

Montreal Protocol on Substances that Deplete the Ozone Layer, 16 September 1987

North-East Atlantic Fisheries Convention, 24 January 1959

Niue Treaty on Cooperation in Fisheries Surveillance and Law Enforcement in the South Pacific Region, 9 July 1992

Nordic Environmental Protection Convention, 19 February 1974

(OECD) Assessment of Projects with Significant Impact on the Environment, 8 May 1979

(OECD) Comprehensive Waste Management Policy, 28 September 1976

(OECD) Control of Air Pollution from Fossil Fuel Combustion, 20 June 1985

(OECD) Control of Eutrophication of Waters, 14 November 1974

(OECD) Control of Transfrontier Movements of Wastes Destined for Recovery Operations, 30 March 1992

(OECD) Declaration of Anticipatory Environmental Policies, 8 May 1979

(OECD) Declaration on Environment Resources for the Future, 20 June 1985

(OECD) Declaration on Environmental Policy, 14 November 1974

(OECD) Energy and the Environment, 14 November 1974

(OECD) Environment and Economics Guiding Principles Concerning International Economic Aspects of Environmental Policies, 26 May 1972

(OECD) Implementation of the Polluter-Pays Principle, 14 November 1974

(OECD) International Conference on Environment and Economics: Conclusions, 21 July 1984

(OECD) Measures to Reduce All Man-Made Emissions of Mercury to the Environment, 18 September 1973

(OECD) Recommendation of the Council on Further Measures for the Protection of the Environment by Control of Polychlorinated Biphenyls, 13 February 1987

(OECD) Recommendation of the Council on the Reduction of Transfrontier Movements of Wastes, 31 January 1991

(OECD) Strategies for Specific Water Pollutants Control, 4 November 1974

(OECD) Waste Paper Recovery, 30 January 1980

(OECD) Protection of the Environment by Control of Polychlorinated Biphenyls, 13 September 1973

Optional Protocol of Signature Concerning the Compulsory Settlement of Disputes Arising out of the United Nations Conference on the Law of the Sea, 29 April 1958

Plant Protection Agreement for the South-East Asia and Pacific Region (as amended), 27 February 1956

Protocol Amending the 1978 Agreement Between the United States of America and Canada on great Lakes Water Quality, 1987

Protocol Amending the International Convention Relating to the Limitation of the Liability of Owners of Sea-Going Ships, 21 December 1979

Protocol Concerning Co-operation in Combating Pollution of the Mediterranean Sea by Oil and Other Harmful Substances in Cases of Emergency, 2 February 1976

Protocol Concerning Cooperation in Combating Oil Spills in the Wider Caribbean Region, 24 March 1983

Protocol Concerning Cooperation in Combating Pollution Emergencies in the South Pacific Region, 25 November 1986

Protocol Concerning Cooperation in Combating Pollution in Cases of Emergency, 21 March 1981

Protocol Concerning Marine Pollution Resulting from Exploration and Exploitation of the Continental Shelf, 29 March 1989

Protocol Concerning Mediterranean Specially Protected Areas, 3 April 1982

Protocol Concerning Regional Cooperation in Combating Pollution by Oil and Other Harmful Substances in Cases of Emergency, 14 February 1982

Protocol for the Prevention of Pollution of the Mediterranean Sea by Dumping from Ships and Aircraft, 16 February 1976

Protocol for the Prevention of Pollution of the South Pacific Region by Dumping, 25 November 1986

Protocol for the Protection of the Mediterranean Sea Against Pollution from Land-Based Sources, 17 May 1980

Protocol for the Suppression of Unlawful Acts against the Safety of Fixed Platforms Located on the Continental Shelf, 10 March 1988

Protocol of 1978 Relating to the International Convention for the Prevention of Pollution from Ships, 17 February 1978

Protocol on Environmental Protection to the Antarctic Treaty, 4 October 1991

Protocol on Protection of the Black Sea Marine Environment Against Pollution from Land Based Sources, 21 April 1992

Protocol Relating to Intervention on the High Seas in Cases of Pollution by Substances Other than Oil, 2 November 1973

Protocol to Amend the Convention on Wetlands of International Importance Especially as Waterfowl Habitat, 3 December 1982

Protocol to Amend the International Convention on Civil Liability for Oil Pollution Damage, 25 May 1984

Protocol to Amend the International Convention on the Establishment of an International Fund for Compensation for Oil Pollution Damage, 25 May 1984

Protocol to the 1979 Convention on Long-Range Transboundary Air Pollution on Long-Term Financing of Cooperative Programme for Monitoring and Evaluation of the Long-Range Transmissions of Air Pollutants in Europe (EMEP), 28 September 1984

Protocol to the International Convention on Civil Liability for Oil Pollution Damage, 19 November 1976

Protocol to the International Convention on the Establishment of an International Fund for Compensation for Oil Pollution Damage, 19 November 1976

Protocol to the Kuwait Regional Convention for the Protection of the Marine Environment Against Pollution from Land-Based Sources, 21 February 1990

Regional Convention for the Conservation of the Red Sea and Gulf of Aden Environment, 14 February 1982

Rio Declaration, 3-14 June 1992

South Pacific Forum Fisheries Agency Convention, 10 July 1979

South Pacific Nuclear Free Zone Treaty, 6 August 1985

Supplementary Protocol to the Agreement on Regional Co-Operation in Combating Pollution of the South-East Pacific by Hydrocarbons or Other Harmful Substances, 22 July 1983

Treaty Banning Nuclear Weapon Tests in the Atmosphere, in Outer Space and Under Water, 10 October 1963

Treaty for Amazonian Cooperation, 3 July 1978

United Nations Conference on Desertification (UNCOD) Plan of Action to Combat Desertification and General Assembly Resolutions, 29-9 August 1977 (200k file!!)

United Nations Convention on the Law of the Sea, 10 December 1982

United Nations Convention to Combat Desertification in those Countries Experiencing Serious Drought and/or Desertification, Particularly in Africa, 12 September 1994

United Nations Framework Convention on Climate Change, 9 May 1992

Vienna Convention for the Protection of the Ozone Layer, 22 March 1985

Vienna Convention on the Law of Treaties, 23 May 1969

World Charter for Nature, 1982

INTERNATIONAL TREATIES RELATING TO THE ENVIRONMENT

Listing for North America

Agreement for Cooperation, Environmental Problems in the Border Areas, U.S.-Mexico

Agreement on Energy Cooperation between Quebec and New York

Agreement on Acid Precipitation

Agreement Regarding Establishment of Joint Pollution Contingency Plans for Spills of Oil

Agreement in Sustainable Development

Agreement Approving IBWC Minute 218 Colorado River Salinity Problem

Agreement on the Exchange of Information on Weather Modification Activities

Agreement on Permanent and Definitive Solution (IBWC Minute 242) Colorado River Salinity Problem

Agreement on Arctic Cooperation

Agreement Regarding Canada-U.S. Committee on Water Quality in the St. John River and its Tributary Rivers

Agreement Regarding the Establishment of a Border Environmental Cooperation Commission and NADBANK

Agreement on Great Lakes Water Quality

Agreement Relating to the Establishment of the Roosevelt Campobello International Park

Agreement on Cooperation Improve Management of Arid and Semi-Arid Lands and Control Desertification

Agreement on Air Quality

Agreement Concerning the Transboundary Movement of Hazardous Waste

Agreement Respecting Cooperation in Radioactive Waste Management

Agreement on the Conservation of Polar Bears

Agreement of Cooperation Re: Pollution of the Marine Environment by Discharges of Hydrocarbons

Agreement on the Conservation of the Porcupine Caribou Herd

Agreement to Track Air Pollution Across Eastern North America (Cross Appalachian Tracer Experiment)
Antarctic Treaty
Arrangement Prohibiting the Importation of Raccoon Dogs
Arctic Cooperation, Agreement on and Exchange of Notes Concerning Transit of Northwest Passage
Binational Childhood Lead Surveillance and Education
Binational Wastewater Treatment (Sanitation) Agreement
Black Bass Task Force of the American Fisheries Society, Northern Division
Boundary Waters Treaty
British Columbia and Alaska Agreement in Natural Resources
British Columbia and United States Collaboration in Economic Development, MOU
British Columbia and Washington Air Management Task Force
British Columbia and Washington Arrangement in Fisheries Management
British Columbia and Washington Memorandum of Understanding
British Columbia and Washington, Cooperative Working Group Written Agreements between
Code of Practice International Transboundary Movement of Radioactive Waste (IAEA)
Columbia River Treaty
Controlling Traffic in Wild Species of Flora and Fauna, U.S.-Mexico Joint Committee Cooperative Agreements
Convention on International Trade in Endangered Species of Wild Fauna and Flora (CITES)
Convention for the Protection of Migratory Birds and Game Mammals
Convention on the Prevention of Marine Pollution by Dumping of Wastes and Other Matter
Convention for the Protection, Preservation and Extension of the Sockeye Salmon Fisheries in Fraser
Convention for the Protection of Migratory Birds
Convention on Great Lakes Fisheries
Convention on Long-Range Transboundary Air Pollution
Convention on Future Multilateral Cooperation in the North-West Atlantic Fisheries (1978)
Convention for the Establishment of a Tribunal Operation of the Smelter at Trail, BC
Convention on Biological Diversity
Convention on the Conservation of Antarctic Marine Living Resources
Convention on the Regulation of Antarctic Mineral Resources Activities
Convention on Nature Protection and Wildlife Preservation in the Western Hemisphere
Convention for the Preservation of Halibut Fishery of the Northern Pacific Ocean and the Bering Sea
Convention for the Preservation of the Halibut Fishery of the Northern Pacific Ocean
Convention on Wetlands (Ramsar, Iran, 1971)
Cooperation in the Field of Climate-Related Programs, Memorandum of Understanding on
Cooperation Relating to the Management of Lake Champlain
Cooperative Agreement with the (SCERP) Southwest Center for Environmental Research and Policy
Cooperative Resources Management
Cooperative Agreement Concerning Plant Protection Against the Mediterranean Fruit Fly
Delineation of Boundaries of Exclusive Fishery Zones in the Gulf of Mexico and the Pacific
Eastern Pacific Ocean Tuna Fishing Agreement
Environmental Health Monitoring and Cooperation, El Paso - Juarez
Energy Cooperation Arrangement
Environmental Cooperation Agreement between Quebec and New York
Environmental Cooperation Agreement between Quebec and the State of Mexico
Environmental Cooperation Agreement between Quebec and Vermont

Environmental Cooperation Agreement between Quebec and Wisconsin

Environmental Cooperation between British Columbia and Washington

Environmental Cooperation between Government of United Mexican States and Government of Canada

FAO International Code of Conduct of the Distribution and Use of Pesticides

Fisheries Agreement

Fisheries Enforcement, Agreement on

Great Lakes Charter

Growth Management Agreement

International Convention for the High Seas Fisheries of the North Pacific Ocean

International Convention for the Conservation of Atlantic Tunas

International Convention for the Regulation of Whaling and 1956 Protocol

International Air Quality Control District Air Pollution Control Program, El Paso - Juarez

Elimination of Toxic Substances from the Great Lakes Environment

Integrated Environmental Plan for the United States-Mexican Border Area

Interlaboratory Cooperation

Interim Convention on Conservation of and Amending Protocols North Pacific Fur Seals

International Tropical Timber Agreement

International Convention for the Prevention of Pollution from Ships 1978

International Convention (with annexes) for the Prevention of Pollution of the Sea by Oil

International Steering Committee, Rainy Lake and Namakan Reservoir Water Level

Loon Habitat Suitability Model/Management Strategy Development

Memorandum of Understanding on Cooperation in the Management and Protection of National Parks

Memorandum of Understanding on the Environmental Cooperation on the Management of Lake Champlain

Memorandum of Understanding between Mexico & U.S. EPA Cooperation on Environmental Programs and Transboundary Problems,

Memoandum of Understanding on Mineral Resource Assessment Data and Services Manitoba and North Dakota

Memorandum of Understanding on the Conservation of the Whooping Crane

Memorandum of Understanding Regarding Accidental and Unauthorized Discharges of Pollutants Along Inland Boundary

Memorandum of Understanding Regional Coordination on Hazardous Waste Management

Memorandum of Understanding Between Maine and Manitoba Regarding St. Croix International Waterway

Memorandum of Understanding Concerning Cooperation in Research and Control of Dryland Salinity

Memorandum of Understanding Research and Development Cooperation in Pollution Measurement and Control

Memorandum of Understanding on Cooperation in Technical Assistance in the General Area of Earth Sciences

Memorandum of Intent Concerning Transboundary Air Pollution

Montreal Protocol on Substances that Deplete the Ozone Layer

Multi-State Fish and Wildlife Information Systems and the Ontario Wildlife Information Management

North American Plant Protection Agreement

North American Waterfowl Management Plan

North American Waterfowl Management Plan—Eastern Habitat Joint Venture

North Dakota and Manitoba Agreement on Consultation and Cooperation

Pacific Salmon Treaty

OECD Council Recommendation on Principles Regarding Transfrontier Pollution
Protection and Improvement of the Environment in the Metropolitan Area of Mexico City
Protocol Regarding Intervention on the High Seas in Cases of Pollution by Substances Other than Oil
Protocol to the Cartagena Convention Regarding Specially Protected Areas and Wildlife
Regional Agreement on the Research and Management of Marine Turtles in the American Pacific
Reciprocal Fisheries Agreement
Salmon Convention for the Protection of in the North Atlantic Ocean
State-to-State Environmental Exchange Program
States-Provinces Agricultural Accord
Treaty on Maritime Boundaries
Treaty and Protocol Between U.S. and Mexico on the Utilization of Colorado and Tijuana Rivers
Vienna Convention for the Protection of the Ozone Layer

Part III

Environmental Standards

6 An Introduction to International Standards Organizations and the Development Process

CONTENTS

Standards are documented agreements containing technical specifications or criteria to be used as guidelines (and possibly legal requirements) to ensure materials, products, processes, or services are appropriate for their purpose. A standardization body is responsible for the development, presentation, and publication of appropriate standards. These organizations can be national bodies, international organizations, or both.

The standards development process begins when a need is expressed to a national standards organization, which may develop and publish a standard or present the issue as a new work item to an international body such as the International Organization for Standardization (ISO). The following standards organizations participate in the development of consensus, industry-wide, voluntary standards. [A complete listing of U.S. and international standards organizations appears in Chapter 12.]

Committees

- European Committee for Standardization (CEN)
- European Committee for Electrical Standardization (CENELEC)
- Technical Committee (TC) 207 Environmental Management (ISO 14000)
- Subcommittees (SC) to TC 207- SC1 through SC6

Associations, Councils, Organizations

- American National Standards Institute (ANSI)
- American Society of Testing and Materials (ASTM)
- British Standards Institution (BSI)
- International Organization for Standardization (ISO)
- American Society of Quality Control (ASQC)

- Canadian Standards Association (CSA)
- Standards Association of Australia (SAA)
- Deutches Insitut for Normung e. V. (DIN)
- National Fire Protection Association (NFPA)

Standards organizations have developed codes, practices, and standards designed to protect property, the environment, and the health and welfare of workers and the public. Also, international standardization resulting from consensus agreements has helped facilitate international trade by helping eliminate technical trade barriers caused by nonharmonized standards. Standards are the key to current and future business strategies designed to reduce costs, improve quality, and secure a larger market share.

Conforming to internationally recognized standards for a company's product or service builds confidence in its customers. Standardization has moved beyond product specifications and service industries to include such diverse areas as environmental management, information processing and communications, textiles, packaging, and financial services.

Standards organizations may be comprised of qualified representatives of industry, research institutes, government authorities, or other standards organizations that provide membership opportunities and serve as international technical and educational organizations. Some standards organizations have been established by the federal government, such as the Occupational Safety and Health Administration (OSHA) in the United States, to develop regulations and standards to protect the health and safety of working men and women. Others, such as the NFPA, have been established to improve methods of fire protection and prevention, electrical safety, and other related safety goals.

Many of these standards are voluntary, but they demonstrate a level of performance or quality for products and services, which is valuable in marketing and product liability actions. Also, many standards are referenced by government agencies as minimum requirements required to meet federal law. As barriers to international trade are eliminated, there are more opportunities for companies to sell products abroad. However, the need to have consistent environmental and safety standards also increases.

National and international organizations develop standards designed to be universally recognized and applied. For example, ANSI developed the *American National Standard for Hazardous Industrial Chemicals—Precautionary Labeling* (Z129.1–1994, revision of ANSI Z129.1–1988). This standard was developed by a technical committee of the Chemical Manufacturers Association (CMA) and is designed to provide guidance for precautionary information on hazardous chemical container labels.

ISO has developed a series of *International Standards* for environmental management. ISO 14001 Environmental Management Systems (EMS) was developed to provide elements of an effective EMS to organizations wanting to demonstrate sound environmental performance.

STANDARDS ORGANIZATIONS IN THE UNITED STATES

Voluntary standards in the United States are developed by a decentralized network of private-sector participants with oversight and coordination provided by ANSI. ANSI, founded in 1918, is a private, not-for-profit membership organization that harmonizes private-sector standards activities within the United States. ANSI consists of approximately 1,300 national and international companies; 30 government agencies; 20 institutional members; and 250 professional, technical, trade, and consumer organizations. ANSI is the vehicle by which a company can become involved in the international standardization process.

Specialized "technical committees" also may develop a particular standard, which is then submitted to ANSI for approval under its accredited procedures. All ANSI standards must be reviewed and reaffirmed, modified, or withdrawn every 5 years in order to provide quality, current standards. ANSI is the sole U.S. representative and dues-paying member of the two major non-treaty standards organizations, ISO and the International Electrotechnical Commission (IEC).

ANSI participates in the technical program of both ISO and IEC and administers many key committees and subgroups. Comments and proposals from Europe to the European standardizing bodies (CEN/CENELEC) must originate through the respective national bodies. In the United States, ANSI is the member body for ISO and the United States National Committee.

ANSI provides its members with a variety of services, including publications such as *Standards Action* and the *ANSI Reporter*. Information databases, research, training and consulting services, and publication sales are also available. Publications available from ANSI include copies of approved American National Standards, ISO and IEC standards, proposals of regional groups tied to the European Union, and specifications from national standards organizations that belong to ISO.

ANSI maintains the Executive Standards Council, which accredits ISO Technical Advisory Groups (TAGs). The United States is represented by TAGs that participate in the development of technical positions relating to international standardization brought to ISO. The TAG is responsible for developing America's positions on standards for the ISO TC and for selecting delegates to the TC meetings.

ANSI appoints a TAG administrator to provide the overall administration of the TAG and ensure that the procedures governing their activities are followed. Standards submitted to ANSI are addressed by the *Executive Standards Council, Board of Standards Review*, and special planning panels to avoid duplicating efforts or developing conflicting standards. Developers of these standards (usually national trade, technical, professional, consumer, or labor organizations) are guided through the approval process by ANSI staff professionals. In summary, ANSI:

- coordinates voluntary standards activities;
- approves American National Standards;
- represents U.S. interests in international standardization; and
- provides information on and access to world standards.

As the U.S. member of ISO for Environmental Management (ISO/TC 207), ANSI works in conjunction with the ASTM and the ASQC on program activities (Figure 4).

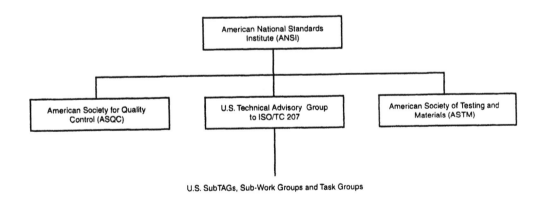

Figure 4 U.S. Organizational Structure for ISO 14000

The organizational structure for ISO/TC 207 in the U.S. is comprised of the following:

- U.S. Technical Advisory Group (U.S. TAG)
- U.S. SubTAGs (ST) and Sub-Work Group (SWG)

    - ST1 - Environmental Management System

        ♦ TG1 - Specification
        ♦ TG2 - General Guidelines

    - ST2 - Environmental Auditing

        ♦ TG1 - General Principles
        ♦ TG2 - Audit Procedures
        ♦ TG3 - Qualifications Criteria
        ♦ TG4 - Environmental Site Assessments

    - ST3 - Environmental Labeling

        ♦ TG1 - Guiding Principles for Practitioner Programs
        ♦ TG2 - Self-Declaration Claims
        ♦ TG3 - Guiding Principles for Environmental Labeling Programs

    - ST4 - Environmental Performance Evaluation
    - ST5 - Life-Cycle Assessment
    - ST6 - Terms and Definitions
    - SWG - Environmental Aspects in Product Standards

- U.S. Task Groups

Each SubTAG has a chairman, and each task group's activities are coordinated by a "leader." The ASTM and the ASQC act as administrators.

INTERNATIONAL STANDARDIZATION ACTIVITIES OF ASTM

Other standardization activities are being pursued by the ASTM. ASTM has developed and published more than 10,000 technical standards. These standards, used by organizations worldwide, have been developed from the work of standards-writing committees made up of technically qualified ASTM members. ASTM was organized in 1898, and is now one of the largest voluntary standards development organizations in the world. ASTM publishes the following principal types of consensus standards:

- Standard Test Methods
- Specifications
- Practices
- Guides
- Classifications
- Terminology

Standards writing is conducted by 132 main ASTM technical committees, which are divided into subcommittees. Subcommittees are further divided into task groups, responsible for preparing a draft standard. The activities of the technical committees are governed by the ASTM Board of Directors. As part of the standards process, an ASTM subcommittee will develop a working document for use by the appropriate committee. Full procedures for the standards development process are found in the *Regulations Governing ASTM Technical Committees and the Form and Style for ASTM Standards*. ASTM standards are published each year in the *Annual Book of ASTM Standards*. This publication includes more than 9,000 standards contained in 71 volumes.

ASTM has also developed the following test methods, practices, and standards for environmental subjects:

- Environmental Site Assessment (ASTM E-50)

  - ASTM E 1527, *Standard Practice for Environmental Site Assessments: Phase I Environmental Site Assessment Process*
  - ASTM E 1528, *Standard Practice for Environmental Site Assessment Transaction Screen Process*

- Environmental Risk Management (ASTM E-51)
- Water and Environmental Technology
- Hazardous Substances and Oil Spill Response
- Environmental Auditing and Management

  - ASTM PS-11, *Provisional Standard Practice for Environmental Regulatory Compliance Audits*
  - ASTM PS-12, *Provisional Standard for the Study and Evaluation of an Organization's Environmental Management System*

Relating to the international standardization of Environmental Management, ASTM is the administrator of the U.S. TAG to ISO TC/207 and SubTAGs 3-6 and SWG on Environmental Aspects in Product Standards (Figure 1).

INTERNATIONAL STANDARDIZATION ACTIVITIES OF ISO

ISO is a worldwide federation of national standards bodies from more than 100 countries. ISO is a nongovernmental organization established in 1947 with its first standard published in 1951. The mission of ISO is to promote the development of standardization activities to promote the exchange of goods and services.

ISO establishes TCs to address specific standards issues. These committees are identified by number with an ISO prefix. (e.g., ISO/TC207: *Environmental Management*). ISO's work results in international agreements, which are published as International Standards. Standards issued by ISO, such as the *Draft International Standard for Environmental Management Systems* (ISO/DIS 14001) are referenced using the issuing body abbreviation first (ISO) and DIS (draft international standard), if appropriate, followed by the standard number (14001).

When a standard has been finalized, the date of the last revision is also listed. Like ANSI, the technical work of ISO is decentralized and implemented by approximately 2,700 technical committees, subcommittees, and working groups. The committee members meet to resolve global standardization issues.

ISO and the IEC established the Strategic Advisory Group on the Environment (SAGE) in 1991. In December 1992, SAGE recommended a TC established to develop international Environmental Management Standards (Figure 5). ISO/TC 207 Environmental Management is made up of six subcommittees (SC) and one working group. Each subcommittee is responsible for developing standards documents for a specific area and is assigned a secretariat. For example, SC1 is responsible for Environmental Management Systems. The secretariat for SC1 is the BSI of the United Kingdom. Each SC is further divided into SC working groups. SC1 contains two working groups (Specification and General Guidelines for EMS).

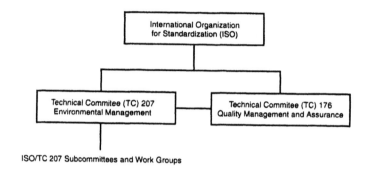

ISO/TC 207 Subcommittees and Work Groups

Figure 5 ISO Organization Structure for ISO TC/207 Environmental Management

To participate in ISO TC/207, member status must be obtained from ISO. Three types of member status include the following:

- Participating (P-members)
- Observing (O-members)
- Liaison (L-members)

Participating members represent those countries who wish to vote, participate actively in discussions, and have access to all relevant documentation. Observing and liaison members participate in discussions, receive relevant information, but do not vote.

RESPONSIBILITIES FOR THE TECHNICAL WORK OF ISO

A Technical Management Board (TMB) is responsible for the overall management of the technical work, including:

- Establishment of technical committees in ISO;
- Appointment of chairmen for technical committees;
- Allocation or reallocation of secretariats of technical committees;
- Approval of titles and programs of work of technical committees;
- Ratification of the establishment of subcommittees by technical committees;
- Ratification of the dissolution of subcommittees by technical committees;
- Coordination of technical work for a particular issue;
- Assignment of responsibility for the development of standards;
- Monitoring the progress of the technical work;
- Reviewing the need for work in new fields of technology;
- Planning the need for work in new fields of technology;
- Maintenance of the ISO/IEC Directives and other rules for the technical work; and
- Consideration of appeals concerning decisions on new work item proposals, on committee drafts, on inquiry drafts, or on Final Draft International Standards.

A TMB may establish advisory groups comprised of experts in the relevant fields. A group having advisory functions may be established by one of the TMBs or jointly by two TMBs. The tasks allocated to such a group may include the making of proposals relating to the drafting or agreement of normative documents (in particular, International Standards and Technical Reports), but does not include the preparation of the documents unless specifically authorized by the TMB(s). The work of advisory groups is presented in the form of recommendations to the TMB(s). The recommendations may include proposals for the establishment of a working group or a joint working group for the preparation of normative documents.

The Joint Technical Programming Committee (JTPC) has the responsibility of avoiding or eliminating overlap in the technical work of ISO and IEC. Decisions of the JTPC are communicated to both organizations for immediate implementation. Joint technical committees may be established by a common decision of the ISO TMB and IEC Council. Technical committees are numbered in sequence in the order in which they are established and must agree on their title and scope after its establishment.

The chief executive officer (CEO) of the respective organization is responsible for implementing the ISO/IEC Directives and other rules for the technical work. The office of the CEO arranges all contacts between the TCs, Council, and TMB. Establishment and dissolution of technical committees is conducted by the TMB in ISO and by the Council of IEC.

A proposal for work in a new field, which appears to require the establishment of a new technical committee, may be made in the respective organization by the following:

- a national body;
- a technical committee or subcommittee;
- the Technical Management Board; and/or
- the CEO.

Subcommittees are established and dissolved by the parent technical committee, subject to ratification by the TMB. A subcommittee may be established only on the condition that a national body has expressed its readiness to undertake the role of secretariat. A subcommittee is comprised of at least five members of the parent technical committee, who are prepared to participate actively in the work of the subcommittee. The title and scope of a subcommittee is determined by the parent technical committee. The secretariat of the parent technical committee informs the office of the CEO of the decision to establish a subcommittee.

All national bodies have the right to participate in the work of technical committees and subcommittees. Each national body must indicate to the office of the CEO if it intends to participate actively in the work or follow the work as an observer (O-members). A national body may choose to be neither P-member nor O-member of a given committee.

National bodies have the responsibility to organize their national input, taking into account relevant interests at their national level. Membership of a technical committee does not imply automatic membership of a subcommittee. If a member of a technical committee is interested in participating in the activities of a particular subcommittee, notification of this interest must be made. If a participating member does not meet their obligations of a P-member, their status can be changed to an O-member by the CEO.

Chairpersons of technical committees shall be nominated by the secretariat of the Technical Committee and appointed by the TMB for a maximum period of 6 years. The chairperson of a technical committee is responsible for the overall management of that technical committee, including any subcommittees and working groups. The chairperson of a technical committee or subcommittee is also responsible for conducting meetings in order to reach an agreement on committee drafts.

Secretariats of technical committees and subcommittees are allocated by ISO. For example, the secretariat for ISO/TC207 has been allocated to Canada (specifically, the Canadian Standards

Association), and the secretariat for SC4 (Environmental Performance Evaluation) to ISO/TC207 has been allocated to ANSI. The secretariat is responsible for monitoring, reporting, and ensuring active progress of the work and ensuring that the ISO/IEC Directives and the decisions of Council and the TMB are followed.

Technical committees or subcommittees may establish working groups for specific tasks. Working groups are comprised of individually appointed experts brought together to deal with specific tasks allocated to the working group. In special cases, a joint working group may be established to undertake a specific task in which more than one ISO and/or IEC technical committee or subcommittee is interested.

CURRENT INTERNATIONAL STANDARDIZATION ACTIVITIES OF ANSI

ANSI hosted a workshop in May 1995 for American organizations and companies to address the need for international standardization of occupational health and safety management systems (OHSMS). The perspectives gained from this meeting will provide the basis for the position of the United States on the issue to be presented to ISO. At the time of this writing, the position of the United States had not been decided, nor had ISO determined whether or not to pursue the development of OHSMS standards.

OSHA in the United States has long recognized that compliance with its standards alone cannot achieve the goals of the statute which initiated it, the *Occupational Safety and Health Act* (OSHAct). Therefore, OSHA developed the Voluntary Protection Program designed to encourage the improvement of employer-provided occupational safety and health programs providing the systematic protection of workers. These programs are composed of management systems for preventing or controlling occupational hazards. An outline of OHSMS issues has been prepared by a task group of the ANSI Board of Directors' International Advisory Committee and is available from ANSI.

Interest in having ISO develop an international standard on OHSMS has been expressed by various organizations in the United States and around the world. Further, the issue has been discussed by ISO TC/176 (Quality Management Systems) and ISO TC/207 (Environmental Management). These two ISO Technical Committees brought the issue to the ISO TMB for input. The TMB initially determined that it did not have sufficient information to recommend proceeding with OHSMS standards development. Because both areas are components of quality systems, many companies have already addressed the need to include occupational safety and health management with their environmental concerns.

Like the ISO/DIS for Environmental Management, a standard for OHSMS would focus on *systems*. These systems would cover a wide range of occupational health and safety and employee-employer issues. One aspect of worker protection in which an international standard exists is personal protective equipment (PPE). The European Committee for Standardization (CEN) has developed PPE standards produced by seven technical committees coordinated by an advisory group. Standards for OHSMS have been proposed in Australia, the United Kingdom, and Norway. The British Standards Institution (BSI) passed BS 8800, Occupational Health and Safety Management Systems in 1996.

DEVELOPMENT OF INTERNATIONAL ENVIRONMENTAL LAWS

Science has provided the necessary evidence and supporting data to evaluate environmental concerns such as global climate change, stratospheric ozone depletion, loss of biodiversity, and air and water contamination. These concerns raise the question whether there is sufficient evidence to develop and implement controls. Pollution has been traditionally thought of as a localized problem. However, it is now known that pollutants travel in air streams and waterways causing negative effects great distances from their source. Pollutants may travel in the tissue of animals or through the import or export of products or even on the vessel itself as has been the case with certain organisms transplanted to the Great Lakes as stowaways on ships. Transboundary effects of pollution, loss of vital resources, and global environmental damage have united the international community to protect important resources through the implementation of international agreements. Treaties or agreements may be between two or more parties that provide the legal obligations necessary to protect the environment. Treaties between two parties are called bilateral agreements and between multiple parties multilateral agreements. A treaty may apply to a "State" even if the State did not participate in its development. Also, a treaty can become binding national law even if it has not been ratified by that country. Although treaty guidelines are part of the Vienna Convention on the Law of Treaties ("Vienna Convention"), no formal process exists for the development of a treaty. Usually, a need arises that motivates a government, organization, or coalition to begin negotiations regarding an environmental issue. For example, the United Nations has convened numerous conventions that produced significant international agreements protecting the environment as well as producing action plans. The following steps are included in the development and implementation of a treaty:

1. Initiation
2. Negotiation
3. Adoption and Signature
4. Ratification
5. Implementation

During the initiation phase, a need is expressed, such as the depletion of the protective ozone layer. Since stratospheric ozone protects us from the harmful rays of the sun, a significant risk to human health and the environment may result if more ultraviolet radiation is allowed to hit the earth. Therefore, parties were motivated to enter into negotiations regarding protection of this valuable ozone layer. Negotiations may go on for years and may be influenced by economics and political circumstances.   Countries may represent themselves during these negotiations or be represented by another party. For example, the European Union represents its member states as a single entity. Countries may unify on an issue, as did the African nations when banning the importation of hazardous waste. Observers to these negotiations may include business and industry, environmental organizations, and the media. As text of the treaty is developed, states must decide if it is acceptable. Adoption of the treaty text must be put to a vote. Following adoption of the treaty text, it must be "authenticated." Authentication is the process where states sign indicating this is indeed the text to which they agree. A state is not bound to the requirements of a treaty, even if signed, until its ratification. Each country follows its own ratification. In the United States, ratification of a treaty is completed when the President presents it to the Senate and if approved, the President ratifies it and deposits a "notice of ratification" with the appropriate organization (e.g., United Nations). Adoption of implementing statutes completes the process in the United States.

DEVELOPMENT OF EUROPEAN UNION DIRECTIVES

Development of legislation applicable to all member states of the EU is conducted by its institutions (Council of Ministers, European Parliament, European Commission, European Court of Justice, and the European Council). Legislative acts may include Council Regulations, Council Decisions, Council Recommendations and Opinions, or Council Directives. Specific environmental issues such as hazardous waste, air and water pollution, or the adoption of an international agreement may require legislative action. The driving force for development of environmental protection directives came in 1972 with the Conference of Heads of States and Governments and the Stockholm Declaration. Efforts may be made to address the issue without legislation, or scientific study may be undertaken to help determine the extent of the problem.   Programs are in place in Europe to collect and analyze pollutant data used in decision making. If it is determined that Community legislation and specifically a directive is required, a proposal for a directive is developed. The European Commission is responsible for developing environmental action plans and drafting legislation, including proposals for environmental protection. Adoption of a directive can be by consultation procedure or cooperation procedure. The procedure used is dependent on which EEC Treaty article was used to develop the proposal. Directive proposals are sent from the Commission to the Council of Ministers. Input on the proposal is requested from the European Parliament and the Economic and Social Committee. Eventually, the Council must come to a common position on the proposal, which is sent to the Parliament. Parliament may accept, reject, or propose amendments to the proposal. Also, the Commission has the "right of amendment" and may do so at this stage. The Council must make a final decision and vote on the proposal. If passed (requires unanimous vote for matters relating to the environment), the directive is published in the Official Journal of the European Communities, which specifies the date of entry into force in the member states. Finally, member states must develop national legislation implementing the provisions of the directive. However, the doctrine of direct effect provides protection in that the directive may be applicable even in the absence of national legislation.

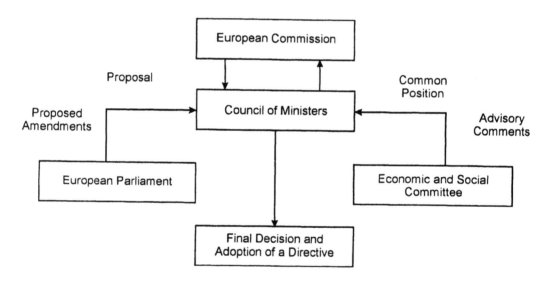

Figure 6 Adoption of an EU Directive

DEVELOPMENT OF NATIONAL LAW

Drivers for the development of national law include the need for environmental or resource protection, international law, EU legislation in the case of EU member states, political pressure, or even economics. In certain cases, an environmental catastrophe such as the Bhopal incident has motivated the passage of national laws. Most countries have designated an agency, ministry, or authority to handle environmental issues, including development of national programs. In the United States, the Environmental Protection Agency (EPA) was assembled in 1970 from existing agencies and took responsibility for programs formerly under the Department of Agriculture, Department of the Interior, and Health, Education and Welfare. Congress creates most environmental laws in the United States Bills (proposed laws) are developed by working committees and subcommittees such as the Water Resources Subcommittee of the Public Works and Transportation Committee in the House of Representatives and the Environmental Pollution Subcommittee of the Environment and Public Works Committee in the Senate. Bills approved by Congress are sent to the President for signature and if signed become law. Upon passage of a law, implementing regulations must be developed by the appropriate agency. The proposed rules developed by the agency (e.g., EPA) are made available for public comment through publication in the Federal Register. After review of public comments, the agency may revise the proposed rules and publish a final rule in the Federal Register. Effective dates for the rule are part of the Federal Register notice. States may be required to develop state law implementing the federal requirements. Like many countries, U.S. states may adopt more stringent requirements but must at least meet the federal requirements. Some countries such as Mexico adopt all international treaties they ratify into national law. The Unites States has been used as a model for regulatory programs for environmental protection. These programs were designed much the same as the media-specific U.S. laws of the 1970s. Of late, it has been realized that integrated programs which provide incentives are more effective.

7 Guiding Principles and Practices for Business

CONTENTS

INTRODUCTION

Tremendous environmental improvements have been realized over the past three decades through national and international legislation. However, those improvements have recently slowed and have been comparatively small. To improve the condition of the environment beyond what the regulatory structure can provide requires economic drivers for business. Companies, with the help of industry groups and strong corporate environmental commitments, have parlayed consumer interest in environmental protection into business opportunities. Guiding principles and codes of practice have been developed over the past eight to ten years by a variety of organizations for use by business and industry to improve environmental performance and promote sustainable development, including associations representing industries such as chemical manufacturing, agricultural, petroleum and energy, paper, travel and tourism, and engineering. Other programs have been developed by nongovernmental and economic organizations; environmental and business groups have also developed recommendations. These voluntary guidelines address environmental management, sustainability, environmental reporting, environmental and product stewardship, pollution prevention, and response to emergencies. They are designed to promote sustainable business practices and have been adopted by hundreds of companies worldwide.

Some of these guiding principles include

- Bellagio Principles
- Berlin Guidelines (mining and environment guidelines)
- CERES Principles
- Charter for Environmental Action in the International Hotel and Catering Industry (The Prince of Wales Business Leaders Forum)
- Charter of Ethics for Tourism and the Environment (Alliance Internationale de Tourisme)
- Code for Environmentally Responsible Tourism (Pacific Asia Travel Association)
- Code of Best Agricultural Practices to Optimize Fertilizer Use (Europe)
- Code of Ethics for the Industry (Tourism Industry Association of Canada)
- Codes of Management Practice (Chemical Industry of Finland)

- Codes of Practice (for workplace injuries and environmental performance indicators, New Zealand Chemical Industry Council)
- Declaration of the Business Council on Sustainable Development
- Environmental Code of Practice (Canadian Petroleum Association)
- Environmental Guidelines (European Petroleum Industry Association)
- Environmental Policy Statement (American Iron and Steel Institute)
- Environmental Principles (International Iron and Steel Institute)
- FAO International Code of Conduct on the Distribution and Use of Pesticides
- Guide for Environmental Practice (The Mining Association of Canada)
- Guidelines for Environmental Care and Protection in Finnish Industry
- ICC Business Charter for Sustainable Development
- Industrial Environmental Policy (Federation of Swedish Industries)
- International Council on Metals and the Environment
- Keidanren Principles
- Mission Statement (Petroleum Association for Conservation of the Environment)
- Mission Statement (The Institute of Petroleum, U.K.)
- PERI Guidelines
- Product Stewardship Program (Lead Industries Association)
- Responsible Care (Chemical Manufacturers Association)
- Statement on the Environment (Business and Industry Advisory Committee to the OECD)
- Statement of Environmental Policy and Code of Conduct (European Committee of Electricity Supply Undertakings)
- Statement of the International Leather Industry Principles for Improved Environmental, Health and Safety Performance (International Council of Tanners)
- Strategies for Today's Environmental Partnership Program (American Petroleum Institute)

BELLAGIO PRINCIPLES

In 1987, the World Commission on Environment and Development (Brundtland Commission) called for the development of new ways to measure and assess progress toward sustainable development. This call has been subsequently echoed in Agenda 21 of the 1992 Earth Summit and through activities that range from local to global in scale. In response, significant efforts to assess performance have been made by corporations, nongovernment organizations, academics, communities, nations, and international organizations. In November 1996, an international group of measurement practitioners and researchers from five continents came together at the Rockefeller Foundation's Study and Conference Center in Bellagio, Italy to review progress to date and to synthesize insights from practical ongoing efforts. The attached principles resulted and were unanimously endorsed.

These principles deal with four aspects of assessing progress toward sustainable development. Principle 1 deals with the starting point of any assessment - establishing a vision of sustainable development and clear goals that provide a practical definition of that vision in terms that are meaningful for the decision-making unit in question. Principles 2 through 5 deal with the content of any assessment and the need to merge a sense of the overall system with a practical focus on current priority issues. Principles 6 through 8 deal with key issues of the process of assessment, and Principles 9 and 10 deal with the necessity for establishing a continuing capacity for assessment.

1. Guiding Vision and Goals

    Assessment of progress toward sustainable development should be guided by a clear vision of sustainable development and goals defining that vision.

2. Holistic Perspective

Assessment of progress toward sustainable development should:

- include review of the whole system as well as its parts,
- consider the well-being of social, ecological, and economic subsystems, their state as well as the direction and rate of change of that state, of their component parts, and the interaction between parts, and
- consider both positive and negative consequences of human activity, in a way that reflects the costs and benefits for human and ecological systems, in monetary and non-monetary terms.

3. Essential Elements

Assessment of progress toward sustainable development should:

- consider equity and disparity within the current population and between present and future generations, dealing with such concerns as resource use, over-consumption and poverty, human rights, and access to services, as appropriate,
- consider the ecological conditions on which life depends, and
- consider economic development and other, non-market activities that contribute to human/social well-being.

4. Adequate Scope

Assessment of progress toward sustainable development should:

- adopt a time horizon long enough to capture both human and ecosystem time scales thus responding to needs of future generations as well as those current to short term decision-making,
- define the space of study large enough to include not only local but also long distance impacts on people and ecosystems, and
- build on historic and current conditions to anticipate future conditions—where we want to go, where we could go.

5. Practical Focus

Assessment of progress toward sustainable development should be based on:

- an explicit set of categories or an organizing framework that links vision and goals to indicators and assessment criteria,
- a limited number of key issues for analysis,
- a limited number of indicators or indicator combinations to provide a clearer signal of progress,
- standardizing measurement wherever possible to permit comparison, and
- comparing indicator values to targets, reference values, ranges, thresholds, or direction of trends, as appropriate.

6. Openness

Assessment of progress toward sustainable development should:

- make the methods and data that are used accessible to all, and
- make explicit all judgments, assumptions, and uncertainties in data and interpretations.

7. Effective Communication

Assessment of progress toward sustainable development should:

- be designed to address the needs of the audience and set of users,
- draw from indicators and other tools that are stimulating and serve to engage decision-makers, and
- aim, from the outset, for simplicity in structure and use of clear and plain language.

8. Broad Participation

Assessment of progress toward sustainable development should:

- obtain broad representation of key grass-roots, professional, technical and social groups, including youth, women, and indigenous people, to ensure recognition of diverse and changing values, and
- ensure the participation of decision-makers to secure a firm link to adopted policies and resulting action.

9. Ongoing Assessment

Assessment of progress toward sustainable development should:

- develop a capacity for repeated measurement to determine trends,
- be iterative, adaptive, and responsive to change and uncertainty because systems are complex and change frequently,
- adjust goals, frameworks, and indicators as new insights are gained, and
- promote development of collective learning and feedback to decision-making.

10. Institutional Capacity

Continuity of assessing progress toward sustainable development should be assured by:

- clearly assigning responsibility and providing ongoing support in the decision-making process,
- providing institutional capacity for data collection, maintenance, and documentation, and
- supporting development of local assessment capacity.

CERES PRINCIPLES

Over the past nine years, the Coalition for Environmentally Responsible Economies (CERES) has emerged as a worldwide leader in standardized corporate environmental reporting and promoting the transformation of environmental management within firms. Formed from a unique partnership between some of America's largest institutional investors and environmental groups, CERES has pioneered an innovative, practical approach toward encouraging greater corporate responsibility on environmental issues. The CERES coalition was formed in 1989. It united fifteen major U.S. environmental groups with an array of socially responsible investors and public pension funds. In the early years (1989-92), the CERES Principles were mainly adopted by companies that already had strong "green" reputations, as exemplified by such firms as The Body Shop, Ben & Jerry's, Seventh Generation, and Aveda. But the momentum behind the CERES concept continued to build. Finally in 1993, following lengthy negotiations, Sun Oil became the first Fortune 500 company to endorse the CERES Principles. Sun's leadership triggered a new round of conversations, leading to endorsements by other large companies such as Arizona Public Service, Bethlehem Steel, Catholic Healthcare West, General Motors, H.B. Fuller, and Polaroid. Today, 46 companies and organizations have endorsed the CERES Principles.

By adopting these Principles, we publicly affirm our belief that corporations have a responsibility for the environment, and must conduct all aspects of their business as responsible stewards of the environment by operating in a manner that protects the Earth. We believe that corporations must not compromise the ability of future generations to sustain themselves. We will update our practices constantly in light of advances in technology and new understandings in health and environmental science. In collaboration with CERES, we will promote a dynamic process to ensure that the Principles are interpreted in a way that accommodates changing technologies and environmental realities. We intend to make consistent, measurable progress in implementing these Principles and to apply them to all aspects of our operations throughout the world.

PROTECTION OF THE BIOSPHERE

We will reduce and make continual progress toward eliminating the release of any substance that may cause environmental damage to the air, water, or the earth or its inhabitants. We will safeguard all habitats affected by our operations and will protect open spaces and wilderness, while preserving biodiversity.

SUSTAINABLE USE OF NATURAL RESOURCES

We will make sustainable use of renewable natural resources, such as water, soils and forests. We will conserve non-renewable natural resources through efficient use and careful planning.

REDUCTION AND DISPOSAL OF WASTES

We will reduce and where possible eliminate waste through source reduction and recycling. All waste will be handled and disposed of through safe and responsible methods.

ENERGY CONSERVATION

We will conserve energy and improve the energy efficiency of our internal operations and of the goods and services we sell. We will make every effort to use environmentally safe and sustainable energy sources.

RISK REDUCTION

We will strive to minimize the environmental, health and safety risks to our employees and the communities in which we operate through safe technologies, facilities and operating procedures, and by being prepared for emergencies.

SAFE PRODUCTS AND SERVICES

We will reduce and where possible eliminate the use, manufacture or sale of products and services that cause environmental damage or health or safety hazards. We will inform our customers of the environmental impacts of our products or services and try to correct unsafe use.

ENVIRONMENTAL RESTORATION

We will promptly and responsibly correct conditions we have caused that endanger health, safety or the environment. To the extent feasible, we will redress injuries we have caused to persons or damage we have caused to the environment and will restore the environment.

INFORMING THE PUBLIC

We will inform in a timely manner everyone who may be affected by conditions caused by our company that might endanger health, safety or the environment. We will regularly seek advice and counsel through dialogue with persons in communities near our facilities. We will not take any action against employees for reporting dangerous incidents or conditions to management or to appropriate authorities.

MANAGEMENT COMMITMENT

We will implement these Principles and sustain a process that ensures that the Board of Directors and Chief Executive Officer are fully informed about pertinent environmental issues and are fully responsible for environmental policy. In selecting our Board of Directors, we will consider demonstrated environmental commitment as a factor.

AUDITS AND REPORTS

We will conduct an annual self-evaluation of our progress in implementing these Principles. We will support the timely creation of generally accepted environmental audit procedures. We will annually complete the CERES Report, which will be made available to the public.

These Principles establish an environmental ethic with criteria by which investors and others can assess the environmental performance of companies. Companies that endorse these Principles pledge to go voluntarily beyond the requirements of the law. The terms "may" and "might" in Principles one and eight are not meant to encompass every imaginable consequence, no matter how remote. Rather, these Principles obligate endorsers to behave as prudent persons who are not governed by conflicting interests and who possess a strong commitment to environmental excellence and to human health and safety. These Principles are not intended to create new legal liabilities, expand existing rights or obligations, waive legal defenses, or otherwise affect the legal position of any endorsing company, and are not intended to be used against an endorser in any legal proceeding for any purpose.

RESPONSIBLE CARE

Codes of Management Practices

- The Community Awareness and Emergency Response Code
- The Pollution Prevention Code
- The Distribution Code
- The Process Safety Code

- The Employee Health and Safety Code
- The Product Stewardship Code

INTERNATIONAL CHAMBER OF COMMERCE (ICC), WORLD BUSINESS ORGANIZATION

The ICC promotes international trade, investment, and the market economy system worldwide; makes rules that govern the conduct of business across borders; and provides essential services, foremost among them the ICC International Court of Arbitration, the world's leading institution of its kind. Members from 63 national committees and more than 7,000 member companies and associations from over 130 countries throughout the world present ICC views to their governments and coordinate with their membership to address the concerns of the business community.

THE BUSINESS CHARTER FOR SUSTAINABLE DEVELOPMENT

Principles for Environmental Management

Foreword

There is widespread recognition today that environmental protection must be among the highest priorities of every business.

In its milestone 1987 report, "Our Common Future," the World Commission on Environment and Development (Brundtland Commission) emphasised the importance of environmental protection to the pursuit of sustainable development.

To help business around the world improve its environmental performance, the International Chamber of Commerce created this Business Charter for Sustainable Development. It comprises sixteen Principles for environmental management which, for business, is a vitally important aspect of sustainable development.

This Charter assists enterprises in fulfilling their commitment to environmental stewardship in a comprehensive fashion, in line with national and international guidelines and standards for environmental management. It was formally launched in April 1991 at the Second World Industry Conference on Environmental Management in Rotterdam, and continues to be widely applied and recognised around the world.[1]

Principles

1. Corporate priority
2. Integrated management
3. Process of improvement
4. Employee education
5. Prior assessment
6. Products and services
7. Customer advice
8. Facilities and operations
9. Research
10. Precautionary approach
11. Contractors and suppliers
12. Emergency preparedness
13. Transfer of technology

---

[1] From the ICC, World Business Charter, Principles for Environmental Management.

14. Contributing to the common effort
15. Openness to concerns
16. Compliance and reporting

THE PERI GUIDELINES

Each reporting organization may decide how, when, and to what extent to present the PERI reporting components listed below. No specific order of presentation is mandatory or encouraged. The recommended content to be included is provided in the following sections.

1. Organizational Profile

   Provide information about the organization that will allow the environmental data to be interpreted in context:

   • Size of the organization (e.g., revenue, employees),
   • Number of locations,
   • Countries in which the organization operates,
   • Major lines of activity, and
   • The nature of environmental impacts of the organization's operations.
   • Provide a contact name in the organization for information regarding environmental management.

2. Environmental Policy

   Provide information on the organization's environmental policy(ies), (e.g., scope and applicability, content, goals and date of introduction or revision, if relevant).

3. Environmental Management

   Summarize the level of organizational accountability for environmental policies and programmes and the environmental management structure, (e.g., corporate environmental staff and/or organizational relationships).

   Indicate how policies are implemented throughout the organization and comment on such items as:

   • Board involvement and commitment to environmental matters,
   • Accountability of other functional units of the organization,
   • Environmental management systems in place (if desired, include references or registration under—or consistency with—any relevant national or international standards),
   • Total Quality Management (TQM), Continuous Improvement or other organization-wide programmes that may embrace environmental performance,
   • Identify and quantify the resources committed to environmental activity (e.g., management, compliance, performance, operations, auditing),
   • Describe any educational/training programmes in place that keep environmental staff and management current on their professions and responsibilities, and
   • Summarize overall environmental objectives, targets and goals, covering the entire environmental management programme.

4. Environmental Releases

Environmental releases are one indicator of an organization's impact on the environment. Provide information that quantifies the amount of emissions, effluents or wastes released to the environment.

Information should be based on the global activity of the organization, with detail provided for smaller geographic regions, if desired.

Provide the base line data against which the organization measures itself each year to determine its progress, and quantify, to the extent possible, the following—including historical information (e.g., last three years, where available) to illustrate trends:

- Emissions to the atmosphere, with specific reference to any,
- Chemical-based emissions (include those listed in any national reportable inventories, (e.g., TRI in the U.S., NPRI in Canada, SEDESOL's Emissions Inventory in Mexico),
- Use and emissions of ozone-depleting substances,
- Greenhouse gas emissions, e.g., carbon dioxide, methane, nitrous oxide and halocarbons,
- Discharges to water (include those considered to be a priority for your organization),
- Hazardous waste, as defined by national legislation. Indicate the percentage of hazardous waste that was recycled, treated, incinerated, deep-well injected or otherwise handled, either on- or off-site. Comment on how hazardous waste disposal contractors (storers, transporters, recyclers or handlers of waste) are monitored or investigated by the organization,
- Waste discharges to land. Include information on toxic/hazardous wastes, as well as solid waste discharges from facilities, manufacturing processes or operations,
- Objectives, targets and progress made regarding the above-listed items, including any information on other voluntary programme activity (e.g., U.S. EPA 33/50 programme), and
- Identify the extent to which the organization uses recommended practices or voluntary standards developed by other organizations, such as the International Chamber of Commerce, the International Standards Organizations, CMA, API, CEFIC, U.S. EPA, Environment Canada, MITI Guidelines, etc.

5. Resource Conservation

- Materials conservation

  Describe the organization's commitment to the conservation and recycling of materials and the use and purchase of recycled materials. Include efforts to reduce, minimize, reuse or recycle packaging.

- Energy conservation

  Describe the organization's activity and approach to energy conservation: commitment made to reduce energy consumption, or to use renewable or more environmentally benign energy sources, energy efficiency programme activities, reductions achieved in energy consumption and the resulting reductions achieved in VOCs, NOX, air toxics and greenhouse gas emissions.

- Water conservation

Describe the organization's efforts in reducing its use of water or in recycling of water.

- Forest, land and habitat conservation

Describe the organization's activities to conserve or reduce/minimize its impact on natural resources such as forest, lands and habitats

6. Environmental Risk Management

Describe the following:

- Environmental audit programmes and their frequency, scope, number completed over the past two years, as well as extent of coverage. Indicate whether the audits are conducted by internal or external personnel or organizations, and to whom and to which management levels the audit findings are reported. Describe follow-up efforts included in the programme to ensure improved performance.
- Remediation programmes in place or being planned, indicating type and scope of activity.
- Environmental emergency response programmes, including the nature of training at local levels, frequency and the extent of the programme. Indicate the degree and method of communications extended to local communities and other local organizations regarding mutual aid procedures and evacuation plans in case of an emergency.
- Workplace hazards. Indicate the approach taken to minimize health and safety risks in the organization's operations, and describe any formal policies or management practices to reduce these risks (e.g., employee and contractor safety training and supervision, statistical reporting).

7. Environmental Compliance

Provide information regarding the organization's record of compliance with laws and regulations. Summary history for the last three years should be given. Additional details should be provided for any significant incidents of non-compliance since the last report, including:

- Significant fines or penalties incurred (define in accordance with local situation, e.g., over $25,000 in the U.S.), and the jurisdiction in which it was applied,
- The nature of the non-compliance issues (e.g., reportable, uncontrolled releases, including oil and chemical spills at both manufacturing and distribution operations),
- The scope and magnitude of any environmental impact, and
- The programmes implemented to correct or alleviate the situation.

8. Product Stewardship

This component defines "product" as the outcome of the organization's activity and is applicable whether an organization manufactures, provides services, advocates, governs, etc. In addition, the section is intended to focus on both the organization's activities in producing its products or services not addressed elsewhere in the guidelines and any activities associated with the "end-of-line" products or service.

Provide information that indicates the degree to which the organization is committed to evaluating the environmental impact of its products, processes and/or services.

Describe any programme activity, procedure, methodology or standard that may be in place to support the organization's commitment to reduce the environmental impacts of its products and services. For example,

- Discuss technical research or design (e.g., new products, services or practices, redesign of existing products or services, practices implemented or discontinued for environmental reasons, design for recyclability or disassembly, or redesign of accounting practices),
- Provide information on waste reduction/pollution prevention programmes from the organization's products, processes or services, including conservation and reuse of materials, and the use of recycled materials,
- Describe the organization's efforts to make its products, processes and services more energy efficient,
- Describe post-consumer materials management, or end-of-life programmes, such as product take-back,
- Detail customer cooperative or partnership programmes and their development (e.g., used oil collection and energy efficiency services),
- Describe supplier programmes and cooperative or partnership activities designed to reduce environmental impacts or add environmental value to the design or redesign of products and services,
- Include information regarding selection criteria for environmentally responsible suppliers and standards to which they must adhere, and
- Identify the scope of the supplier certification process (e.g., all suppliers, major suppliers or those in specific sectors).

Other components

- Specify product stewardship targets and goals, and comment on established procedures to monitor and measure company performance, and
- Provide any baseline data against which the organization can measure its progress.

9. Employee Recognition

Include information regarding employee recognition and reward programmes that encourage environmental excellence. Comment on other education and information programmes that motivate employees to engage in sound environmental practices.

10. Stakeholder Involvement

Describe the organization's effort to involve other stakeholders in its environmental initiatives.

Indicate any significant work undertaken with research or academic organizations, policy groups, nongovernmental organizations, and/or industry associations on environmentally preferable technologies.

Describe how the organization relates to the communities in which it operates, and provide a description of its activities. For example, indicate the degree to which the organization shares pertinent facility-specific environmental information with the communities in which it has facilities.

PROFESSIONAL CODES

Many professional organizations both national and international have established codes of conduct, codes of ethics, or mission statements to which members are to adhere in the course of their professional and personal lives. The codes cover issues of honesty, integrity, fair practice, conflicts of interest, professionalism, promotion of safety, and protection of human health and the environment. Some require notification of environmental risks, mandate pollution prevention and environmental protection. Some recognize the impact the profession may have on the environment and that proper planning and management are important components of environmental protection. The following associations from a variety of occupations have incorporated these types of codes:

- American Chemical Society
- American Society for Biochemistry and Molecular Biology
- American Society of Safety Engineers
- Canadian Council of Professional Engineers
- International Federation of Consulting Engineers
- National Association of Environmental Professionals
- Water Quality Association
- World Engineering Partnership for Sustainable Development
- World Federation of Engineering Organizations

However, even some environmental professional organizations have not established codes of conduct promoting environmental protection or they have not been adequately addressed. Regardless of the number of environmental standards or treaties that are adopted, environmental protection is best achieved by the efforts of individuals. Application of environmentally sound philosophies at work and home provide an opportunity for all of us to contribute to improved environmental conditions worldwide. The following organizations promote environmentally responsible codes of conduct and ethics and provide their members with guidance in this area.

The Chemists Code of Conduct established by the American Chemical Society were prepared by the Council Committee on Professional Relations. The code serves as a guidance document and framework for behavior for members. One of the responsibilities of members is to the environment and reads: "Chemists should understand and anticipate the environmental consequences of their work. Chemists have responsibility to avoid pollution and to protect the environment."

The Council of the American Society for Biochemistry and Molecular Biology has approved a Code of Ethics for the Society and it is expected that "investigators will follow government and institutional requirements regulating research such as those ensuring the welfare of human subjects, the comfort and humane treatment of animal subjects and the protection of the environment."

The American Society of Safety Engineers (ASSE) has not only established a code of ethics that incorporates environmental concerns, they have identified their actions indeed have an impact on the environment. "As a member of the American Society of Safety Engineers, I recognize that my work has an impact on the protection of people, property and the environment." Also, ASSE places great importance on environmental protection and reporting of unacceptable environmental risk: "Hold paramount the protection of people and property and the environment" and "Advise employers, clients, employees or appropriate authorities when my professional judgment indicates that the protection of people, property or the environment is unacceptably at risk."

International Federation of Commercial, Clerical, Professional and Technical Employees, Code of Professional, Social and Ethical Responsibility for Professional and Managerial Staff: "In the pursuit of their professional activities, professional and managerial staff shall take into account not merely the scientific, technical and economic considerations, but also the social, environmental and ethical implications of their work. The responsibility of professional and managerial staff for the sustainable welfare of the community is an integral part of their professional responsibility. Professional and managerial staff shall ensure that their activity contributes to an equitable distribution of world resources." ..."Professional and managerial staff shall take all steps to maintain sustainable systems of work and to avoid dangers which may cause death, injury or ill-health to any person. They shall also avoid damage to nature and goods by any act or omission as a consequence of the execution of their duties. Professional and managerial staff shall take all steps to safeguard public interest in matters of health and safety."

The National Association of Environmental Professionals has developed a Code of Ethics and "Guidance for Practice as an Environmental Professional." The code of ethics states "As an Environmental Professional, I will incorporate the best principles of the environmental sciences for the mitigation of environmental harm and enhancement of environmental quality." The guidance for practice states "As an Environmental Professional, I will incorporate the best principles of design and environmental planning when recommending measures to reduce environmental harm and enhance environmental quality."

WORLD FEDERATION OF ENGINEERING ORGANIZATIONS, CODE OF ETHICS

"Professional Engineers shall uphold the values of truth, honesty and trustworthiness and safeguard human life and welfare and the environment. In keeping with these basic tenets, Professional Engineers shall hold paramount the safety, health and welfare of the public and the protection of the environment and promote health and safety within the workplace."

In addition to industries and individuals pledging to abide by a code of ethics and environmentally responsible behavior, countries have made a similar commitment. Thirty-four nations, including the United States, the Soviet Union, Canada, and 31 European nations have signed the "Charter of Paris for a New Europe," 1990.

"...We pledge to intensify our endeavors to protect and improve our environment in order to restore and maintain a sound ecological balance in air, water, and soil... We commit ourselves to promoting public awareness and education on the environment as well as the public reporting of the environmental impact of policies, projects and programs...

We stress the need for new measures providing for the systematic evaluation of compliance with existing commitments and, moreover, for the development of more ambitious commitments with regard to notification and exchange of information about the state of the environment and potential environmental hazards."

REFERENCES

American Chemical Society, The Chemists Code of Conduct.

American Society of Safety Engineers, Code of Professional Conduct.

Council of the American Society for Biochemistry and Molecular Biology, ASBMB Code of Ethics.

National Association of Environmental Professionals, Code of Ethics and Standards of Practice for Environmental Professionals.

World Federation of Engineering Organizations, Code of Ethics.

Environmental Change and International Law, edited by Edith Brown Weiss, United Nations University Press, 1992.

Environmental Managers Guide to ISO 14000, Business and Legal Reports, 1995.

International Environmental Law and Policy, David Hunter, James Salzman, Durwood Zaelke, Foundation Press, 1998.

International Environmental Auditing, David D. Nelson, Government Institutes, 1998.

Precautionary Legal Duties and Principles of Modern International Environmental Law, Harold Hohmann, Graham & Trotman/Martinus Nijhoff, 1998.

Environmental Strategies Handbook, A Guide to Effective Policies & Practices, Rao V. Kolluru, McGraw-Hill, Inc. 1994.

8 Programs of the United Nations

CONTENTS

INTRODUCTION

The world offices of the United Nations (UN) include the United Nations Office at Geneva (UNOG) and the United Nations Office at New York (headquarters). The UN is an international organization consisting of 126 countries that was formed to promote international peace, security, and cooperation. This was to be done under a charter signed in San Francisco in 1945 with the intent of coordinating activities of member states. Although not officially a legislative body, the UN has adopted declarations of law much like a legislature. The UN consists of political bodies and agencies or "organs" that carry out its work, including the initiation, negotiation, and adoption of multilateral and intergovernmental treaties. Member states maintain the right to participate in the treaty-making process. Decisions by the UN are made through voting by the member states. In addition to dealing with problems relating to health, food, education, and human rights, the UN addresses pollution and environmental issues. As a result of the 1972 Stockholm Conference on the Human Environment, the United Nations Environment Program (UNEP) was created by the UN General Assembly, Resolution 2997 (XXVII). For the first time, an agency of the UN had a specific environmental agenda. Their mission

> "Facilitate international cooperation in the environmental field; to keep the world environmental situation under review so that problems of international significance receive appropriate consideration by Governments; and to promote the acquisition, assessment and exchange of environmental knowledge."
>
> United Nations, Everyone's United Nations 168, 1978

UNITED NATIONS CONFERENCE ON ENVIRONMENT AND DEVELOPMENT (UNCED)

INTRODUCTION

UNCED originated from the 1972 United Nations Conference on the Human Environment that met in Stockholm, Sweden. This conference was the first major, modern, international gathering on human activities relating to the environment. The conference produced the Stockholm Declaration, which is a set of principles "...to inspire and guide the peoples of the world in the preservation and enhancement of the human environment." Chapter 4 contains the entire text of the declaration.

UNCED COLLECTION

The UNCED collection is a set of documents produced by the United Nations Conference on Environment and Development (Earth Summit). Documents from the preconference activities are also included. UNCED was held June 3 through 14, 1992, in Rio de Janeiro, Brazil. Delegations from 178 countries, heads of state of more than 100 countries, and representatives of more than 1,000 nongovernmental organizations attended. Four preparatory committees ("prepcoms") met in the two years prior to UNCED and produced the texts of major UNCED agreements. The UNCED collection includes

- Draft versions of the agreements
- Reports by experts
- Materials from the negotiating process.
- 100 national environmental reports submitted to UNCED
- Speeches of national leaders at the conference

UNCED major agreements

1. Agenda 21—40-chapter statement of goals and potential programs related to sustainable development
2. The Rio Declaration - principles on sustainable development
3. The Biodiversity Treaty - a binding international agreement aimed at strengthening national control and preservation of biological resources
4. The Statement of Forest Principles - a nonbinding agreement on development, preservation, and management of the Earth's remaining forests
5. The Framework Convention on Climate Change - a binding international agreement relating to global warming that seeks to limit or reduce emissions of greenhouse gases, mainly carbon dioxide and methane. This convention was negotiated prior to Rio but, as planned, was opened for signature at the conference.

A CD-ROM that includes all official documents and a complete collection of national reports is available from the UN. The final UNCED documents in English, French, and Spanish are available through a gopher server at the UN.

AGENDA 21

Introduction

Follow-up on official UN implementation of Agenda 21 and the non-binding UNCED agreements has formally been assigned to the United Nations Commission on Sustainable Development (UNCSD), an institution organized as a result of UNCED.

Agenda 21 Contents

Chapter 1 Preamble

Section I. Social and Economic Dimensions
Chapter 2 International Cooperation for Sustainable Development
Chapter 3 Combating Poverty
Chapter 4 Changing Consumption Patterns
Chapter 5 Demographic Dynamics & Sustainability
Chapter 6 Human Health

UNITED NATIONS ENVIRONMENT PROGRAM

UNEP uses a variety of instruments to address environmental issues including promotion of international and intergovernmental treaties, sponsorship of conferences designed to facilitate negotiation, guidelines or "soft law", technical support information exchange, and environmental monitoring systems. In the 1981Conclusions and Recommendations of Montevideo, the UNEP Governing Council determined that eleven areas required action:

1. Coastal zone management
2. Environmental impact assessment

3. General development of environmental law
4. International cooperation in environmental emergencies
5. International trade in potentially harmful chemicals
6. Legal and administrative mechanisms for the prevention and redress of pollution damage
7. Marine pollution from land-based sources
8. Protection of rivers and other inland waters against pollution
9. Protection of the stratospheric ozone layer
10. Transboundary air pollution
11. Transport, handling and disposal of toxic and dangerous waste

The UNEP procedures recognize that action must be taken at the global, regional, and national level. For its part, UNEP initiates and promotes international agreements some of which they also serve as secretariat. Also, UNEP assists member states in developing national legislation by providing legislative frameworks from other countries and environmental data and information from around the world. Information is collected and disseminated through the following databases and satellite monitoring systems:

• Global Environmental Monitoring System
• Infoterra Information System
• International Register of Potentially Toxic Chemicals

Where to find more information

The United Nations Scholars' Workstation is available on-line and contains a wide variety of research information on such topics as disarmament, economic and social development, environment, human rights, international relations, international trade, peacekeeping, and population and demography. The workstation was developed by the Yale University Library and the Social Science Statistical Laboratory. Yale University maintains the Yale Center for International and Area Studies, which is supported by the workstation. The Library of the United Nations Office at Geneva is available to students, researchers, diplomats, journalists, UN, staff, and scholars. Some UNOG Library information is available on the internet along with some helpful links (see list of internet sites).

List of Geneva-Based UN Agencies

• Economic Commission for Europe
• International Computing Center
• International Labor Organization
• International Telecommunications Union
• International Trade Center UNCTAD/WTO
• United Nations Conference on Trade and Development
• United Nations Convention to Combat Desertification
• United Nations High Commissioner for Human Rights
• United Nations High Commissioner for Refugees
• United Nation Institute for Training and Research
• United Nations Research Institute for Social Development
• World Health Organization
• World Intellectual Property Organization
• World Meteorological Organization

Programs of the United Nations

- Centre on Transnational Corporations
- Agenda for the 21st Century
- Environment Program (UNEP)
- International Program on Chemical Safety (with ILO and WHO)
- INFOTERRA (a system to exchange environmental information)
- Environmentally Sound Management of Inland Water (UNEP)
- Global Environment Monitoring System UNEP, Earthwatch Programme
- International Tropical Timber Agreement (UNEP)
- Regional Seas Programme (UNEP)
- Intergovernmental Panel on Climate Change (UNEP, WHO)
- Cairo Guidelines and Principles for the Environmentally Sound Management of Hazardous Wastes (adopted by UNEP)
- Basel Convention on the Control of Transboundary Movements of Hazardous Wastes and Their Disposal (1989)
- Environmental Perspective for the Year 2000 and Beyond (UNEP)
- UN Development Programme

UNEP Environmental Guidelines

- Environmentally sound management of hazardous wastes (1987)
- Exchange of information on chemicals in international trade (1989)
- Goals and principles of environmental impact assessment (1987)
- Offshore mining and drilling, especially for oil (1982)
- Pollution of the marine environment from land-based sources (1985)
- Shared natural resources (1978)
- Weather modification (1980)

THE WORLD HEALTH ORGANIZATION (WHO)

Regional and Other Offices

- Regional Office for Africa
- Regional Office for the Americas/Pan American Health Organization
- Regional Office for the Eastern Mediterranean
- Regional Office for Europe
- Regional Office for Southeast Asia
- Regional Office for the Western Pacific

International Agency for Research on Cancer (IARC)

- Office for the Organization of African Unity
- Center for Health Development in Kobe
- Office at the European Union
- Officed at the United Nations
- Liaison Office in Washington
- Office for the United Nations System in Vienna
- Onchocerciasis Control Program

CONVENTIONS FOR WHICH UNEP PROVIDES SECRETARIAT

- Basel Convention on Transboundary Movements of Hazardous Wastes and their Disposal
- Bonn Convention on Migratory Species (CMS)
- Convention on Biological Diversity (CBD)
- Convention on International Trade in Endangered Species (CITES)
- Vienna Convention for the Protection of the Ozone Layer
- Montreal Protocol on Substances that Deplete the Ozone Layer
- Lusaka Agreement on Cooperative Enforcement Operation Directed at Legal Trade in Wild Fauna and Flora
- Regional Seas Conventions
- Barcelona Convention (Mediterranean Action Plan)

SOURCES OF ADDITIONAL INFORMATION

The Rio Declaration, Volume I
UNCED Report, Volume I
UNCED Report, Volume II
Forest Principles, Volume III
UNCED Report, Volume III
A Guide to UNCED: Process and Documentation, Shanna Halpern, 1992
Thematic Guide to Political Institutions and Global Environmental Change, Consortium for International Earth Science Information Networks (CIESIN)
Register of International Treaties and Other Agreements in the Field of the Environment, United Nations Environment Programme (UNEP)
World Treaties for the Protection of the Environment (1992), Instituto per L'Ambiente of Milan, Italy
International Environmental Law, Emerging Trends and Implications for Transnational Corporations, United Nations Environment Series, 1993 (out of print)
The UN Commission on Sustainable Development (UNCSD): dpcsd@igc.apc.org
    Lowell Flanders
    Department for Policy Coordination & Sustainable Development
    1 United Nations Plaza, Room DC1-868
    New York, NY 10017, U.S.
    Telephone: 1-212-963-0251
    Fax: 1-212-963-2180 or 1-212-963-0377
The Interim Secretariat of the Framework Convention on Climate Change
    16 Ave Jean Trembley, CH-1209
    Geneva, Switzerland
    Telephone: 1-41-22-798-8400
    Fax: 1-41-22-788-3823
    e-mail: unfccc@igc.apc.org
Gopher directory at gopher.undp.org

REFERENCES

Environmental Change and International Law, edited by Edith Brown Weiss, United Nations University Press, 1992.

International Environmental Law and Policy, David Hunter, James Salzman, Durwood Zaelke, Foundation Press, 1998.

International Environmental Auditing, David D. Nelson, Government Institutes, 1998.

Precautionary Legal Duties and Principles of Modern International Environmental Law, Harold Hohmann, Graham & Trotman/Martinus Nijhoff, 1998.

Environmental Strategies Handbook, A Guide to Effective Policies & Practices, Rao V. Kolluru, McGraw-Hill, Inc. 1994.

9 Environmental Management Systems

CONTENTS

"Organisations of all kinds are increasingly concerned to achieve and demonstrate sound environmental performance. They do so in the context of increasingly stringent legislation, the development of economic and other measures to foster environmental protection, and a general growth of concern about environmental affairs."

BS 7750 EMS Standard

INTRODUCTION TO ENVIRONMENTAL MANAGEMENT SYSTEMS (EMS)

With scientific data supporting a greater demand on environmental protection and increased public awareness and participation, instruments to protect the environment have steadily grown. Much progress has been achieved. However, as Lester Brown writes in State of the World 1998: "While economic indicators such as investment, production and trade are consistently positive, the key environmental indicators are increasingly negative." To accomplish additional improvements in environmental quality, instruments beyond regulatory requirements are necessary. The EMS can provide the change industries need to manage their environmental impacts and realize not only benefits for the environment but for business as well. For example, improved environmental performance results in reduced operating and compliance costs, improved company image, and reduced company liability. A compliant EMS can help gain an advantage over competitors and may even be required to qualify for contracts or sales opportunities. Environmental aspects of a business operation have an impact on the bottom line and must be integrated with other business planning functions. Environmental management is fundamental to business strategy. Its role in business operations is continually growing. Total Quality Management (TQM) applied to environmental programs (Total Quality Environmental Management, TQEM) is relatively new but has proven effective in helping organizations meet business objectives. Environmental performance is not a separate operational function; rather it is a contributor to productivity. Once viewed as a necessary evil and cost burden, environmental programs are now seen as providing business opportunities and increased profits.

The core of strategic environmental planning is the EMS. The EMS is the perfect tool to integrate environmental issues with business concerns, and the standards are a simply stated set of provisions easily adapted to any organization regardless of size, function, or location. The ISO 14001 standard defines an EMS as

> "That part of the overall management system which includes the organizational structure, planning activities, responsibilities, practices, procedures, processes and resources for developing, implementing, achieving, reviewing and maintaining the environmental policy."

Eco-Management and Audit Scheme (EMAS) similarly defines an EMS. Clearly, the EMS is built from an organization's environmental policy and is the key to the system's success. Since the environmental policy is defined by top management, any successful program requires their support and involvement. In certain circumstances, it has proven difficult to educate top management and secure necessary support. Also, the EMS must include provisions for communication relating to the importance of compliance with the environmental policy and objectives to company personnel. In the words of G.B. Shaw, "The greatest problem with communication is the illusion that it has been accomplished." Herein lies two key problems with successful development and implementation of an EMS that must be recognized and overcome from the beginning. The EMS standards are designed to accomplish the same basic goal: improve environmental performance through the implementation of an environmental management system.

These standards include

BS 7750 (1994)	British National Standard
EMAS (1993)	EU Regulation
I.S. 310 (1994)	Irish National Standard
ISO 14001 (1996)	International Standard
UNE 77-801(2) (1994)	Spanish National Standard

All the standards are performance oriented and most are open to virtually any organization from airports to manufacturing facilities. Only EMAS specifies applicability of the standard to one type of site (industrial). Notice that each of the above national standards preceded ISO 14001 and in response to the passage of ISO 14001 and ratification by CEN (the European standards organization) were scheduled to be withdrawn. Therefore, Member States of the EU may seek EMS registration through EMAS 1836/93 or ISO 14001. The standards include more than just the requirement for an EMS. They require environmental policies, objectives and programs. They require top management support and review, audits, operational controls, emergency planning, training, and communication. They require an initial or general environmental review and the use of pollution prevention practices. Note, ISO 14001 and BS 7750 do not explicitly require an initial review. However, one must be performed to complete the other requirements of the standard. Improving environmental performance is dependent on other factors such as pollution control technology. All activities performed by an organization to reduce environmental impacts must be integrated to achieve maximum improvements in environmental performance.

ISO 14001 ENVIRONMENTAL MANAGEMENT SYSTEMS

The International Standards Organization (ISO) established a Strategic Advisory Group on the Environment (SAGE) in 1991 to consider the need to develop international standards relating to protection of the environment. Since that time, ISO technical committees have developed and passed draft standards pertaining to environmental management systems (ISO 14000). ISO 14000 is a series of international environmental standards focused on management systems rather than specific goals or product characteristics. It is expected that the 14000 series will become a prerequisite for doing business worldwide and even within the United States. Also, the USEPA is very interested in ISO 14000 and has closely followed the development of the standards providing comments on draft documents.

Work on a revision of ISO 14001 may begin in 1999. Also, ISO has announced there is to be a joint work program involving both TC 207 (EMS standards technical committee) and TC 176 (quality management system standards technical committee) designed to improve the compatibility of 14001 and the ISO 9000 series of quality management system standards.

TECHNICAL COMMITTEE 207 (TC 207) ENVIRONMENTAL MANAGEMENT

The scope of activity for the committee is standardization in the field of environmental management tools and systems. The committee is further divided into subcommittees (SC) and working groups (WG). Environmental management systems and tools include

- EMS (SC 1)
- Environmental auditing (SC 2)
- Environmental performance evaluation (SC 3)
- Environmental labeling
- Life cycle assessment (SC 4)
- Environmental aspects in product standards (SC 5)
- Terms and definitions (SC 6)

As you may expect, TC 207 is to have a close working relationship with ISO/TC 176, environmental systems and audits. Other environmental Technical Committees (TC) have responsibility for test methods for pollutants, including ISO/TC 146 for air quality, ISO/TC 147 water quality, ISO/TC 190 soil quality, and ISO/TC 43 acoustics.

The Standard

"Organisations of all kinds are increasingly concerned to achieve and demonstrate sound environmental performance by controlling the impact of their activities, products or services on the environment, taking into account their environmental policy and objectives. They do so in the context of increasingly stringent legislation, the development of economic policies and other measures to foster environmental protection, and a general growth of concern from interested parties about environmental matters including sustainable development."

Introduction to ISO 14001

ISO 14001 Conformance: A Recommended Approach
Jennifer L. Kraus, MPH, Principal, Global Environmental Consulting Company, Inc.

Introduction

ISO 14000 is a series of voluntary standards designed to promote environmental protection while spurring international trade and commerce by establishing a common worldwide approach to environmental management systems. The standards represent a new approach for companies to manage and track their environmental performance. It is a movement away from the "command-and-control" regime to an encouragement of strategic environmental management.

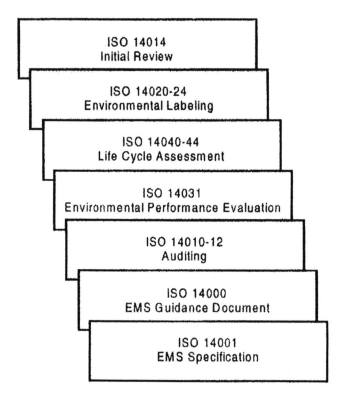

ISO 14014
Initial Review

ISO 14020-24
Environmental Labeling

ISO 14040-44
Life Cycle Assessment

ISO 14031
Environmental Performance Evaluation

ISO 14010-12
Auditing

ISO 14000
EMS Guidance Document

ISO 14001
EMS Specification

A company demonstrates its commitment to the environment by adopting the ISO 14000 standard to show the public that it is an environmentally responsible company. The word is that these standards may soon become a requirement for doing business in Europe and other regions. The result? There has been a tremendous move within businesses in the U.S. and internationally to begin positioning themselves to become registered under these standards. Efforts to become certified to ISO 14001, or simply design a program that conforms to the standard will vary from company to company. The flexibility built into the standard allows for some creativity with regard to implementation. However, one can identify a consistent methodology to use from site to site for implementation. A recommended approach that can be applied to any type of company, large or small, is herein presented.

ISO 14001: A RECOMMENDED APPROACH

This recommended approach consists of eight major steps

1. Training
2. Gap Analysis
3. Assessment of Environmental Aspects
4. Objectives and Targets Identification
5. Environmental Documentation
6. Employee Awareness Campaign
7. Registrar Selection
8. Preassessment

Although each of these steps is individually addressed below, they are not meant to be separately completed and managed. Overlap and integration is encouraged and necessary throughout conformance activities. The more holistic of an approach to conformance, the better.

Step 1—Training

The purpose of this first step is to ensure that there is adequate knowledge of the ISO 14000 standards. A solid understanding of the issues and cognizance of the basic elements of an EMS are critical. The training can also establish a common starting point and uniform knowledge base for the team of individuals within a facility that will be actively working to conform to the standard.

Training can take place at different levels and target different individuals. It may be based upon roles and responsibilities within a facility, and vary in terms of sophistication and length depending upon the background of persons being trained and purpose for training. General awareness training for senior managers, for instance, may consist of brief presentations over a one to two hour period. More technical training, however, may be required for those intimately involved in the ISO 14001 conformance efforts and could last from one day to one week.

Step 2—Gap Analysis

The purpose of the gap analysis is to compare a facility's current EMS to the EMS elements presented in the ISO 14001 standard. In other words, the analysis identifies conformance gaps and their approximate extent. The gap analysis is an essential tool to enable the facility to fully understand the level of effort that must be applied to conform to the standard. Once this step is completed, additional steps and activities may proceed.

When performing a gap analysis, it is recommended that a simple spreadsheet or checklist be utilized. This enables a systematic review of all standard elements while documenting findings in a clear, concise, and easy-to-manage format. The findings can be used to prepare an action plan that lays out what the facility must do in order to achieve conformance. How this is to be accomplished and who will be playing what role in conformance-related activities should also be included in the action plan.

Information that should be documented when performing a gap analysis includes

- Standard Section Number
- Element (e.g., Planning)
- Description
- Status
- Relevant Facility Documentation
- Priority

- Responsible Party
- Estimated Completion Date

Step 3—Assessment of Environmental Aspects

An environmental aspect is defined in ISO 14001 as an "element of an organization's activities, products, and services which can interact with the environment." A clear understanding of a facility's environmental aspects and those which are significant is essential to developing or updating a facility's environmental policy, setting environmental objectives and targets, identifying additional training requirements, establishing operational controls, and developing an appropriate monitoring and measurement system. It is important to remember that aspects can apply within, or external to, a facility's property boundaries.

Typically, an assessment of environmental aspects will focus on two broad categories of aspects:

- Aspects directly regulated by applicable laws and regulations, and
- Unregulated environmental aspects.

An initial starting point to enable aspect identification is a process or block flow diagram that identifies resource inputs and emission outputs at the site. Next, consideration should be given to the following issues:

- Location of the activity,
- Frequency (or potential of occurrence), and
- Severity.

ISO 14001 does not require that a facility use any enumerated criteria in determining significance. Therefore, a facility's definition of significance will depend on a variety of considerations related to the business and environmental issues.

Documenting the definition of significance and the overall assessment of environmental aspects is important to demonstrate how thoroughly the facility has investigated the potential impacts on the environment. Spreadsheets, tables, lists, and graphs are an easy-to-read and user-friendly way of documenting the facility's review and findings. The key is keeping the methodology as systematic as possible in order to prevent the facility from "missing" any issues.

Step 4—Environmental Objectives and Targets

Developing environmental objectives and targets is a critical part of the ISO 14001 planning process. During the consideration of objectives and targets the following should be considered:

- Legal and other requirements,
- Significant environmental impacts,
- Technological options,
- Financial and business requirements, and
- The views of other interested parties.

Objectives and targets must be consistent with the facility's environmental policy and commitment to pollution prevention. One suggestion is to hold a brainstorming session with key people within a facility to think about the following with respect to setting objectives:

```
OBJECTIVE: REDUCE WASTE
FROM MATERIALS USED IN
RESEARCH AND DEVELOPMENT
ACTIVITIES.

TARGET: REDUCE LAB-PACKED
WASTE BY 15 PERCENT BY
MARCH 31, 1998.

TARGET: BY JANUARY 31, 1998,
COMPLETE AN ANALYSIS
OF ALL VENDOR SUPPLIED R&D
MATERIALS AND IDENTIFY
SPECIFIC OPPORTUNITIES FOR
RETURNING MATERIALS TO
VENDOR.
```

- Where does the facility want to invest financial resources?
- Are there any outstanding or crucial compliance issues?
- Is the facility committed to pollution prevention and if so, how?
- Is there the potential for process and design innovations?
- What are current and future training needs?
- What are the significant environmental aspects?
- Consider information technology and how this might help with objective setting and tracking.

The key is tying environmental objectives to the facility's existing business objectives. This is the best way to elevate environmental performance to the same level as other business functions. It will also encourage sharing limited resources within a facility that will enable meeting objectives.

Once the objectives have been set, there should be at least one measurable target for each. For example, if a facility's objective is to reduce energy consumption, measurable targets might include reducing energy usage in lighting by 15 percent by a certain date, or reducing energy usage in compressed air applications by 10 percent by a certain date.

Step 5—Environmental Documentation

This step entails the review of the facility's documentation structure. Depending upon the findings associated with this review, this step may also include the development of procedures, specifications, work instructions, best management practices, and/or an environmental manual.

Here are some steps to help a facility through this documentation stage.

1. First, determine what is needed. This information should already be noted in the results of the gap analysis.
2. Determine the format for all documents. Assess whether there is an existing format requirement at your facility or whether there is any flexibility with respect to this issue.

3. Consider prototyping (visualize the end from the beginning) each document in an attempt to generate documentation that is easy to read and use.
4. Assign writing tasks. For instance, procurement personnel should be involved in the establishment of written procedures for procuring equipment, hazardous materials, or environmental services.
5. Set up a document tracking system. Ensure that within the format chosen there is a way to track document revisions, page numbers, and authorization signatures.
6. Set up a master list or "library" of environmental documents.
7. Finally, don't forget to use the documents.

> THE FACILITY'S ENVIRONMENTAL
> DOCUMENTATION MUST
> ENCOMPASS THE SITE'S EMS
> AND THE REQUIRED EMS
> COMPONENTS:
>
> - POLICY;
> - PLANNING;
> - IMPLEMENTATION AND
>   OPERATION;
> - CHECKING AND
>   CORRECTIVE ACTION;
>   AND
> - MANAGEMENT REVIEW

Another recommendation for addressing documentation requirements is to create a manual describing and documenting the facility's EMS. The manual may serve as a level one, or simple document which anyone can pick up and use, should the plant manager or environmental manager be unavailable in the event of an inspection or other incident. The manual can also serve as a tool to provide the third party auditor with a solid overview of the facility's multimedia environmental management efforts.

Step 6—EMS Awareness Campaign

One of the most important steps in implementing an EMS is making stakeholders not only aware of the EMS but achieving their buy-in to the system. It will be important that at a minimum the facility's commitment to the environment be communicated. Sharing the facility's environmental policy on a large-scale level is a good start. Next is the implementation of creative ways to communicate with all facility employees how their job responsibilities could potentially impact the environment and why their support of the facility's EMS is so important.

Communication methodologies will vary from facility to facility; however, as long as the key concepts are communicated and internalized by the receptors, the task will be effectively completed. Communication methodologies a facility may want to consider include

- Producing fact sheets,
- Holding site tours,
- Holding public meetings,

- Establishing phone lines to answer questions or record concerns, and
- Going to local schools or community colleges that may provide a focal point of interest about a facility.

Step 7—Registrar Selection

A facility should choose a registrar with whom a positive professional relationship may be established. Steps to facilitate this selection could include soliciting statements of qualifications and interviewing. It is important that the registrar understand and be sensitive to the facility's business priorities and culture.

Step 8—Preassessment

An excellent way to prepare a facility for the conformity assessment or to verify ISO 14001 conformance is to conduct a preassessment. The purpose of the preassessment is to identify areas requiring additional work prior to the certification audit or to ensure conformance.

The preassessment should be comprehensive and systematic. The focus should be on how ISO 14001 is used and implemented and how well the facility's products, processes, services, and management systems meet the specifications defined in the standard.

IMPORTANT TIPS

The following important tips are presented based upon this author's experience with the development and implementation of numerous EMSs that conform to ISO 14001.

- Documentation is crucial. The only way to truly verify conformance of a facility's EMS to ISO 14001 is to show documentation that it does conform. Unless something is in writing, it is difficult to prove that it actually exists.
- The gap analysis can set the tone for the rest of the EMS development and implementation efforts. If the gap analysis is hurried and random, it is likely that subsequent activities will be the same. However, if the facility engages in simple strategic planning to assess how it is going to get from point A to point B, the process will be smooth and efficient.
- Don't just adopt, but adapt when developing and implementing an EMS.
- Be prepared. Prior to the conformance assessment, ensure that personnel and resources are available during the audit process. Consider reserving a conference room for the duration of the audit. If practical, have all of the environmental documentation available in one place for review—the conference room would make it easy.

OTHER ISO STANDARDS

In addition to the ISO 14000 environmental management series of standards, ISO has developed other standards relating to environmental protection and worker health and safety.

ISO 13.020 – Environmental Protection
ISO 13.030 – Solid Wastes
ISO 13.040 – Air Quality
    General aspects
    Ambient atmospheres
    Workplace atmospheres
    Stationary emissions
    Transport exhaust emissions

ISO 13.060 – Water Quality
ISO 13.080 – Soil Quality. Pedology
ISO 13.100 – Occupational Safety. Industrial Hygiene
ISO 13.180 – Ergonomics
ISO 13.200 – Accident and Disaster Control
ISO 13.220 – Protection Against Fire and Explosion

ECO-MANAGEMENT AND AUDIT SCHEME, EU DIRECTIVE

The Eco-Management and Audit Scheme, 1836/93/EEC, 1993 is a European Union Regulation open to companies operating a site where industrial activities are performed. It is a program or "scheme" designed to promote continuous improvements in the environmental performance of industrial activities. Registration to EMAS may be done within any of the member states, while verification of conformance is available for sites outside the member states. Although limited to specified industrial sites, a provision exists for the inclusion of other organizations on an experimental basis. The scheme was to be reviewed by the Commission after five years and, in fact, a proposal for a new version of EMAS has been published by the European Commission with possible implementation in 2000. The new version is out for comment and contains provisions to include all organizations in the scheme and additional requirements designed to bridge the gap between EMAS and ISO 14001. The ISO 14001 requirement clauses are in the new EMAS. EMAS as stated in Article 3—Participation in the Scheme includes eight elements.

1. Environmental policy
2. Initial review of environmental effects
3. Compliance with environmental legislation
4. Continual improvement in environmental performance
5. An environmental management system
6. An environmental program
7. Internal environmental auditing
8. A validated and publicly available environmental statement

The EMAS Regulation provides a specification for an EMS that is more prescriptive than some of the other standards. For example, ISO 14001 provides general guidelines. Also, requirements to establish and operate accreditation and verification schemes by member states are included. The European Commission has evaluated the other four EMS standards in relation to EMAS and published European Directives detailing the correspondence between them.

BS 7750   Directive 96/150/EC
I.S. 310   Directive 96/149/EC
ISO 14001   Directive 97/265/EC
UNE 77-801(2) Directive 96/151/EC

EMAS is designed so that sites meeting the requirements can be registered with the member state and the European Commission. This involves the submission of a validated environmental statement to the member state. Each member state is required to establish a "competent body" to perform registrations and develop procedures for suspension or cancellation of a sites registration. Also, member states are required to develop a program for accredited environmental verifiers who evaluate sites for compliance with EMAS and validate environmental statements.

Outline of Key Requirements of EMAS

Article 1 – EMAS Objectives

"The Objective of the scheme shall be to promote continuous improvement in the environmental performance of industrial activities by:

a. the establishment and implementation of environmental policies, programmes and management systems, in relation to their sites;
b. the systematic, objective and periodic evaluation of the performance of such elements;
c. the provision of information of environmental performance to the public."

Article 2 – Definitions

This article defines each of the important terms for compliance with the regulation including environmental review, environmental programme, environmental objectives, environmental audit and environmental management system

"...environmental management system shall mean that part of the overall management system which includes the organizational structure, responsibilities, practices, procedures, processes and resources for determining and implementing the environmental policy;"

Article 3 – Participation in the Scheme

The key elements that must be met for registration to the scheme are contained in this article. In addition to the eight items above, this article contains provisions for the establishment of environmental objectives set at the highest appropriate management level.

Article 4 – Auditing and Validation

This article contains the requirements for site auditing including items that must be reviewed and frequency of audits. Environmental policies, programs, management systems, reviews and audits must be verified against the requirements of the regulation. The accredited environmental verifier and the site auditor cannot be the same person. The article contains the items the verifier is to check.

Article 5 – Environmental Statement

This requirement of EMAS is one of the most important and not shared by ISO 14001 or BS 7750. The statement must be prepared and include information about the site's environmental performance. The statement must also be validated, submitted to the member state and disseminated as appropriate to the public.

BRITISH NATIONAL STANDARD, BS 7750, BRITISH STANDARDS INSTITUTION (BSI)

BSI represents British interests in the development of international standards. BSI standards such as BS 7750 are developed by committees made up of representatives, including manufacturers, users, research organizations, government departments, and consumers. BSI provides administrative assistance to these committees and coordinates the work. Before publication, British standards are made available for public comment. Distributed to members, *Business Standards* is the magazine of the British Standards Institution and is published on a monthly basis. The BSI Board is responsible for making BSI policies. Under the BSI board are Industry Sector Boards that authorize work on new standards projects. The Industry Sector Boards are Building and Civil Engineering, Engineering, Healthcare & Environment, Materials & Chemicals, Electrotechnical/BEC Executive, Consumer Products & Services, Management Systems, and DISC Board. BS 7750 (1994) Environmental Management Systems is a British national standard from the Environmental Management Standards Policy Committee that serves as a specification for an environmental management system but does not establish absolute requirements for environmental performance. However, organizations must comply with applicable environmental legislation and must commit to continual improvement. Although a national standard, registration is internationally possible. Although similar to ISO 14001, some of the requirements and definitions are different. For example, BS 7750 uses the term "environmental effects" as opposed to "environmental impacts" and defines continual improvement with a time frame: "year on year enhancement." BS 7750 has additional definitions for the terms environmental effects evaluation, environmental effects register, environmental management manual, environmental management program, and environmental management review. Unlike ISO 14001, this standard specifies that an EMS manual be developed for the whole organization that outlines the program and serves as a permanent reference.

BS 7750 Table of Contents

Specification
1 – Scope
2 – Informative References
3 – Definitions
4 – Environmental Management System Requirements
    4.1 Environmental Management System
    4.2 Environmental Policy
    4.3 Organization and Personnel
        4.3.1 Responsibility, authority, and resources
        4.3.2 Verification of resources and personnel
        4.3.3 Management representative
        4.3.4 Personnel, communication, and training
        4.3.5 Contractors
    4.4 Environmental Effects
        4.4.1 Communications
        4.4.2 Environmental effects evaluation
        4.4.3 Register of regulatory requirements
    4.5 Environmental Objectives and Targets
    4.6 Environmental Management Program
    4.7 Environmental Management Manual
        4.7.1 Manual
        4.7.2 Documentation
    4.8 Operational Control
        4.8.1 General
        4.8.2 Control
        4.8.3 Verification, measurement, and testing
        4.8.4 Non-compliance and corrective action
    4.9 Environmental Records
    4.10 Environmental Management Records
        4.10.1 General
        4.10.2 Audit program

4.10.3 Audit protocols and procedures
4.11 Environmental Management Reviews

ENVIRONMENTAL AUDITING

An important component of an EMS is the requirement that the system be audited or reviewed. The ISO 14001 auditing requirements are supported by ISO 14010, 14011, and 14012. EMAS contains detailed requirements for auditing, including a maximum of three-year cycle between audits. I.S. 310 (Irish Standard) also specifies a maximum three-year audit cycle. Remember, this is an audit of the EMS and not necessarily of specific systems. Overall, advantages of environmental auditing include

- Compliance with an EMS standard
- Better compliance record
- Fewer penalties and citations
- Improved public image
- Improved image with regulators
- Reduced overall costs
- Improved communication and awareness
- Improved relations (internal and external)

Auditing does not come without potential disadvantages as well, including

- Requires resources and commitment
- May disrupt company operations
- May provide negative information
- Potential increased liability
- Potential conflict with company personnel

Program objectives must be established for the audit which may include assurance of compliance, management of liabilities, accountability, tracking of compliance costs, and EMS conformance. Information collected during an environmental audit process must be sufficient, reliable, relevant, and useful. An auditing program requires

- Top management support
- Independent audit team
- Adequate resources and tools
- Program objectives
- Information collection and processing
- Written report of findings
- Quality assurance

U.S. REGULATORY POLICY TOWARD EMSS AND ISO 14001

By John W. Grosskopf, PE, DEE
President and Owner of Environmental Resources Engineering (ERE), Inc.
Co-founding Principle of the ISO Network, LLC

U.S. REGULATORY SETTING, BACKGROUND, AND HISTORY

Very few observers question that the movement toward environmental management systems in this country is both a relatively recent phenomenon and one that was largely driven by the more proactive environmental leaders in industry. Although an increasingly overused term, the phrase "command and control" is a fairly accurate one that describes the United States Environmental Protection Agency (USEPA) general approach to environmental management since they were placed in the environmental drivers seat in this country in the early 1970s. Although USEPA remains in that driver's seat, the car is now filled with other drivers, including environmental management systems (EMSs). The question seems to be, and now EPA's (apparent) struggle, is how they evolve toward, or perhaps blend EMS and other advanced approaches with their traditional command and control strategies.

Even EPA admits this is a vastly different country today than when the first wave of environmental statutes and their attendant rules and regulations were introduced over 25 years ago. It is difficult to argue the need for command and control strategies for those organizations that need the kind of encouragement command and control and current enforcement strategies dictate. One would hope the numbers of organizations in this category of environmental nonperformers would swiftly decline. Aside from this class of corporate environmental citizenry, however, there is an increasing need for EPA and other federal and state regulators to develop and encourage innovative approaches for the more environmentally proactive organizations. An increasing number of organizations have demonstrated a desire to comply, or even go beyond compliance, and improve their overall environmental management and performance.

This search, if one would call it that, has been helped along by the government's attempts to find better ways to get the job done. In addition, it may well be that the environmental regulatory agencies, particularly EPA, are undergoing a paradigm shift in both the way they think and act with respect to protecting the health and safety of its citizenry and the environment.

Some in U.S. industry have undergone this shift and have practiced proactive environmental management and have embraced environmental management systems for going on two decades. Some of these demonstrated leaders include 3M, Ciba Geigy, Monsanto, General Dynamics and others. The ranks of those recognized as environmental leaders in this country is small but is on the rise. Most, but not all of these leaders have come from the more resource-rich larger firms.

THE IMPACT OF USEPA'S REGULATORY REINVENTION EFFORTS

In 1995, President Clinton announced a number of government "reinvention" initiatives to be implemented by EPA as part of their efforts to achieve greater public health and environmental protection at a more reasonable cost. Among these initiatives EPA prioritized "Project XL," "eXcellence and Leadership". Project XL is a national pilot program that has attempted to find, develop, and test ways of achieving better public health and environmental protection. Although EMSs were not central to the strategies initially developed for Project XL, they have found their way into the mix of alternatives offered and pursued in this initiative. On June 23, 1998 EPA issued a Federal Register notice that outlined its intention to improve the three-year old project, in two broad category areas. One addresses EMSs and the EPA's role with ISO 14000 initiatives.

Through the application of comprehensive EMSs, EPA intends that participating organizations would be asked to collect information and report on implementation of the EMS in such key areas as compliance, pollution prevention, EMS costs and benefits, and overall environmental performance improvement. EPA states (as it always has) that the EMS must achieve compliance, but since Project XL attempts to explore and test new approaches, EPA will consider streamlining or modifying existing environmental requirements. Participants will be required to demonstrate their EMS achieves superior environmental performance based on the objectives of the EMS. EPA requires proposals for regulatory relief be linked to how the EMS may create ways for transferable improvements in EPA's regulatory system. EPA also encourages public and private sector facilities, sectors, states, local governments and communities to sponsor projects that truly reinvent the way they currently conduct environmental management.

Observers looking for examples of regulatory reinvention (although the driver is perhaps more closely associated with creative enforcement rather than regulatory reinvention) and actual work around EMSs within EPA, will be surprised to find it in the enforcement branch, of the National Investigations Investigation Center (NEIC) located in Denver, Colorado.

Although certain regions and offices of the USEPA have been interested in EMSs, such as Region IX's Merit Partnership ISO 14000 Demonstration Project, or the previously mentioned Project XL and ELP (both of which have features that include EMSs), these three programs are pilot projects. The NEIC and Department of Justice, on the other hand, are applying their unique pressure to bear to install a full-scope EMS that is designed to achieve increased compliance at a multi-state, multi-facility organization. The NEIC has studied and has begun to apply some of the principles common to all EMSs including the importance of the environmental policy, EMS auditing, application of performance based metrics, management reviews and the concept of continual improvement, particularly the improvement of the overall EMS.

THE USEPA/NEIC COMPLIANCE FOCUSED EMS

The NEIC has been dubbed by some as the "A-Team" of the enforcement branch of USEPA. The NEIC typically will take on the tougher, larger (multi-regional or national in scope), and more complex regulatory enforcement cases within EPA. Over the past several years, senior investigators within the NEIC began to recognize certain patterns of noncompliance of those firms they were investigating. They began to apply root-cause analysis to determine the fundamental causes for firms' observed noncompliant behaviors.

Recently, NEIC began to study and learn about the move by the more proactive firms relative to development and application of environmental management systems. NEIC began to marry the concepts of root-cause failure analysis and concepts contained in EMSs and reasoned that the root cause for noncompliance could be traced to a deficiency in, failure of, or the complete absence of an EMS.

For example, a company under investigation may have been discovered to have an inordinately high incidence of certain water quality violations from facilities operating in several regions of the country. Applying root-cause analyses, they theorize the company has a faulty corporate methodology in determining its legal requirements with respect to determining which controls may be required to control or eliminate contaminant discharges. Having discovered this, the company can be compelled, NEIC reasons, to improve their methodology of determining their legal requirements, and applying this new learning. Or a root-cause analysis may reveal that key individuals within the company require more in-depth, or more frequent training regarding how to properly prepare a complete listing of regulatory and legal requirements the firm faces. As more learn of these new approaches, they will discover that comprehensive failures in the compliance arena usually can be traced back to multiple organizational failures.

This type of thinking began to crystallize in the mid-1990s, when the New York-based smelting and mining operator, ASARCO was under investigation by the NEIC. As part of a landmark consent decree lodged in federal court in Montana and Arizona in February 1998, USEPA and ASARCO agreed that the company would establish a court ordered EMS that is applicable to all its active facilities nationwide (38 facilities, 6,000 employees in 7 states). ASARCO committed to improve its environmental controls nationwide by developing and implementing an internal EMS to identify and correct the root causes of the company's alleged, observed noncompliance.

Perhaps the most telling of EPA's EMS efforts and ISO 14000 in particular, is the NEIC Compliance Focused EMS (CFEMS), which was developed coincident with the NEIC-ASARCO consent decree. Some might argue the CFEMS and its application in the USEPA-ASARCO consent decree reveals more than their formal position toward EMSs and ISO 14001 in the March 4, 1998 Federal Register.

ASARCO and EPA agreed that ASARCO would install an EMS at all the facilities affected by the consent decree and further agreed to the general form of the EMS and specific elements contained therein. The basis of the EMS model utilized by the NEIC was derived from a combination of NSF's EMS standards, the Global Environmental Management Initiative (GEMI), and ISO 14000. After a year in development, NEIC released the Compliance Focused EMS (CFEMS) in 1997. The CFEMS is fundamentally based on the principles of continual improvement and performance-based management, as are GEMI, NSF's EMS, and ISO 14000.

The CFEMS contains the following 12 elements:

1. Management policies/procedures
2. Oversight, personnel, oversight of EMS
3. Accountability and responsibility
4. Environmental requirements
5. Assessment, prevention and control
6. Environmental incident/noncompliance investigations
7. Environmental training, awareness and competence
8. Planning for environmental matters
9. Maintenance of records and documentation
10. Pollution prevention program
11. Continual program evaluation and improvement
12. Public involvement/community outreach

After studying ASARCO's prior record of noncompliance and applying the root-cause analysis methods discussed earlier, NEIC determined that ASARCO's environmental management program (destined to become a system) required an emphasis in particular areas where the ASARCO EMS was currently weak or missing. These included, among others, the determination of the environmental requirements (element 4), assessment prevention and control (element 5), environmental training, awareness and competence (element 7), and continual program evaluation and improvement (element 11).

The consent decree included these elements and others in an attempt to improve ASARCO's overall EMS to improve its environmental compliance, while at the same time improving its environmental performance.

The CFEMS significantly differs from other EMS models, particularly ISO 14000. First, the CFEMS is written in a prescriptive style so that it can be used in other settlement agreements (NEIC contemplates continued application of the CFEMS and its elements in future actions). Unlike ISO 14000, the CFEMS doesn't explicitly require that emergency preparedness and response plans are a part of the EMS. Another important difference is that the CFEMS does not require that top management conduct the periodic review of the EMS, as do some of the other EMS models currently in existence. The CFEMS does require the EMS be reviewed, but it does not specifically state it be done by top management.

MAJOR FEDERAL AGENCY EMS/ISO INITIATIVES

THE DEPARTMENT OF ENERGY (DOE)

The Department of Energy (DOE) had been analyzing and using ISO 14001 even before ratification of ISO 14001 and acceptance by the American National Standards Institute (ANSI) in this country in the fall of 1996. At the time of this writing, a number of DOE facilities have received ISO 14001 certification and a number of others are undergoing the certification process, including the DOE-Allied Signal Federal Manufacturing and Technology Site in Kansas City, (DOE's first ISO 14001 certification), the Westinghouse Electric Co. at the Waste Isolation Pilot plant (WIPP) near Carlsbad, NM, and the Savannah River Site in SC. Among other prominent DOE facilities that are seeking certification is the Oak Ridge Operations in Tennessee, and officials at the Hanford Nuclear cleanup in the state of Washington.

Behind the ISO 14001 certification activities is an overall strategy embraced by DOE; the integration of EMSs within existing safety and health programs. The intended result is to mesh the EMS with DOE's integrated safety management system to develop a fully integrated environmental, safety and health (ES&H) management system. Fundamental to this strategy is the way DOE defines safety: protecting the worker, the public, and the environment.

This strategy was articulated in a major DOE report introduced by Energy Secretary Pena, entitled, Accelerating Cleanup: Paths to Closure, in the summer of 1998. The report addresses the many challenges DOE faces in cleaning up radioactive, chemical, and other hazardous waste. The report also outlines how DOE plans to carry out the world's largest environmental cleanup program. The report states how DOE's program is developing formal integrated management systems for the major DOE functions of planning, budgeting, and management functions. In effect, the report formalizes DOE's management systems approach as the way they will conduct themselves, in their evolving mission of production to cleanup. This approach includes the integrated ES&H management systems already in place, being implemented or planned at all of their facilities.

DOE's serious intentions regarding ISO 14001 are illustrated by requiring an EMS that conforms to ISO 14001 as a contractual requirement and performance incentive in Fluor Daniel's winning bid for the Hanford site contract in 1995.

Having ISO 14001 conforming programs in place, management can demonstrate DOE's capability and commitment to continuous improvement, which is consistent with the USEPA's CEMP. The overarching reason for the integrated ES&H management systems approach is DOE's desire to improve its overall business management system, not just that of environmental, health and safety.

DEPARTMENT OF DEFENSE (DOD)

The Department of Defense (DOD) has taken a different tack regarding its pursuit of ISO 14001 initiatives and its integration into DOD facilities nationwide. One goal is to meet the requirements of Executive Order 12856, which is designed to recognize and reward outstanding environmental performance at federal agencies and facilities. In a memorandum to the chief environmental officers of the Department of the Army, Navy, Air Force, and Marine Corps, Peter Walsh, Assistant Under Secretary of Defense (Environmental Quality) states in part that it is advantageous for DOD to adopt ISO 14001 instead of CEMP. In support for this more broad-based approach, the DOD reasons that ISO 14001 has less of a compliance focus and more of a quality management approach, it is better known internationally and is more readily adaptable to DOD's diverse operations.

The DOD followed through with this more broad-based approach by notifying the USEPA on February 29, 1996 that DOD intends to begin voluntary adoption of ISO 14001 on a pilot basis at the installation level in lieu of endorsing the CEMP. They reason that adoption of the ISO 14000 will enable federal agencies, such as DOD, to achieve meaningful improvements from an EMS. Further, it will allow federal agencies to consider combining a comprehensive EMS with ongoing regulatory reinvention efforts such as DOD's ENVVEST and EPA's Project XL and ELP programs.

Furthermore, in September 1997, DOD kicked off its ISO 14001 Pilot Program in which select facilities would undergo a two-year pilot of ISO 14001. DOD had already determined that many potential benefits can be derived from adoption of ISO 14001. This two-year self evaluation and development program was fundamentally designed to determine if the added benefit of its adoption is worth the associated cost. If the benefits are worth the cost, the pilot study will help DOD make the decision on whether to mandate ISO 14001, encourage ISO14001, or leave it as an option where the benefits make sense.

Assistant Deputy Under Secretary of Defense, Walsh, in his February 13, 1996 memorandum regarding voluntary adoption of ISO 14001, states "ISO 14001 has more of a quality management approach and less of a compliance focus." One might easily construe the DOD surmises that the path to improved environmental performance is through improved quality management and not that of compliance. The EPA, on the other hand, through its NEIC, CFEMS published in 1997 is highly suggestive of matters of import to them.

INDUSTRY RESPONSE TO ISO 14000 IN THE UNITED STATES

At the time of this writing, just over two years after ratification of ISO 14001, a number of statistics are highly suggestive that in the U.S., industry remains firmly in the category of "wait and see" toward ISO 14001. As of August 1, 1998, there are 182 ISO 14001 certifications in the U.S. versus nearly 1,100 in Japan, 650 in Great Britain, 630 in Germany, and 292 in Sweden. The U.S., which is the most advanced and largest economic power in the world is ranked ninth in the world in terms of numbers of registrations, barely ahead of much lesser economic powers Korea (160) and Taiwan (152). All told, there are nearly 5,400 registrations worldwide, so the U.S. possesses less than 3.5% of the overall total.

The status of the U.S. industry response has been summarily described by many, including this author, as one of "wait and see" in the two years since the standard has been ratified. Indications suggest that there has been more activity going on in this country than meets the eye in terms of industry's response. Many, if not most of the Fortune 500 and corporate 1000 firms have done much more than "wait and see" with respect to ISO 14000. The larger firms are leading the charge in this country toward ISO 14000. That "charge" might be more accurately described as "examine, analyze and then utilize," meaning that many have used the standard as a benchmark by which to measure their existing EMS. Most of these larger firms already have mature environmental programs, including fairly well articulated and fully functioning EMSs. Their EMSs typically are hybrids of various EMS models that currently exist, or that have evolved over the years from individual programs that were successively improved and ultimately formed into a system. Despite this, those that have conducted a comparison of their EMS to ISO 14001 discover areas in which they are deficient or could improve. Many have proceeded to analyze the "gaps" between their EMS and the ISO standard and then have addressed those gaps. Many choose to improve those aspects of their EMS where they will receive direct benefit or where it simply makes sense.

EMERGING TRENDS IN STRATEGIC ENVIRONMENTAL MANAGEMENT

So what does this all mean? Posed differently, we might ask, where are we going?

Although it is perhaps too early to tell with much certainty, according to long time observers of EPA's command and control strategies and industries response to them, some patterns are beginning to emerge.

From industry's perspective, today's industry leaders have clearly placed its environmental responsibilities among the highest corporate priorities. As a priority, they have begun to apply the latest and more advanced management techniques toward the environmental function, just as they have in other priority business functions. So for an increasing number, the principles of performance-based management (indicators, metrics, etc.), TQM (now TQEM Total Quality Environmental

Management), employee incentive and reward systems, and other advanced management methods are included as important features or adjuncts to their EMS. The results are undeniably impressive for nearly all that have learned to effectively apply these new approaches, and more and more in industry and within the agencies begin to take notice.

But the leaders are not those at issue here; it is the smaller and mid-size firms, and those larger firms that have not yet seen fit to apply these advances that need the attention. EPA continues its long-standing policy of targeting the intransigent violators and setting an example. To that end, the author notes that in 1997 EPA collected a record number of environmental fines from industry. However, this is a resource intensive and painstakingly slow process for the agencies, which is not "lost" on EPA policy makers. It is partly because of this that EPA applies alternative enforcement strategies to the simple slap on the wrist (now rising to slaps in the face) "you violate-you pay" penalty schemes.

Just as industry leaders have learned there is a better way (ala EMSs) to manage their environmental impacts, so the agencies too begin to learn. This writer predicts that agencies will continue to experience success in the early application of EMS approaches. Progress will be predictably slow as they undergo the "stops and starts" in applying a new way of doing things, just as leaders in industry did. But through time, success will beget success, just as early efforts in pollution prevention (picking the low hanging fruit) made organizations realize these strategies work and should be driven deeper into the organization, yielding even greater results (evolving into life cycle analysis (LCA) and design for environment (DfE)). As more in industry apply these approaches, more environmental progress will be demonstrated, more will take notice and begin to apply them, and so on. Undoubtedly this will have a continuing positive impact on the agencies, particularly EPA.

Within EPA itself, those that promote EMS and alternative enforcement paths utilizing the CFEMS, will slowly begin to win converts as positive results are demonstrated through their pilot programs and projects and through results from the innovative enforcement strategies as exhibited by NEIC in the ASARCO matters.

In one of the strongest signals yet, the multi-state working group (MSWG) has formed in which ten of the most proactive states in the union have decided to collaborate and conduct a series of pilot and demonstration projects utilizing EMSs and ISO 14001. It will be wise for us all to watch these efforts closely. Regardless of the early results from these efforts, just by their existence, they will have an enormous impact on awareness of what EMSs and ISO 14000 are and what they can do. The federal and state EMS/ISO 14001 pilot projects, the MSWG, the CFEMS, and the EMS features in Project XL, suggest we are truly entering a new era of environmental management in this country.

EPA POSITION STATEMENT ON ENVIRONMENTAL MANAGEMENT SYSTEMS AND ISO14001 AND A REQUEST FOR COMMENTS ON THE NATURE OF THE DATA TO BE COLLECTED FROM ENVIRONMENTAL MANAGEMENT SYSTEM ISO14001 PILOTS

[Federal Register: March 12, 1998 (Volume 63, Number 48)]
[Notices]
[Page 12094-12097]
From the Federal Register Online via GPO Access [wais.access.gpo.gov]
[DOCID:fr12mr98-64]

ENVIRONMENTAL PROTECTION AGENCY

[FRL-5976-6]

EPA Position Statement on Environmental Management Systems and ISO14001and a Request for Comments on the Nature of the Data To Be Collected From Environmental Management System ISO14001 Pilots

AGENCY: Environmental Protection Agency.

ACTION: Position statement; request for comment on information gathering.

SUMMARY: This document communicates the EPA's position regarding Environmental Management Systems (EMSs), including those based on the International Organization for Standardization.

[[Page 12095]]

Standardization (ISO)14001 standard. This document also describes the evaluative stage EPA is entering concerning EMSs. Further, it solicits comments on proposed categories of information to be collected from a variety of sources that will provide data for a public policy evaluation of EMSs.

FOR FURTHER INFORMATION CONTACT:

Office of Reinvention--EMS, Environmental Protection Agency, 401 M St., SW, mail code 1803, Washington, D.C. 20460, Telephone: (202) 260-4261. E-mail: reinvention@epamail.epa.gov.

SUPPLEMENTARY INFORMATION:

I. Background

A diverse group of organizations, associations, private corporations and governments has been developing and implementing various EMS frameworks for the past thirty years. For example, the Chemical Manufacturers Association created its own framework called Responsible Care. In addition, the French, Irish, Dutch, and Spanish governments developed their own voluntary EMS standards. The possibility that these diverse EMS frameworks could result in barriers to international trade led to a heightened interest in formulating an international voluntary standard for EMSs. To that end, the International Organization for Standardization (ISO), consisting of representatives from industry, government, non-governmental organizations (NGOs), and other entities, finalized the ISO14001 EMS standard in September 1996. The intent of this standard is to produce a single framework for EMSs, which can accommodate varied applications all over the world. ISO14001 is unique among the ISO 14000 standards because it can be objectively audited against for internal evaluation purposes or for purposes of self-declaration or third-party certification of the system. EPA participation in the development of voluntary standards, including the ISO 14000 series of standards, is consistent with the goals reflected in section 12(d) of the National Technology Transfer and Advancement Act of 1995 (NTTAA) (Pub. L. No. 104-113, s. 12(d), 15 U.S.C. 272 note). The NTTAA requires federal agencies to use voluntary consensus standards in certain activities as a means of carrying out policy objectives or other activities determined by the agencies, unless the use of these standards would be inconsistent with applicable law or otherwise impractical. In addition, agencies must participate in the development of voluntary standards when such participation is in the public interest and is compatible with an agency's mission, authority, priority, and budget resources. Agency participation in the development of EMS voluntary standards does not necessarily connote EPA's agreement with, or endorsement of, such voluntary standards. On December 16, 1997, EPA Deputy Administrator Fred Hansen asked EPA's newly chartered Office of Reinvention "to take lead responsibility for policy coordination of all EMS pilots, programs, and communications." (Full text of memo available at www.epa.gov/reinvent.) This notice initiates the Office of Reinvention's effort to ensure public input in that endeavor.

II. Statement

Implementation of an EMS has the potential to improve an organization's environmental performance and compliance with regulatory requirements. EPA supports and will help promote the development and use of EMSs, including those based on the ISO 14001 standard, that help an organization achieve its environmental obligations and broader environmental performance goals. In doing so, EPA will work closely with all key stakeholders, especially our partners in the States. EPA encourages the use of EMSs that focus on improved environmental performance and compliance as well as source reduction (pollution prevention) and system performance. EPA supports efforts to develop quality data on the performance of any EMS to determine the extent to which the system can help bring about improvements in these areas. EPA also encourages organizations that develop EMSs to do so through an open and inclusive process with relevant stakeholders, and to maintain accountability for the performance outcomes of their EMSs through measurable objectives and targets. EPA encourages organizations to make information on the actual performance of their environmental management systems available to the public and governmental agencies. In addition, through initiatives such as Project XL and the Environmental Leadership Program, EPA is encouraging the testing of EMSs to achieve superior environmental performance. At this time, EPA is not basing any regulatory incentives solely on the use of EMSs, or certification to ISO 14001. The Commission for Environmental Cooperation (CEC) Council issued on June 12, 1997, a resolution (#97-05) signed by EPA Deputy Administrator Fred Hansen on behalf of the United States concerning "future cooperation regarding environmental management systems and compliance." The CEC Council was formed pursuant to the North American Agreement on Environmental Cooperation, an environmental side agreement to the North American Free Trade Agreement, and is comprised of the environmental ministers for Canada, Mexico and the United States. The declarative and directive paragraphs of the Council's resolution #97-05 read as follows:

The Council * * * Declares That: Governments must retain the primary role in establishing environmental standards and verifying and enforcing compliance with laws and regulations. Strong and effective governmental programs to enforce environmental laws and regulations are essential to ensure the protection of public health and the environment. Voluntary compliance programs and initiatives developed by governments can supplement strong and effective enforcement of environmental laws and regulations, can encourage mutual trust between regulated entities and government, and can facilitate the achievement of common environmental protection goals; Private voluntary efforts, such as adoption of Environmental Management Systems (EMSs) such as those based on the International Organization Standardization's Specification Standard 14001(ISO14001), may also foster improved environmental compliance and sound environmental management and performance. ISO14001 is not, however, a performance standard. Adoption of an EMS pursuant to ISO14001 does not constitute or guarantee compliance with legal requirements and will not in any way prevent the governments from taking enforcement actions where appropriate; Hereby Directs: The Working Group to explore (1) the relationship between the ISO14000 series and other voluntary EMSs to government programs to enforce, verify and promote compliance with environmental laws and regulations, and (2) opportunities to exchange information and develop cooperative positions regarding the role and effect of EMSs on compliance and other environmental performance. The Working Group shall, no later than the 1998 Council Session, report its results to the Council and provide recommendations for future cooperative action in this area. The review and recommendations shall recognize and respect each Party's domestic requirements and sovereignty.

III. Evaluative Phase

EPA is working in partnership with a number of states to explore the utility of EMSs, especially those based substantially on ISO14001, in public policy innovation. The goal of this partnership is to gather credible and compatible information of known quality adequate to address key public policy issues. The primary mechanism to generate this information will be pilot projects.

[[Page 12096]]

Valid, compatible data from other sources will also be used whenever possible. To make efficient use of resources, and to ensure more robust research, EPA and states will work together on the creation of a common data base. The data base will be open and usable, while recognizing the need to insure the appropriate level of confidentiality for participants. A group of federal and state officials involved in EMS pilot projects have been working together to set up a common national database of information gathered through the pilot projects. As part of that process, EPA and states are developing a series of data protocols which provide instructions and survey instruments to guide the actual collection of data for the data base. That document will be available at http://www.epa.gov/reinvent. This document will serve to solicit comments on the categories of information to be collected. From the following general categories of information (and possibly others), EPA and participating states will develop the above mentioned protocols. The following categories are designed to provide a general idea as to the types of information that EPA believes should be collected to evaluate the effectiveness of EMSs from the perspective of regulators. EPA further believes that collection of data in all categories will allow the fullest understanding and evaluation of the benefits of an EMS. The data categories which appear in this document were, to the extent possible, developed around the kinds of data we believe will or could be generated by an ISO 14001EMS.

1. Environmental Performance

The impact a facility has on the environment is of paramount importance to regulators' assessment of EMSs. Thus, it is critical to measure any change in a facility's environmental performance that might be attributable to implementation of an EMs. Information would be collected as to the types, amounts, and properties of pollutants that are released to air, surface water, groundwater, or the land. Information on these pollutants would need to be normalized to a facility's production levels. Information relating to recycling, reuse, and energy requirements could also be included. This inquiry could include both regulated and non-regulated pollutants.

2. Compliance

Implementation of an EMS has the potential to improve an organization's environmental compliance with regulatory requirements. The goal of collecting compliance information is to be able to measure the relationship between an EMs and compliance with local, state and federal environmental regulations. The types of data to be collected would include: information on whether the facility has a recent history of regulatory violations; the number and seriousness of the violations; how quickly violations were discovered and corrected; and measurements of any changes in regulatory compliance status.

3. Pollution Prevention

Pollution prevention is a significant goal for both federal and state regulators. Therefore, better understanding the relationship between an organization's overall performance and the role of pollution prevention in the organization's EMs is important to regulators. In the federal context, pollution prevention is defined as "* * * any practice which--(1) reduces the amount of any hazardous substance, pollutant, or contaminant entering any waste stream, or otherwise released into the environment (including fugitive emissions) prior to recycling, treatment, or disposal; and (ii) reduces the hazards to public health and the environment associated with the release of such substances, pollutants, or contaminants." \1\ This definition will likely serve as a basis for helping an organization identify measures that it might have taken toward pollution prevention. Data collected would include a description of the type of pollution prevention and source reduction techniques used, including good operating practices, inventory control, spill and leak prevention, raw material modification/substitution, process modification, and product reformulation or redesign.

Pollution Prevention Act of 1990 Section 6603, 42 U.S.C. 13102 (1990).

4. Environmental Conditions

In order to understand the impact of an EMs on the environment, it is necessary to know something about the status of the ambient environment surrounding the facility prior to implementation of an EMS. An analysis of this nature will not only help regulators evaluate EMs, it should also help facility mangers prioritize their environmental aspects and shape the policies and objectives of their EMSs. Environmental conditions data will assist all parties in determining the sustainability of certain human activities from an environmental, economic and social perspective. It is difficult, of course, to collect accurate and comparable information about environmental conditions. The time and expense needed for a facility to collect and report such data could be prohibitive. Also, the selection of an appropriate geographic focus—local, regional, or global—will be challenging. One way to minimize this burden would be to utilize available governmental or other surveys (e.g., the 1990 U.S. Census, hydrogeologic reports). Nevertheless, to the degree that these obstacles can be overcome, the analysis conducted by federal and state regulators will benefit.

5. Costs/Benefits to Implementing Facilities

There has been much speculation and assertion about the relative costs and benefits associated with the implementation of an EMS. Data collected in this category should help provide answers to questions concerning possible net financial benefits that might accompany improved compliance and increased environmental performance, or that might result from being able to achieve compliance in less costly ways. The data may also shed light on the costs associated with higher levels of environmental performance. It is important to recognize some of the limitations inherent in traditional approaches to cost/benefit analysis. To address these limitations, organizations could be encouraged to identify intangible costs and benefits associated with the implementation of an EMS, even if they are difficult to quantify. Also, a list of usually "hidden" costs and benefits could be used to help organizations identify and understand costs and benefits that are traditionally overlooked.

6. Stakeholder Participation and Confidence

Community participation has become an increasingly important component of federal and state efforts to increase environmental performance and protect human health. Both federal and state regulators are interested in understanding the involvement of local communities and other stakeholders in the EMS process. Data could be collected to assess the amount and degree of stakeholder participation in both the development and implementation of an organization's EMS, or the effect that such participation has on the public credibility of the facility's EMS implementation. More information concerning the pilot projects as well as other federal, state and international initiatives relating to

[[Page 12097]]

EMSs and ISO14000 can be found in the ISO14000 Resource Directory (copies can be obtained through EPA's Pollution Prevention Information Clearinghouse at 202-260-1023, e-mail: ppic@epa-mail.epa.gov).

Dated: March 6, 1998.
Fred Hansen,
Deputy Administrator.
[FR Doc. 98-6389 Filed 3-11-98; 8:45 am]
BILLING CODE 6560-50-M

IMPORTANT ORGANIZATIONS AND TERMS FOR ENVIRONMENTAL MANAGEMENT SYSTEMS (TERMS FROM ISO 14001)

- ANSI American National Standards Institute

A private not-for-profit organization that coordinates the standards system in the United States and approves American National Standards. ANSI represents the U.S. interests in international standardization and provides information on those standards.

- ASTM: American Society for Testing and Materials

Organized in 1898, ASTM provides a forum for producers, users, and consumers to agree on and write standards for materials, products, systems, and services. ASTM publishes standard test methods, specifications, practices, guides, classifications, and terminology.

- CEN: European Committee for Standardization

Members are national standards bodies from 18 European countries. The committee develops standards for the European Community and is under contract with the Commission of the EC to define the technical specifications for products that will meet the requirements of an EEC directive.

- CENELEC: European Committee for Electrotechnical Standardization

CENELEC is a regional standards organization that develops electrical standards for the EC. This committee is often referenced together with CEN because of similar functions performed by each for the Commission.

- Commission of the European Communities ("The Commission")

One of the EC permanent bodies within the EC. The Commission is the executive branch of the EC.

- "CE" Marking

The symbol that is required to be placed on machinery and other products meeting the requirements of a particular EC directive. The marking of this symbol on a product indicates conformance to the directive.

- Declaration of Conformity

The statements made by a manufacturer of a product declaring the product meets the requirements of a particular directive. Specific information such as the type of equipment, manufacturer's name, and address is required on the declaration.

- Directive

  Standards or laws developed in the EEC to provide for a consistent set of regulations recognized by member states. These directives detail requirements for products, processes, and services within the EC. They are required to be incorporated into each member country's national laws.

- DIS: Draft International Standard

  An international standard developed by ISO in a draft form that is circulated for comment and vote by members.

- EC: European Community

  A term used to describe the association that exists in Europe. The term has now been replaced by the term European Union (EU). The EU is a collective designation of three organizations, including the EEC, European Coal and Steel community, and the European Atomic Energy Community.

- Harmonized Standards

  These standards define the technical specifications needed by professionals for designing and manufacturing products conforming to a particular directive. Harmonized standards are developed by CEN/CENELEC.

- IEC: International Electrotechnical Commission

  This commission is a nongovernmental organization with national participation from 40 countries that develop and publish international standards in the electrical and electronics fields. Works in collaboration with ISO.

- ISO: International Organization for Standardization

  ISO is a worldwide federation of national standards bodies from more than 100 countries. They are a nongovernmental organization established in 1947. ISO's goal is to develop international standards to promote the exchange of goods and services.

- continual improvement

  Process of enhancing the environmental management system to achieve improvements in overall environmental performance in line with the organization's environmental policy.

- environment

  Surroundings in which an organization operates, including air, water, land; natural resources, flora, fauna, humans, and their interrelation.

- environmental aspect

  Element of an organization's activities, products, or services that can interact with the environment.

- environmental impact

  Any change to the environment, whether adverse or beneficial, wholly or partially resulting from an organization's activities products or services.

- environmental management system

  That part of the overall management system which includes organizational structure, planning activities, responsibilities, practices, procedures, processes, and resources for developing, implementing, achieving, reviewing, and maintaining the environmental policy.

- environmental management system audit

  A systematic and documented verification process to objectively obtain and evaluate evidence to determine whether an organization's environmental management system conforms to the environmental management system criteria set by the organization, and communication of the results of this process to management.

- environmental objective

  Overall environmental goal, arising from the environmental policy, that an organization sets itself to achieve, and which is quantified where applicable.

- environmental performance

  Measurable results of the environmental management system, related to an organization's control of its environmental aspects, based on its environmental policy, objectives, and targets.

- environmental policy

  Statement by the organization of its intentions and principles in relation to its overall environmental performance which provides a framework for action and for the setting of its environmental objectives and targets.

- environmental target

  Detailed performance requirement, quantified where practicable, applicable to the organization or parts thereof, that arises from the environmental objectives and that needs to be set and met in order to achieve those objectives.

* interested party

Individual or group concerned with or affected by the environmental performance of an organization.

* organization

Company, corporation, firm, enterprise, or institution, in part or combination thereof, whether incorporated or not, public or private, that has its own functions and administration.

* prevention of pollution

Use of processes, practices, materials, or products that avoid, reduce, or control pollution, which may include recycling, treatment, process changes, control mechanisms, efficient use of resources, and material substitution.

AGENCIES AND ORGANIZATIONS CONDUCTING EMS ACTIVITIES

ENVIRONMENTAL MANAGEMENT SECRETARIAT FOR LATIN AMERICA AND THE CARIBBEAN

* Commission for Environmental Cooperation (CEC)
* Organisation for Economic Cooperation and Development (OEDC)
* United Nations Conference on Trade and Development (UNCTAD), Training Programs for Officials Interested in ISO in Developing Countries
* United Nations Environment Program (UNEP)
* United Nations Industrial Development Organization (UNIDO)
* U.S.-AID, EMS Development for Industry – Electric Power Sector
* World Bank, Informal Working Group on ISO 14000

Individual Countries

* Austria— EU Environmental Management and Audit Scheme
  Institute for Ecological Research in Economics
* Bolivia— Sustainable Development Networking Program (SDNP)
* Canada— The Health Sciences Centre (HSC), ISO 14001 Pilot Project
  Canadian Departments of Environment and industries' National Environmental Training Initiative
  EMS Accreditation Program
* Ireland— I.S. 310 (1994) Irish National Standard
* Spain— UNE 77-801 (2) (1994),

FEDERAL AGENCIES OF THE UNITED STATES, INITIATIVES FOR ENVIRONMENTAL MANAGEMENT

USEPA

The Voluntary Standards Network was established in 1993 by EPA Administrator Carol Browner. The program is responsible for coordinating all of the Agency's ISO 14000 activities. The Network develops and coordinates Agency policies, including participation in ISO 14001. Over 130 members in the Network represent most of EPA's offices and regions. Lead points of contact called standards coordinators have been established in each office and region.

- The Voluntary Standards Network
- Natural Gas STAR Program
- Environmental Leadership Program (ELP)
- ISO 14001/EMS Task Group
- EMS Audit Procedural Guidelines
- Indiana Small Business Pilot Project
- Environmental Accounting Project
- Environmentally Preferable Public Purchasing
- Expanding the Use of Environmental Information by the Banking Industry Through ISO 14000
- Environmental Technology Verification Program
- Implementing EMS in the Metal Finishing Industry
- EMS Demonstration Project
- EMS Implementation Guide for Small and Medium Sized Organizations
- Using EMS to Meet Watershed Protection Goals
- EMS Implementation by Municipal Governments
- OW EMS Implementation Workgroup
- Code of Environmental Management Principles for Federal Agencies (CEMP)
- EMS Primer for Federal Agencies
- Compliance-Focused EMS
- Project XL
- Star Track (Region 1)
- Environmental Leadership Program-New England (Region 1)
- Compliance Leadership Through Environmental Audits and Negotiation (CLEAN) (Region 1)
- ISO 1400 Project XL (Region 3)
- Using ISO 1400 in the Paper Industry (Region 4)
- Life Cycle Assessment Methodology (Region 5)
- ISO 14000: A National Dialogue (Region 6)
- EMS for Federal Facilities (Region 8)
- The Merit Partnership for Pollution Prevention (Region 9)
- Evaluation of Policy Implications of ISO 14000 and Other EMS Standards and State of Washington Department of Ecology ISO 14000 Leadership Project both from Region 10

Department of Commerce

- National Institute of Standards and Technology (NIST) – Informational Paper on ISO 14000 series
- Manufacturing Extension Partnership (MEP)
- ISO 14000 Workgroup for the National P2 Roundtable

- ISO 14000/EMS Gap Analysis Tool Suite
- Greenscore TM
- Environmentally Conscious Manufacturing (ECM)
- ISO 14000 Awareness for Maryland Manufacturers
- Vermont Manufacturing Extension Center

Department of Defense, Environmental Management Systems Committee

- U.S. Army
  - Adoption of ISO 14000 Methodologies for Environmental LCA being conducted on weapon systems and materials
  - Total Quality Environmental Management (TQEM) – Green Initiatives
  - ISO 14001 Feasibility Initiative
- Air Force ISO 14001 Workshops
- U.S. Navy EMS Evaluation
- Naval Surface Warfare Center (NSWC) Carderock ISO 14000 Implementation and Certification

Department of Energy

- Energy Facilities Contractors Group (EFCOG) ISO 14000 Working Group
- EMS Fact Sheets
- Environmental Management Systems at DOE
- EMS Primer for Federal Facilities
- Implementation of ISO 14001 at Westinghouse-Managed DOE Sites
- Strategic and Program Planning for EMS Initiatives

Department of State

- U.S.-Asia Environmental Partnership (U.S.-AEP)
- U.S.-EAP Clean Technology and Environmental Management Initiative
- U.S.-EAP Clean Technology and Environmental Management Information Centers
- U.S.-EAP Environmental Exchange Program (EEP)

Food and Drug Administration

- Standards Policy Committee

Federal Trade Commission

- Environmental Marketing Claims Guidelines

U.S. Postal Service

- Development of ISO 9000/14000 Protocol for Fleet Maintenance Activities

U.S. States (Includes Pollution Prevention Programs)

Environmental Council of States (EPA's ISO 14000/EMS Task Group)
Multi-State Work Group (comprised of 40 states divided between participants and observing states)

Alabama	Department of Environmental Management
Arkansas	ISO 14000 Infrastructure Development
California	CAL/EPA ISO 14000 Pilot Project
	San Francisco Bay Area Green Business Program
Colorado	Pollution Prevention Program
Connecticut	Common Sense Initiative and StarTrack Pilot Project
Delaware	Department of Natural Resources
Florida	FL Department of Environmental Protection (FDEP)
Georgia	Pollution Prevention Assistance Division (P2AD)
Idaho	Idaho Manufacturing Alliance
Indiana	Small Business Pilot Project
Iowa	Iowa Waste Reduction Center (IWRC) EMS Assistance Program
	Waste Reduction Assistance Program
Kansas	Environmental Management System
Kentucky	KY Pollution Prevention Center's (KPPC) ISO 14000 Awareness
Louisiana	Department of Environmental Quality
	Environmental Leadership/ISO 14000
	EL Pollution Prevention Program
Maine	Department for Environmental Protection (DEP)
	Center for Technology Transfer (CTT)
	EMS Development for Industry – Electric Power Sector
Maryland	MD Department of the Environment (MDE)
	ISO 14000 Awareness for Maryland Manufacturers
Michigan	Clean Corporate Citizen Program
Minnesota	Office of Attorney General
	EMS Training
	Banking/Insurance Initiative
Missouri	MO Environmental Improvement & Energy Resources Authority
	ISO 14000 Cooperation Project
Nebraska	Department of Environmental Quality (NDEQ)
New Hampshire	NH Pollution Prevention Program (NHPPP)
New Mexico	Energy, Minerals, and Natural Resources Department
	Green Zia Environmental Excellence Program
New York	Department of Environmental Conservation
	ISO 1400 Regulatory Integration Pilot Program
North Carolina	NC Department of Environment, Health and Natural Resources
	Environmental Management Systems
	ISO 14000 Workgroup for the National P2 Roundtable
North Dakota	ND State Water Commission
	Wetland Conservation Strategy
Ohio	Ohio EPA, Office of Pollution Prevention
	ISO 14000 Information Gathering
Oklahoma	Department of Environmental Quality (DEQ), Pollution Prevention Program

Oregon	ODEQ Pollution Prevention Core Committee
	Environmental Action Agreement Project
Pennsylvania	Department of Environmental Management
	Strategic Environmental Management: Beyond Compliance
	Market-Based Audits of EMS: Implementing ISO 14000
	Pennsylvania Environmental Council
Tennessee	Department of Environment and Conservation (TDEC)
Texas	Texas Natural Resource Conservation Commission
	Office of Pollution Prevention and Recycling
Utah	DEQ, Pollution Prevention Program
Vermont	VT Manufacturing Extension Center (VMEC)
Virginia	VDEQ
Washington	Department of Ecology
	Compliance Assurance and Environmental Audits
	ISO 14000 Leadership Project
	EPA ISO 14000 Task Group
	Performance Based Permit System
Wisconsin	WI Department of Natural Resources
	WI ISO 14000 Working Group
	Wharton/LaFollette Joint Research Effort

Industry Associations

- Air & Waste Management Association (AWMA), Intercommittee Task Force on ISO 14000
- American Petroleum Institute (API), Strategies for Today's Environmental Partnership (STEP)
- American Society for Quality Control (ASCQ), Energy and Environment Division
- Electronic Industries Association
- The Associated Industries of Massachusetts/Massachusetts Manufacturing Partnership ISO 14000 Collaborative
- National Association of Environmental Professionals (NAEP), ISO 14000 Working Group
- Industrial Designers Society of America (IDSA), Environmental Responsibility Section
- Northeast Business Environmental Network (NBEN), The Forum for Best Management Practices
- National Center for Manufacturing Science (NCMS)

List of Nongovernmental Organizations

- Alliance for Environmental Innovation
- American Institute for Pollution Prevention (AIPP)
- Coalition for Environmentally Responsible Economies (CERES)
- Community Nutrition Institute (CNI), Joint Policy Dialogue on Trade and the Environment
- The Global Environmental Management Initiative (GEMI)
- The Good Neighbor Project for Sustainable Industries
- Green Seal Environmental Partners Program
- ANSI/Global Environment & Technology Foundation (GETF) ISO 14000 Integrated Solutions (IIS)

- Green Mountain Institute for Environmental Democracy (GMIED)
- ISO 14000 Legal Issues Forum
- Management Institute for Environment and Business (MEB), Industrial Products, Inc.: Measuring Environmental Performance (Case Study)
- ISO 14000 workgroup for the National P2 Roundtable
- New England Environmental Network, Nothing to Waste Initiative
- The Pacific Institute
- The Rainforest Alliance, The Smart Wood Program
- The Sierra Club

Academic/Education Institutions Providing Training in EMSs

- Advanced Interactive Training Course, Bucharest, Romania
- Brown University, Brown is Green (BIG) Program
- Chalmers University of Technology, Managing for Environmental Opportunities, Gothenburg, Sweden
- Denver University
- Georgia Institute of Technology, Economic Development Institute (EDI)
- Montana State University, Extension Service, MT Pollution Prevention Program
- University of Maryland, Environmental Finance Center
- Red Rocks Community College, Environmental Training Center, Rocky Mountain Education Center, Lakewood, Colorado
- Salt Lake Community College, Environmental Training Center
- University of St. Gall, Institute for Management Technology, Switzerland
- SUNY Buffalo, The Science and Engineering Library
- Tulane Institute for Environmental Law and Policy, ISO 14000 From a Public Interest Perspective
- The United States Environmental Training Institute (USETI)
- Vanderbuilt Center for Environmental Management Studies

SOURCES OF ADDITIONAL INFORMATION

Achieving Environmental Management Standards, by Mike Gilbert and Rick Gould for BSI, second edition.

The BSI Environmental Management Systems Handbook, a guide to the BS EN ISO 14000 series, second edition.

ISO 14001 and Beyond, Environmental management systems in the real world, edited by Christopher Sheldon, published by Greenleaf.

ISO 14000 Answer Book, by Dennis R Sasseville, W. Gray Wilson and Robert Lawson.

DOCUMENTS AND PUBLICATIONS

- Accessing European Parliament Documentation
- The Documentation of the European Communities: A Guide
- The Official Journal of the European Communities
- European Access
- Index to International Statistics (IIS)
- Publications of the European Communities
- EC Index
- Documents Annual Catalog

- Index of COM Documents
- EUROSTAT Index
- Guide to EC Court Decisions
- Guide to EC Legislation
- Directory of Community Legislation in Force
- Panorama of EU Industry 1997

INTERNET SITES FOR EUROPEAN UNION INSTITUTIONS

EUROPA—European Commission
http://europa.eu.int

Council of the European Union
http://ue.eu.int

European Court of Justice
http://europa.eu.int/cj/index.htm

European Parliament
http://www.europarl.eu.int

European Ombudsman
http://www.euro-ombudsman.eu.int

European Court of Auditors
http://www.eca.eu.int

Economic and Social Committee—in French only
http://europa.eu.int/ces/ces.html

European Investment Bank
http://www.eib.org

Committee of the Regions of the European Union
http://www.cor.eu.int

European Monetary Institute
http://www.ecb.int

European Union Services:
Statistical Office of the European Communities
http://europa.eu.int/eurostat.html

Cedefop
http://www.cedefop.gr

European Agency for the Evaluation of Medicinal Products
http:/www.eudra.org/emea.html

European Training Foundation
http://www.etf.it

European Environment Agency
http://www.eea.dk/

CELEX
http://www.europa.eu.int/celex

RAPID
http://www.europa.eu.int/en/comm/spp/rapid.html

I'M Europe (Information Market)
http://www2.echo.lu

Information Society Project Office (ISPO)
http://www.ispo.cec.be

Community Research and Development Information Service (CORDIS)
http://www.cordis.lu

The United States Web Site of the European Union
http://www.eurunion.org

Alphabetical Index to Topics on EU Web Sites
http://www.eurunion.org/infores/euindex.htm

REFERENCES

Trade and the Environment, Law, Economics and Policy, edited by Durwood Zaelke, Paul Orbuch, Robert F. Housman, Center for International Environmental Law, 1993.

Pollution Control in the United States, Evaluating the System, J. Clarence Davies and Jan Mazurek, Resources for the Future, 1998.

Direct Effect of European Law and the Regulation of Dangerous Substances, Christopher J. M. Smith, Gordon and Breach Publishers, 1995.

Environmental Change and International Law, edited by Edith Brown Weiss, United Nations University Press, 1992.

Environmental Managers Guide to ISO 14000, Business & Legal Reports, 1995.

International Environmental Law and Policy, David Hunter, James Salzman, Durwood Zaelke, Foundation Press, 1998.

Environmental Management Systems, Jay G. Martin and Gerald J. Edgley, Governments Institutes, 1998.

International Environmental Auditing, David D. Nelson, Government Institutes, 1998.

Environmental Management in European Companies, Success Stories and Evaluation, edited by Jobst Conrad, Gordon and Breach Science Publishers, 1998.

Environmental Strategies Handbook, A Guide to Effective Policies & Practices, Rao V. Kolluru, McGraw-Hill, Inc. 1994.

REFERENCES FOR U.S. REGULATORY POLICY TOWARD EMSS AND ISO 14001, BY JOHN GROSSKOPF

EPA Position Statement on Environmental Management Systems and ISO 14001 and a Request for Comments on the Nature of the Data to Be Collected From Environmental Management System/ISO 14001 Pilots, Federal Register, March 12, 1998, PP 12094-12097.

Department of Justice Sentencing Guidelines, Part D – Commitment to Environmental Compliance.

A Presentation given by the Office of the Deputy Under Secretary of Defense (Environmental Security), ISO 14001 Environmental Management Systems, August 4, 1998.

Memorandum for the Deputy Assistant Secretaries of the Army, Navy, Air Force and Defense Logistics Agency, Voluntary Adoption of International Organization for Standardization (ISO) 14001, August 13, 1996.

A letter from Peter Walsh, Assistant Deputy Under Secretary of Defense (Environmental Quality) to Barry Breen, Director Office of Federal Facilities Enforcement, EPA, February 29, 1996.

Memorandum for the Assistant Secretaries of the Army, Navy, Air Force, The National Security Agency and Defense Logistics Agencies entitled, "Participation in DOD Component ISO 14001 Environmental Management System Pilot Cost/Benefit Study" prepared by the Office of the Under Secretary of Defense, February 26, 1997.

A presentation given by Larry Stirling, DOE Headquarters at the DOE Pollution Prevention Conference XII entitled, "Changing the Way We DO Business: ISO 14001 Environmental Management Systems."

ISO 14000: International Environmental Management Standards—How Will They Impact DOE, Circular prepared by ANSI/GETF ISO 14000 Integrated Solutions.

A letter from Sherri Goodman, Deputy Under Secretary of Defense (Environmental Security) to Steve Herman, Assistant Administrator Office of Enforcement and Compliance Assurance USEPA, November 5, 1998 endorsing the CEMP by DOE.

ISO 14001 and the Law, by S. Wayne Rosenbaum, 1998 by AQA Co.

The ISO 14001 International Environmental Standards; 14001: 1996 (E) References.

The ISO 14001 International Environmental Standards; 14001: 1996 (E).

EPA Guidance Document: Compliance-Focused Environmental Management System, EPA 330/FB-GD/97-006.

10 The European Union

CONTENTS

The European Union (EU) is a community of 15 countries and 372 million citizens. The EU was originally an economic association (European Community) formed in 1958 (Treaty of Rome) with six member states. Component organizations date back to 1951. On November 1, 1993, with the Treaty of the European Union (Maastrict, The Netherlands), the EC became the European Union (EU). Official community languages include Spanish, German, Greek, English, French, Italian, Dutch, Portugese, Finnish, and Swedish. The EU is the collective designation of three organizations with common membership:

- European Economic Community (EEC) (common market)
- The European Coal and Steel Community
- The European Atomic Energy Community (Euratom)

In addition, the European Economic Area (EEA) is comprised of the 15 member countries in the EU as well as Iceland, Liechtenstein, and Norway. As of January 1, 1995, the fifteen member states (countries) of the EC included

Austria	Belgium	Denmark	Finland
France	Germany	Greece	Ireland
Italy	Luxembourg	Netherlands	Portugal
Spain	Sweden	United Kingdom	

Goals of the European Union

- Integrate the EU economics
- Coordinate social development
- Accomplish political union of the democratic states of Europe

Organizational Structure of the EU

- European Parliament Council of the European Union
- Commission of the European Union
- The Council of Ministers
- European Court of Justice
- Court of Auditors
- Economic and Social Committee
- Committee of the Regions

- European Environment Agency
- As well as Other Community Bodies

EU DIRECTIVES AND ENVIRONMENTAL PROGRAMS

The EC directives create a consistent set of requirements that must be adopted into national law for all EC and EEA member states. The organization of these member countries into an association with common economic, social, and political issues has eliminated restrictions to the free movement of goods, services, capital, workers, and tourists. The EEC is the West's largest trading entity and is comprised of more than 300 million people. A "new approach" policy toward international trade began in 1985. The new approach strategy was to allow the free movement of goods through the technical harmonization of laws and regulations. Many of the directives in the 1970s and 1980s were focused on Europe's goal to improve the living and working conditions of its citizens. Many of the early environmental directives were focused on testing and labeling of dangerous chemicals, drinking and surface water protection, and the control of air pollutants (NO_x, SO_2, particulates) from power plants and motor vehicles. These priorities were shared by regulators in the United States in the 1970s when the Clean Air Act, Clean Water Act, and Safe Drinking Water Act were born.

There are three types of directives, including framework directives, daughter directives, and specific directives. Framework directives provide the general principles or goals of a directive, and daughter directives are more specific and address a particular issue of the framework directive. For example, the directive relating to water pollution control (76/464/EEC) has numerous daughter directives such as Directive 84/491/EEC specific to discharges of hexchlor-hexane. Regulations may also be implemented and are applicable in member states without the State taking action (directly applicable). States must adopt national legislation to implement the provisions of a directive. However, directives are directly effective (doctrine of direct effect), which means even if a member state does not develop national legislation or does not meet the deadlines for implementation, the directive is still considered applicable. Member states must consider the following when implementing directives:

- National laws already in place
- Effective dates for the directive
- Doctrine of direct effect
- Adequacy of national laws
- State authority for implementation

Other legislative acts of the EU include Council Decisions, which are binding decisions, and Council Recommendations and Opinions, which are nonbinding and serve to motivate member states to take action such as adopt certain provisions.

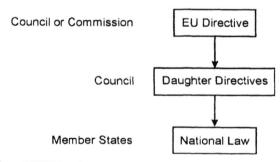

Figure 7 Implementation of EU Directives

Proposals for directives may come from member states, meetings of the Council of Ministers, or the EU Commission. Even though these directives have been adopted to protect the environment, legislative harmonization is based more in eliminating trade barriers and distorted competition than environmental protection.

EXAMPLES OF EU DIRECTIVES

Nitrate Directive (1991)—requires member states to develop codes of good agricultural practice designed to reduce the amount of nitrates that enter surface and groundwater from runoff. Affected or potentially affected waterways must be identified. Like Best Management Practices (BMPs) used for stormwater management in the United States, the strategy of using "codes of good practice" instead of legislation allows each area or site to determine what would be appropriate to control nitrate pollution. This provides freedom to develop workable, cost effective strategies without regulatory restraints. In addition to the codes of good practice, action programs will be developed that contain legally enforceable constraints on agricultural practices and the spreading of organic manure.

Directive 90/313/EEC—access to environmental information. Member states must publish information on the state of the environment. Also, information regarding environmental impacts of industrial activities and government decisions must be available. Although broader in scope, this requirement is similar to a public information law in the United States called "Emergency Planning and Community Right-to-Know" also called SARA Title III.

EU ENVIRONMENTAL PROGRAMS

Important Environmental Issues in the EU

- Climate change
- Acid rain
- Air and water pollution (including coastal areas)
- Depletion of natural resources
- Biodiversity
- Waste and industrial risk
- Deterioration of the urban environment

A priority in environmental legislation in the EU has been reclamation and protection of water resources. A problem, especially for rural communities, is having the funding available to upgrade and operate treatment systems. The EU has made funding available for the construction of water and wastewater treatment facilities. Similarly, in the United States, the U.S. Environmental Protection Agency (EPA) has provided funding to each state under the Clean Water Act.

Agenda 21

This is the global agenda for sustainable development from the United Nations Conference on Environment and Development that took place in 1992 in Rio de Janeiro.

EUROPE AND SUSTAINABLE DEVELOPMENT

"Toward Sustainability" (1992)

This program sets up a proactive strategy designed to integrate environmental policy with economic interests by the year 2000. See the EU's fifth environment action programme. The programme is focused on five main economic sectors that impact the environment and deplete natural resources, including industry, transportation, energy, agriculture, and tourism. Targets have been established for each of these sectors.   The World Commission on Environment and Development defines sustainable development as "development which meets the needs of the present without compromising the ability of future generations to meet their own needs."

The 1992 Maastricht Treaty formally established the concept of sustainable development in EU law. The Amsterdam Treaty of 1997 made sustainable development one of the overriding objectives of the EU.

The six elements of the European Union's sustainable development actions include

1) Integration of environmental considerations into other policy areas
2) Partnership between the EU, member states, the business world, and the public, and shared responsibility
3) Broadening the range of environmental policy instruments to include taxes, subsidies, and voluntary agreements
4) Changing patterns of consumption and production
5) Implementation and enforcement of legislation
6) International cooperation within the framework of Agenda 21 and the fifth environmental action programme

EUROPEAN CONSULTIVE FORUM ON THE ENVIRONMENT AND SUSTAINABLE DEVELOPMENT

A dialogue group was established in 1994 under the fifth action program. The goal of the forum was to bring EU environmental policies and laws closer to the people.

IMPEL NETWORK OF AUTHORITIES IN THE MEMBER STATES

They are responsible for implementing and enforcing national and EU environmental law.

ENVIRONMENTAL POLICY REVIEW GROUP

Top environment officials from the Commission and the member states.

Europe's Environment: The Dobris Assessment (1995) – a report produced by the European Environment Agency (EEA) about the state of the environment in Europe.

Environment in the European Union (1995) – an update report to Europe's Environment: The Dobris Assessment. This report was prepared as part of the progress review for the fifth action program.

EU STANDARDS FOR THE SAFETY OF MACHINERY

In addition to environmental directives, the EU has established directives for the safety of products being imported into member countries. These directives cover such items as construction products, toys, pressure vessels, machines, personal protective equipment, nonautomatic weighing machines, medical devices, gas appliances, boilers, and explosives. Companies exporting these and other products to locations within the EEA must comply with safety directives (standards) that have been established for the product or machinery. The following is an example for machinery. Types of machinery covered by EU directives include lifting platforms, woodworking machines, lasers, packaging machines, tractors, pumps, conveyor belts, compressors, and power tools.

European Community Legislation on Machinery

A specific directive has been developed called the machinery safety directive. The EEC adopts directives which establish essential health and safety requirements and refer to optional harmonized standards as a means of complying with the directive. The machinery safety directive was first officially published in 1989 and assigned number 392. This directive is referred to as Council Directive 89/302/EEC—the machinery safety directive. Directives are identified and referenced by using a consistent format which begins with the date of publication (e.g., 89) followed by the directive number (302) and finally the letters EEC (European Economic Community). During the implementation of a new directive like 89/392/EEC, transition periods are allowed for member countries to implement the legislation and manufacturers to comply with the new requirements. Member countries had to adopt the machinery safety directive by January 1, 1992. The date in which the directive was to be enforced was January 1, 1993; however, another directive which modified 89/392/EEC extended the transition period ending on December 31, 1994. During the transition period, manufacturers may comply with the directive or comply with the requirements of the country of installation. Revisions to the directive were again due in 1996.

> The objective of the directive is to "harmonize the design and manufacture of machinery so as to preserve the health and ensure the safety of people using machinery and in particular workers."

The directive details the essential health and safety requirements for machinery to which the manufacture must confirm compliance.

The directives requirements have been grouped according to the hazards they address and include design, construction, and technical aspects of machinery. The directive is designed for machinery users and people in the proximity. Safety components sold separately and machinery sold dismantled for ease of transport are also included in the scope of the directive. Although machinery that is totally enclosed in a housing presents no hazards, it is still covered by the directive because without the housing the hazards still exist. The machinery safety directive does not apply to machinery if it is specifically covered by another directive. For example, certain types of machinery that present one main hazard such as electrical (e.g., resistance welding) are covered by directive 73/23/EEC which takes precedence over the machinery safety directive. Further, repairs such as the replacement of worn parts to machinery in compliance with the directive that do not modify the characteristics of the machine are not covered.

Main Categories of the Essential Health and Safety Requirements of the Machinery Safety Directive

- Principles of safety integration (hazard identification, risk assessment)
- Controls (e.g., emergency stop, unexpected start-up, and control devices)
- Protection against mechanical hazards (e.g., crushing, shearing, and cutting)
- Required characteristics of guards and protection devices (fixed and moveable guards, and interlocking)
- Protection against other hazards (e.g., electrical, fire and explosion, extreme temperatures, and noise)
- Maintenance (e.g., operation intervention and energy isolation)
- Indicators, warnings, markings, and instruction material in the language of the country

Technical Construction File

This file must be created and maintained for 10 years from the date of manufacture for each piece of machinery to ensure conformance with the directive. The file contains approximately 26 items such as detailed drawings of the machine, accident history of the machine, and operating instructions. In the event a manufacturer is using parts or components from another company, the manufacturer of the machine cannot pass his responsibility on to the supplier or manufacturer of the part or component. The choice of parts or components must be justified in the technical construction file. The manufacturer typically maintains the file on its own premises or with the authorized representative. The authorized representative is the organization approved by the manufacturer to act on their behalf within the EC. An authorized representative, however, cannot alter machinery to comply with the directive. This representative is required to be established in the community and typically handles administrative obligations relating to the directive. The file must be made available in response to substantiated requests by member states. Member states must allow an appropriate amount of time for the manufacturer to produce the file. It is realized some information will be stored electronically and require additional time to retrieve. However, if the file is not made available by the manufacturer or the authorized representative within the specified time period, doubt is cast upon the claims of conformity with the directive. It is important this file be readily accessible to avoid such doubt and possible delays.

Declaration of Conformity

The directive is a "self certification" process called the EC Declaration of Conformity. This declaration is the procedure by which the manufacturer declares the machinery being placed on the market complies with all the essential health and safety requirements. This means the manufacturer certifies themselves as having met the requirements of the directive. The declaration must list the directives requiring the affixing of the 'CE' mark to which the machinery conforms. Currently, ten (10) directives require the affixing of the 'CE' mark (e.g., construction products, simple pressure vessels, appliances burning gaseous fuels). Signature of the EC declaration of conformity authorizes the manufacturer to affix the 'CE' mark to the machinery.

Certain machines covered by the directive require a "third party" review. The third party is a separate organization called a competent or notified body that is recognized by the supervisory authorities of the member states. The notified body assists the manufacturer with certifying compliance with the directive.

The 'CE' marking is a compliance symbol placed directly on the machinery or placed on the packaging or product documentation indicating the machine meets the requirements of the directive. The 'CE' mark consists of the CE symbol followed by the last two digits of the year in which the mark was affixed.

Harmonized European Standards

To provide additional information to the essential health and safety requirements found in the Machinery Safety Directive, harmonized standards or EN (European Norm) have been developed by a separate standardization body (CEN/CENELEC). The standards are technical specifications that cover specific safety requirements and provide additional guidance to companies on how to comply with the directive. Harmonized standards are not compulsory but are needed by professionals for designing and manufacturing products conforming to the requirements established by the directive. Products are assumed to comply with the requirements of the directive if they have been designed and manufactured according to the harmonized standards. The three main categories of the safety standards are Type A, B and C standards. Type A standards provide designers and manufactures with guidance to enable them to produce safer machines and consist of basic concepts and general principles for design. An example of an 'A' standard is EN292–Safety of Machinery (basic concepts). 'B' type standards are specific to one safety aspect (e.g., electrical) or type of safety related device. 'B' standards are further divided into two categories ('B1' and 'B2'). Type 'B1' standards are on particular safety aspects such as safety distances (EN 294). Type 'B2' standards are on safety-related devices such as two-hand controls, interlocking devices, and guards. 'C' type standards give detailed safety requirements for a particular type of machinery or group of machines. 'B1' and 'B2' standards are the most useful for the purpose of the machinery safety directive.

More than 31 European technical committees are currently working on standards for specific types of machinery and as of September 1994, European standards for machinery safety and related items include

- Directive 89/392/EEC (Machinery Safety Directive)
- Directive 91/368/EEC (Safety of Machinery Presenting Hazards Due to Their Mobility or Their Lifting Loads)
- Directive 93/68/EEC (Affixing and Using the CE Marking)
- Directive 85/374/EEC (Liability for Defective Parts)
- Adopted EN standards (17)
- Draft standards (prEN) (114)
- Other work items being addressed (203)

The draft standards that have been developed by CEN/CENELEC and ISO relating to the safety of machinery are expected to be adopted by the EEC. Two committees have been established to handle problems encountered with the application of the directive. One committee deals with standardization issues submitted to CEN or CENELEC. The second acts as an advisory committee to the Commission on ambiguities in the directive. The Commission, with assistance from the advisory committee, can help establish a common and uniform reading of the directive. However, only the Court of Justice of the EC has the authority to interpret the articles or essential health and safety requirements of the directive.

Potential penalties for not complying with the directive include
- Shipment may be stopped at the border and not allowed into a member country.
- Manufacturer would not be allowed to ship product to the EC for five years.
- Withdrawal of product from the European market.
- Possible jail term for the company's European representative.

Member states can question the manufacturers' claims and reverse the assumption of conformity with the directive. The state is required to notify the Commission immediately if they have prohibited the placing on the market, putting into service or use or restricted free movement of machinery. Enforcement will be at the national level and be conducted in a variety of ways. For example, incoming goods may be inspected at the EC/EEA border crossings, or market surveillance may be conducted by regulatory authorities of member countries. In the event of complaints of suspected noncompliance, authorities may conduct investigation activities. Unsafe machinery can be removed from the market even if it is purported to comply with the directive. In fact, member states have an obligation to do so. The authority to remove machinery in this manner is contained in Article 7 of the directive and is called the safeguard clause. Prior to invoking the safeguard clause, a member state must present the grounds for doing so. Arbitrary action is prohibited. Actions taken by a member state must be proportionate to the danger and confirmed by the commission. This means a product may not necessarily be banned outright. Also, whatever measures are taken are not permanent and can be lifted at any time. In the event action is taken against a manufacturer, they should take appropriate legal action and collect as much documentation on the machinery as possible. The Commission will examine the case and consult with all parties. If the Commission decides the measures taken by the member state are justified, the action is extended to the entire EC. Manufacturers may appeal this decision to the courts, but as is typical with legal proceedings, they are lengthy and expensive. Also, during the appeal process, whatever actions had been taken remain in effect until the case is resolved.

SUMMARY OF THE ESSENTIAL HEALTH AND SAFETY REQUIREMENTS RELATING TO THE DESIGN AND CONSTRUCTION OF MACHINERY

Directives are published by the Office for Official Publications of the European Community. The publications are available in the United States from companies that specialize in documents, including international standards (Chapter 14). The machinery safety directive is organized as follows:

- Foreword
- Recitals
- Corpus of Machinery Directive (divided into 14 Articles)
- Annex I— Essential health and safety requirements
- Annex II—Declaration of Conformity
- Annex III—CE Mark
- Annex IV—Procedures for certain types of machinery and safety components
- Annex V—Declaration of Conformity
- Annex VI—EC-type examination of machinery
- Annex VII—Minimum criteria to be taken into account by member states for the notification of bodies.

The recitals are not necessarily part of national legislation that is developed in response to the directive by each member country. However, in the case of litigation, the courts may take them into consideration. The recitals are important because they clarify certain ambiguities in the directive and clarify the meaning of certain words. Also, the order of information appearing in the directive does not imply any hierarchy. The fourteen articles contained in the Corpus of the Machinery Directive detail specific actions, responsibilities, and authority granted member states and manufacturers of machinery. Article 8 is one of the most important for manufacturers because it specifies the procedures that must be followed to certify their machinery conforms to the directive. The seven annexes following the articles provide supplementary and explanatory information. Annex I contains the essential health and safety requirements that must be met by the manufacturers of machinery and make up the major portion of the directive.

Annex I—Essential health and safety requirements relating to the design and construction of machinery and safety components.

1) The essential health and safety requirements are mandatory. The directive imposes an obligation on the manufacturer to use available means to build safe machinery. However, it may not be technically or economically feasible to meet the objectives set by the directive.

2) Summary of the essential health and safety requirements
   a. General
      ◆ Definitions
      ◆ Principles of safety integration
      ◆ Materials and products
      ◆ Lighting
      ◆ Design of machinery to facilitate its handling
   b. Controls
      ◆ Safety and reliability of control systems
      ◆ Control devices
      ◆ Starting
      ◆ Stopping device
      ◆ Emergency stop

- ◆ Mode selection
- ◆ Failure of power supply
- ◆ Failure of control circuit
- ◆ Software

c. Protection against mechanical hazards
- ◆ Stability
- ◆ Risk of break-up during operation
- ◆ Risks due to falling or ejected objects and surfaces, edges, or angles
- ◆ Risks related to combined machinery
- ◆ Risks related to variations in the rotational speed of tools
- ◆ Prevention of risks related to moving parts
- ◆ Choice of protection against risks related to moving parts

d. Required characteristics of guards and protective devices
- ◆ General requirements
- ◆ Special requirements for guards (fixed, moveable, adjustable guards restricting access)
- ◆ Special requirements for protection devices

e. Protection against other hazards
- ◆ Electrical supply
- ◆ Static electricity
- ◆ Energy supply other than electricity
- ◆ Errors of fitting
- ◆ Extreme temperatures
- ◆ Fire and explosion
- ◆ Noise
- ◆ Vibration
- ◆ Risk of slipping, tripping, or falling
- ◆ Radiation
- ◆ External radiation
- ◆ Laser equipment
- ◆ Emissions of dusts, gases, etc.
- ◆ Risk of being trapped in machine

f. Maintenance
- ◆ Machinery maintenance
- ◆ Access to operation position and servicing points
- ◆ Isolation of energy sources
- ◆ Operator intervention
- ◆ Cleaning of internal parts

g. Indicators (warning, marking, instructions)
- ◆ Information and warning devices
- ◆ Warning of residual risk
- ◆ Marking
- ◆ Instructions

DIRECTORY OF COMMUNITY LEGISLATION IN FORCE RELATING TO THE ENVIRONMENT

15—Environment, consumers and health protection: education and culture

    15.10 Environment

        15.10.10   General provisions and programmes

15.10.10—General provisions and programmes, Secondary legislation

75/65/EEC: Commission recommendation of 20 December 1974 to Member States concerning the protection of the architectural and natural heritage

75/436/Euratom, ECSC, EEC: Council recommendation of 3 March 1975 regarding cost allocation and action by public authorities on environmental matters

Council resolution of 3 March 1975 on energy and the environment

Council resolution of 15 July 1975 on the adaptation to technical progress of Directives or other Community rules on the protection and improvement of the environment

76/161/EEC: Council Decision of 8 December 1975 establishing a common procedure for the setting up and constant updating of an inventory of sources of information on the environment in the Community

Resolution of the ECSC Consultative Committee on the Community's Environmental Protection Policies

Council Directive 85/337/EEC of 27 June 1985 on the assessment of the effects of certain public and private projects on the environment

86/479/EEC: Commission Decision of 18 September 1986 establishing an Advisory Committee on the protection of the environment in areas under serious threat (Mediterranean basin)

Council resolution of 3 May 1988 on the close of the European Year of the Environment

Council Regulation (EEC) No. 1210/90 of 7 May 1990 on the establishment of the European Environment Agency and the European Environment Information and Observation Network

Council resolution of 28 January 1991 on the Green Paper on the urban environment

Council Regulation (EEC) No. 1973/92 of 21 May 1992 establishing a financial instrument for the environment (LIFE)

Council resolution of 25 February 1992 on the future Community policy concerning the European coastal zone

Special report No. 3/92 concerning the environment together with the Commission's replies

Council resolution of 3 December 1992 concerning the relationship between industrial competitiveness and environmental protection

Council Regulation (EEC) No. 793/93 of 23 March 1993 on the evaluation and control of the risks of existing substances

Council Regulation (EEC) No. 1836/93 of 29 June 1993 allowing voluntary participation by companies in the industrial sector in a Community eco-management and audit scheme

Special report No. 4/94 on the urban environment together with the Commission's replies

95/365/EC: Commission Decision of 25 July 1995 establishing the ecological criteria for the award of the Community eco-label to laundry detergents

95/533/EC: Commission Decision of 1 December 1995 establishing the ecological criteria for the award of the Community eco-label to single-ended light bulbs (text with EEA relevance)

96/13/EC: Commission Decision of 15 December 1995 establishing the ecological criteria for the award of the Community eco-label to indoor paints and varnishes

96/149/EC: Commission Decision of 2 February 1996 on the recognition of the Irish standard IS310: First Edition, establishing specifications for environmental management systems, in accordance with Article 12 of Council Regulation (EEC) No. 1836/93 (text with EEA relevance)

96/150/EC: Commission Decision of 2 February 1996 on the recognition of the British standard BS7750: 1994, establishing specifications for environmental management systems, in accordance with Article 12 of Council Regulation (EEC) No. 1836/93 (text with EEA relevance)

96/151/EC: Commission Decision of 2 February 1996 on the recognition of the Spanish standard UNE 77-801(2)-94, establishing specifications for environmental management systems, in accordance with Article 12 of Council Regulation (EEC) No. 1836/93 (text with EEA relevance)

96/160/EC: Commission Decision of 8 February 1996 on the appointment of members of the General Consultative Forum on the Environment (text with EEA relevance)

96/304/EC: Commission Decision of 22 April 1996 establishing the ecological criteria for the award of the Community eco-label to bed linen and T-shirts (text with EEA relevance)

96/337/EC: Commission Decision of 8 May 1996 establishing the ecological criteria for the award of the Community eco-label to double-ended light bulbs (text with EEA relevance)

96/461/EC: Commission Decision of 11 July 1996 establishing ecological criteria for the award of the Community eco-label to washing machines (text with EEA relevance)

96/733/EC: Commission Recommendation of 9 December 1996 concerning Environmental Agreements implementing Community directives (text with EEA relevance)

Commission Decision of 24 February 1997 on the setting up of a European consultative forum on the environment and sustainable development (text with EEA relevance)

97/264/EC: Commission Decision of 16 April 1997 on the recognition of certification procedures in accordance with Article 12 of Council Regulation (EEC) No. 1836/93 of 29 June 1993, allowing voluntary participation by companies in the industrial sector in a Community eco-management and audit scheme (text with EEA relevance)

97/265/EC: Commission Decision of 16 April 1997 on the recognition of the international standard ISO 14001:1996 and the European standard EN ISO 14001:1996, establishing specification for environmental management systems, in accordance with Article 12 of Council Regulation (EEC) No. 1836/93 of 29 June 1993, allowing voluntary participation by companies in the industrial sector in a Community eco-management and audit scheme (text with EEA relevance)

97/872/EC: Council Decision of 16 December 1997 on a Community action programme promoting non-governmental organizations primarily active in the field of environmental protection

Council Regulation (EC) No. 722/97 of 22 April 1997 on environmental measures in developing countries in the context of sustainable development

Decision of 21 March 1997 on public access to European Environment Agency documents

Council Resolution of 7 October 1997 on the drafting, implementation and enforcement of Community environmental law

Council Resolution of 7 October 1997 on environmental agreements

98/22/EC: Council Decision of 19 December 1997 establishing a Community action programme in the field of civil protection

Agreement of the Representatives of the Governments of the Member States meeting in Council of 5 March 1973 on information for the Commission and for the Member States with a view to possible harmonization throughout the Communities of urgent measures concerning the protection of the environment

Declaration of the Council of the European Communities and of the representatives of the Governments of the Member States meeting in the Council of 22 November 1973 on the programme of action of the European Communities on the environment

Resolution of the Council and of the representatives of the Governments of the Member States of the European Communities, meeting within the Council of 3 October 1984, on the link between the environment and development

Resolution of the Council of the European Communities and of the representatives of the Governments of the Member States, meeting within the Council of 19 October 1987 on the continuation and implementation of a European Community policy and action programme on the environment (1987-1992)

Resolution of the Council and the Ministers for Health, meeting within the Council of 11 November 1991 on health and the environment

Resolution of the Council and the Ministers for Health, meeting within the Council of 11 November 1991 on the treatment and rehabilitation of drug addicts serving sentences for criminal offences

Conclusions of the Council and the Ministers of Education meeting within the Council of 1 June 1992 on the development of environmental education

Resolution of the Council and the Representatives of the Governments of the Member States, meeting within the Council of 1 February 1993 on a Community programme of policy and action in relation to the environment and sustainable development—A European Community programme of policy and action in relation to the environment and sustainable development

Resolution of the Council and the Ministers for Health, meeting within the Council of 27 May 1993 on future action in the field of public health

15.10.20.20 Water Protection and Management

External relations

Convention on the International Commission for the Protection of the Elbe (EEC Translation)

Protocol to the Convention of 8 October 1990 between the Governments of the Federal Republic of Germany and the Czech and Slovak Federal Republic and the European Economic Community on the International Commission for the Protection of the Elbe

Cooperation Agreement for the protection of the coasts and waters of the north-east Atlantic against pollution

Secondary legislation

Council Directive 75/440/EEC of 16 June 1975 concerning the quality required of surface water intended for the abstraction of drinking water in the Member States

Council Directive 76/160/EEC of 8 December 1975 concerning the quality of bathing water

Council Directive 76/464/EEC of 4 May 1976 on pollution caused by certain dangerous substances discharged into the aquatic environment of the Community

77/795/EEC: Council Decision of 12 December 1977 establishing a common procedure for the exchange of information on the quality of surface fresh water in the Community

Council resolution of 26 June 1978 setting up an action programme of the European Communities on the control and reduction of pollution caused by hydrocarbons discharged at sea

Council Directive 79/869/EEC of 9 October 1979 concerning the methods of measurement and frequencies of sampling and analysis of surface water intended for the abstraction of drinking water in the Member States

Council Directive 79/923/EEC of 30 October 1979 on the quality required of shellfish waters

80/686/EEC: Commission Decision of 25 June 1980 setting up an Advisory Committee on the control and reduction of pollution caused by hydrocarbons discharged at sea

Council Directive 80/68/EEC of 17 December 1979 on the protection of groundwater against pollution caused by certain dangerous substances

Council Directive 80/778/EEC of 15 July 1980 relating to the quality of water intended for human consumption

Council Directive 82/176/EEC of 22 March 1982 on limit values and quality objectives for mercury discharges by the chlor-alkali electrolysis industry

Council Directive 83/513/EEC of 26 September 1983 on limit values and quality objectives for cadmium discharges

Council resolution of 7 February 1983 concerning the combating of water pollution

Council Directive 84/491/EEC of 9 October 1984 on limit values and quality objectives for discharges of hexachlorocyclohexane

86/85/EEC: Council Decision of 6 March 1986 establishing a Community information system for the control and reduction of pollution caused by the spillage of hydrocarbons and other harmful substances at sea

Council Directive 86/280/EEC of 12 June 1986 on limit values and quality objectives for discharges of certain dangerous substances included in List I of the Annex to Directive 76/464/EEC

Council Directive 88/347/EEC of 16 June 1988 amending Annex II to Directive 86/280/EEC on limit values and quality objectives for discharges of certain dangerous substances included in List I of the Annex to Directive 76/464/EEC

Council resolution of 19 June 1990 on the prevention of accidents causing marine pollution

Council Directive 91/271/EEC of 21 May 1991 concerning urban waste-water treatment

Council Directive 91/676/EEC of 12 December 1991 concerning the protection of waters against pollution caused by nitrates from agricultural sources

Council Directive 91/692/EEC of 23 December 1991 standardizing and rationalizing reports on the implementation of certain Directives relating to the environment

92/446/EEC: Commission Decision of 27 July 1992 concerning questionnaires relating to Directives in the water sector

Council resolution of 25 February 1992 on the future Community groundwater policy

93/481/EEC: Commission Decision of 28 July 1993 concerning formats for the presentation of national programmes as foreseen by Article 17 of Council Directive 91/271/EEC

Council Directive 93/75/EEC of 13 September 1993 concerning minimum requirements for vessels bound for or leaving Community ports and carrying dangerous or polluting goods

Council Resolution of 20 February 1995 on groundwater protection

Resolution of the Council and of the representatives of the Governments of the Member States of the European Communities, meeting within the Council of 3 October 1984, on new forms of cooperation in the sphere of water

15.10.20.30—Monitoring of atmospheric pollution

External relations

Protocol to the 1979 Convention on long-range transboundary air pollution on long-term financing of the cooperative programme for monitoring and evaluation of the long-range transmission of air pollutants in Europe (EMEP)

Amendment to the Montreal protocol on substances that deplete the ozone layer

Protocol to the 1979 Convention on long-range transboundary air pollution concerning the control of emissions of nitrogen oxides or their transboundary fluxes

Amendment to the Montreal Protocol on substances that deplete the ozone layer

United Nations Framework Convention on Climate Change—Declarations

Secondary legislation

Council Directive 72/306/EEC of 2 August 1972 on the approximation of the laws of the Member States relating to the measures to be taken against the emission of pollutants from diesel engines for use in vehicles

Commission Directive 77/102/EEC of 30 November 1976 adapting to technical progress Council Directive 70/220/EEC of 20 March 1970 on the approximation of the laws of the Member States relating to measures to be taken against air pollution by gases from positive ignition engines of motor vehicles

Council Directive 77/537/EEC of 28 June 1977 on the approximation of the laws of the Member States relating to the measures to be taken against the emission of pollutants from diesel engines for use in wheeled agricultural or forestry tractors

Council Directive 80/779/EEC of 15 July 1980 on air quality limit values and guide values for sulphur dioxide and suspended particulates

Council Directive 80/1268/EEC of 16 December 1980 on the approximation of the laws of the Member States relating to the fuel consumption of motor vehicles

Council Resolution of 15 July 1980 on transboundary air pollution by sulphur dioxide and suspended particulates

82/459/EEC: Council Decision of 24 June 1982 establishing a reciprocal exchange of information and data from networks and individual stations measuring air pollution within the Member States

Council Directive 82/884/EEC of 3 December 1982 on a limit value for lead in the air

Council Directive 84/360/EEC of 28 June 1984 on the combating of air pollution from industrial plants

Council Directive 85/203/EEC of 7 March 1985 on air quality standards for nitrogen dioxide

Council Directive 85/210/EEC of 20 March 1985 on the approximation of the laws of the Member States concerning the lead content of petrol

86/277/EEC: Council Decision of 12 June 1986 on the conclusion of the Protocol to the 1979 Convention on long-range transboundary air pollution on long-term financing of the cooperative programme for monitoring and evaluation of the long-range transmission of air pollutants in Europe (EMEP)

Council Regulation (EEC) No. 3528/86 of 17 November 1986 on the protection of the Community's forests against atmospheric pollution

Commission Regulation (EEC) No. 1696/87 of 10 June 1987 laying down certain detailed rules for the implementation of Council Regulation (EEC) No. 3528/86 on the protection of the Community's forests against atmospheric pollution (inventories, network, reports)

Council Directive 88/76/EEC of 3 December 1987 amending Directive 70/220/EEC on the approximation of the laws of the Member States relating to measures to be taken against air pollution by gases from the engines of motor vehicles

Council Directive 88/77/EEC of 3 December 1987 on the approximation of the laws of the Member States relating to the measures to be taken against the emission of gaseous pollutants from diesel engines for use in vehicles

Council Directive 88/436/EEC of 16 June 1988 amending Directive 70/220/EEC on the approximation of the laws of the Member States relating to measures to be taken against air pollution by gases from engines of motor vehicles (Restriction of particulate pollutant emissions from diesel engines)

Council Directive 88/609/EEC of 24 November 1988 on the limitation of emissions of certain pollutants into the air from large combustion plants

Council resolution of 14 October 1988 for the limitation of use of chlorofluorocarbons and halons

Council Directive 89/369/EEC of 8 June 1989 on the prevention of air pollution from new municipal waste incineration plants

Council Directive 89/429/EEC of 21 June 1989 on the reduction of air pollution from existing municipal waste-incineration plants

Council resolution of 21 June 1989 on the greenhouse effect and the Community

90/437/EEC: Commission recommendation of 27 June 1990 on the reduction of chlorofluoro-carbons used by the Community's foam plastics industry

90/438/EEC: Commission recommendation of 27 June 1990 on the reduction of chlorofluoro-carbons used by the Community's refrigeration industry

Council Directive 91/441/EEC of 26 June 1991 amending Directive 70/220/EEC on the approx-imation of the laws of the Member States relating to measures to be taken against air pollution by emissions from motor vehicles

Council Directive 92/72/EEC of 21 September 1992 on air pollution by ozone

93/389/EEC: Council Decision of 24 June 1993 for a monitoring mechanism of Community CO_2 and other greenhouse gas emissions

Council Directive 93/12/EEC of 23 March 1993 relating to the sulphur content of certain liquid fuels

Council Directive 93/59/EEC of 28 June 1993 amending Directive 70/220/EEC on the approx-imation of the laws of the Member States relating to measures to be taken against air pollution by emissions from motor vehicles

Council Directive 93/76/EEC of 13 September 1993 to limit carbon dioxide emissions by improving energy efficiency (SAVE)

Directive 94/12/EC of the European Parliament and the Council of 23 March 1994 relating to measures to be taken against air pollution by emissions from motor vehicles and amending Directive 70/220/EEC

European Parliament and Council Directive 94/63/EC of 20 December 1994 on the control of volatile organic compound (VOC) emissions resulting from the storage of petrol and its distribution from terminals to service stations

Commission Regulation (EC) No. 1091/94 of 29 April 1994 laying down certain detailed rules for the implementation of Council Regulation (EEC) No. 3528/86 on the protection of the Com-munity's forests against atmospheric pollution

Council Regulation (EC) No. 3093/94 of 15 December 1994 on substances that deplete the ozone layer

96/511/EC: Commission Decision of 29 July 1996 concerning the questionnaires provided for in Council Directives 80/779/EEC, 82/884/EEC, 84/360/EEC and 85/203/EEC (text with EEA relevance)

96/604/EC: Commission Decision of 8 October 1996 allocating quotas for placing hydrochlo-rofluorocarbons on the market in the Community for the period 1 January to 31 December 1996 (Only the German, Greek, English, French, Italian and Dutch texts are authentic) (text with EEA relevance)

96/737/EC: Council Decision of 16 December 1996 concerning a multiannual programme for the promotion of energy efficiency in the Community—SAVE II

Commission Directive 96/44/EC of 1 July 1996 adapting to technical progress Council Directive 70/220/EEC on the approximation of the laws of the Member States relating to measures to be taken against air pollution by emissions from motor vehicles (text with EEA relevance)

Council Directive 96/62/EC of 27 September 1996 on ambient air quality assessment and management

Council Decision of 27 January 1997 establishing a reciprocal exchange of information and data from networks and individual stations measuring ambient air pollution within the Member States (97/101/EC)

Commission Directive 97/20/EC of 18 April 1997 adapting to technical progress Council Directive 72/306/EEC on the approximation of the laws of the Member States relating to the measures to be taken against the emission of pollutants from diesel engines for use in vehicles (text with EEA relevance)

Directive 97/68/EC of the European Parliament and of the Council of 16 December 1997 on the approximation of the laws of the Member States relating to measures against the emission of gaseous and particulate pollutants from internal combustion engines to be installed in non-road mobile machinery

98/27/EC: Commission Decision of 16 December 1997 allocating import quotas for the fully halogenated chlorofluorocarbons 11, 12, 113, 114 and 115, other fully halogenated chlorofluoro-carbons, halons, carbon tetrachloride, 1,1,1-trichloroethane, hydrobromofluorocarbons and methyl bromide for the period 1 January to 31 December 1998, and in addition, allocating placing market quotas for hydrochlorofluorocarbons for the period 1 January to 31 December 1998 (Only the Dutch, English, French, German, Greek, Italian and Portuguese texts are authentic) (text with EEA relevance)

98/67/EC: Commission Decision of 16 December 1997 on the allocation of quantities of controlled substances allowed for essential uses in the Community in 1998 under Council Regulation (EC) No. 3093/94 on substances that deplete the ozone layer (Only the Dutch, English, Finnish, French, German, Italian and Spanish texts are authentic) (text with EEA relevance)

98/352/EC: Council Decision of 18 May 1998 concerning a multiannual programme for the promotion of renewable energy sources in the Community (Altener II)

15.10.20.50—Chemicals, industrial risk and biotechnology

Secondary legislation

Council Directive 73/404/EEC of 22 November 1973 on the approximation of the laws of the Member States relating to detergents

Council Directive 73/405/EEC of 22 November 1973 on the approximation of the laws of the Member States relating to methods of testing the biodegradability of anionic surfactants

Council Directive 76/769/EEC of 27 July 1976 on the approximation of the laws, regulations and administrative provisions of the Member States relating to restrictions on the marketing and use of certain dangerous substances and preparations

Council resolution of 30 May 1978 on fluorocarbons in the environment

79/3/EEC: Council recommendation of 19 December 1978 to the Member States regarding methods of evaluating the cost of pollution control to industry

80/372/EEC: Council Decision of 26 March 1980 concerning chlorofluorocarbons in the environment

82/795/EEC: Council Decision of 15 November 1982 on the consolidation of precautionary measures concerning chlorofluorocarbons in the environment

Council Directive 82/242/EEC of 31 March 1982 on the approximation of the laws of the Member States relating to methods of testing the biodegradability of nonionic surfactants and amending Directive 73/404/EEC

Council Directive 82/501/EEC of 24 June 1982 on the major-accident hazards of certain industrial activities

Council Directive 82/883/EEC of 3 December 1982 on procedures for the surveillance and monitoring of environments concerned with waste from the titanium dioxide industry

85/71/EEC: Commission Decision of 21 December 1984 concerning the list of chemical substances notified pursuant to Council Directive 67/548/EEC on the approximation of laws, regulations and administrative provisions relating to the classification, packaging and labelling of dangerous substances

Council Directive 87/216/EEC of 19 March 1987 amending Directive 82/501/EEC on the major-accident hazards of certain industrial activities

Council Directive 88/320/EEC of 9 June 1988 on the inspection and verification of Good Laboratory Practice (GLP)

Council resolution of 25 January 1988 on a Community action programme to combat environmental pollution by cadmium

Council resolution of 16 June 1988 concerning export from and import into the Community of certain dangerous chemicals

89/569/EEC: Council Decision of 28 July 1989 on the acceptance by the European Economic Community of an OECD decision/recommendation on compliance with principles of good laboratory practice

Council resolution of 16 October 1989 on guidelines to reduce technological and natural hazards

Council Directive 90/219/EEC of 23 April 1990 on the contained use of genetically modified micro-organisms

Council Directive 90/220/EEC of 23 April 1990 on the deliberate release into the environment of genetically modified organisms

91/274/EEC: Commission Decision of 21 May 1991 concerning a list of Community legislation referred to in Article 10 of Council Directive 90/220/EEC

91/448/EEC: Commission Decision of 29 July 1991 concerning the guidelines for classification referred to in Article 4 of Directive 90/219/EEC

91/596/EEC: Council Decision of 4 November 1991 concerning the Summary Notification Information Format referred to in Article 9 of Directive 90/220/EEC on the deliberate release into the environment of genetically modified organisms

Council Directive 91/692/EEC of 23 December 1991 standardizing and rationalizing reports on the implementation of certain Directives relating to the environment

92/146/EEC: Commission Decision of 11 February 1992 concerning the summary notification information format referred to in Article 12 of Council Directive 90/220/EEC

Council Directive 92/112/EEC of 15 December 1992 on procedures for harmonizing the programmes for the reduction and eventual elimination of pollution caused by waste from the titanium dioxide industry

93/572/EEC: Commission Decision of 19 October 1993 concerning the placing on the market of a product containing genetically modified organisms pursuant to Article 13 of Council Directive 90/220/EEC

93/584/EEC: Commission Decision of 22 October 1993 establishing the criteria for simplified procedures concerning the deliberate release into the environment of genetically modified plants pursuant to Article 6 (5) of Council Directive 90/220/EEC

Commission Directive 93/67/EEC of 20 July 1993 laying down the principles for assessment of risks to man and the environment of substances notified in accordance with Council Directive 67/548/EEC

Commission Directive 93/72/EEC of 1 September 1993 adapting to technical progress for the nineteenth time Council Directive 67/548/EEC on the approximation of the laws, regulations and administrative provisions relating to the classification, packaging and labelling of dangerous substances

94/385/EC: Commission Decision of 8 June 1994 concerning the placing on the market of a product consisting of a genetically modified organism, seeds of herbicide-resistant tobacco variety ITB 1000 OX, pursuant to Article 13 of Council Directive 90/220/EEC

94/505/EC: Commission Decision of 18 July 1994 amending the Decision of 18 December 1992 concerning the placing on the market of a GMO containing product, the vaccine Nobi-Porvac Aujeszky live (gI,tk), pursuant to Article 13 of Council Directive 90/220/EEC

94/730/EC: Commission Decision of 4 November 1994 establishing simplified procedures concerning the deliberate release into the environment of genetically modified plants pursuant to Article 6.5 of Council Directive 90/220/EEC (Only the Spanish, Danish, German, English, French, Italian, Dutch, and Portuguese texts are authentic)

94/783/EC: Commission Decision of 14 September 1994 concerning the prohibition of PCP notified by the Federal Republic of Germany (Only the German text is authentic)

Council Directive 94/55/EC of 21 November 1994 on the approximation of the laws of the Member States with regard to the transport of dangerous goods by road

Commission Regulation (EC) No. 1179/94 of 25 May 1994 concerning the first list of priority substances as foreseen under Council Regulation (EEC) No. 793/93

Commission Regulation (EC) No. 1488/94 of 28 June 1994 laying down the principles for the assessment of risks to man and the environment of existing substances in accordance with Council Regulation (EEC) No. 793/93 (text with EEA relevance)

Commission Regulation (EC) No. 2268/95 of 27 September 1995 concerning the second list of priority substances as foreseen under Council Regulation (EEC) No. 793/93

96/158/EC: Commission Decision of 6 February 1996 concerning the placing on the market of a product consisting of a genetically modified organism, hybrid herbicide-tolerant swede-rape seeds (Brassica napus L. oleifera Metzq. MS1Bn × RF1Bn), pursuant to Council Directive 90/220/EEC (text with EEA relevance)

96/211/EC: Commission Decision of 26 February 1996 concerning the prohibition of pentachlorophenol (PCP) notified by Denmark (Only the Danish text is authentic) (text with EEA relevance)

96/281/EC: Commission Decision of 3 April 1996 concerning the placing on the market of genetically modified soya beans (Glycine max L.) with increased tolerance to the herbicide glyphosate, pursuant to Council Directive 90/220/EEC (text with EEA relevance)

96/424/EC: Commission Decision of 20 May 1996 concerning the placing on the market of genetically modified male sterile chicory (Cichorium intybus L.) with partial tolerance to the herbicide glufosinate ammonium pursuant to Council Directive 90/220/EEC (text with EEA relevance)

Council Directive 96/82/EC of 9 December 1996 on the control of major-accident hazards involving dangerous substances

97/98/EC: Commission Decision of 23 January 1997 concerning the placing on the market of genetically modified maize (Zea mays L.) with the combined modification for insecticidal properties conferred by the Bt-endotoxin gene and increased tolerance to the herbicide glufosinate ammonium pursuant to Council Directive 90/220/EEC (text with EEA relevance)

97/392/EC: Commission Decision of 6 June 1997 concerning the placing on the market of genetically modified swede-rape (Brassica napus L. oleifera Metzg. MS1, RF1), pursuant to Council Directive 90/220/EEC (text with EEA relevance)

97/393/EC: Commission Decision of 6 June 1997 concerning the placing on the market of genetically modified swede-rape (Brassica napus L. oleifera Metzg. MS1, RF2), pursuant to Council Directive 90/220/EEC (text with EEA relevance)

97/549/EC: Commission Decision of 14 July 1997 concerning the placing on the market of T102-test (Streptococcus thermophilus T102) pursuant to Council Directive 90/220/EEC (text with EEA relevance)

Commission Regulation (EC) No. 142/97 of 27 January 1997 concerning the delivery of information about certain existing substances as foreseen under Council Regulation (EEC) No. 793/93 (text with EEA relevance)

Commission Regulation (EC) No. 143/97 of 27 January 1997 concerning the third list of priority substances as foreseen under Council Regulation (EEC) No. 793/93 (text with EEA relevance)

98/291/EC: Commission Decision of 22 April 1998 concerning the placing on the market of genetically modified spring swede rape (Brassica napus L. ssp. oleifera), pursuant to Council Directive 90/220/EEC (text with EEA relevance)

98/292/EC: Commission Decision of 22 April 1998 concerning the placing on the market of genetically modified maize (Zea mays L. line Bt-11), pursuant to Council Directive 90/220/EEC (text with EEA relevance)

98/293/EC: Commission Decision of 22 April 1998 concerning the placing on the market of genetically modified maize (Zea mays L. T25), pursuant to Council Directive 90/220/EEC (text with EEA relevance)

98/294/EC: Commission Decision of 22 April 1998 concerning the placing on the market of genetically modified maize (Zea mays L. line MON 810), pursuant to Council Directive 90/220/EEC (text with EEA relevance)

15.10.30.20—Conservation of wild fauna and flora

External relations

Convention on the conservation of migratory species of wild animals
Convention on the conservation of Antarctic marine living resources

Secondary legislation

75/66/EEC: Commission recommendation of 20 December 1974 to Member States concerning the protection of birds and their habitats

Council Directive 78/659/EEC of 18 July 1978 on the quality of fresh waters needing protection or improvement in order to support fish life

Council Directive 79/409/EEC of 2 April 1979 on the conservation of wild birds

Council resolution of 2 April 1979 concerning Directive 79/409/EEC on the conservation of wild birds

Council Regulation (EEC) No. 348/81 of 20 January 1981 on common rules for imports of whales or other cetacean products

Council Directive 83/129/EEC of 28 March 1983 concerning the importation into Member States of skins of certain seal pups and products derived therefrom

Commission Regulation (EEC) No. 3418/83 of 28 November 1983 laying down provisions for the uniform issue and use of the documents required for the implementation in the Community of the Convention on international trade in endangered species of wild fauna and flora

Commission Regulation (EEC) No. 526/87 of 20 February 1987 laying down certain detailed rules for the application of Council Regulation (EEC) No. 3528/86 on the protection of the Community's forests against atmospheric pollution

Council Directive 92/43/EEC of 21 May 1992 on the conservation of natural habitats and of wild fauna and flora

Council Regulation (EEC) No. 2158/92 of 23 July 1992 on protection of the Community's forests against fire

Commission Regulation (EEC) No. 1170/93 of 13 May 1993 laying down certain detailed rules for the application of Council Regulation (EEC) No. 2158/92 on protection of the Community's forests against fire

Commission Regulation (EC) No. 804/94 of 11 April 1994 laying down certain detailed rules for the application of Council Regulation (EEC) No. 2158/92 as regards forest-fire information systems

Council Regulation (EC) No. 3062/95 of 20 December 1995 on operations to promote tropical forests

96/15/EC: Commission Opinion of 18 December 1995 on the intersection of the Peene Valley (Germany) by the planned A 20 motorway pursuant to Article 6 (4) of Council Directive 92/43/EEC on the conservation of natural habitats and of wild fauna and flora

97/266/EC: Commission Decision of 18 December 1996 concerning a site information format for proposed Natura 2000 sites

Council Regulation (EC) No. 338/97 of 9 December 1996 on the protection of species of wild fauna and flora by regulating trade therein

Commission Regulation (EC) No. 939/97 of 26 May 1997 laying down detailed rules concerning the implementation of Council Regulation (EC) No. 338/97 on the protection of species of wild fauna and flora by regulating trade therein

Commission Regulation (EC) No. 2551/97 of 15 December 1997 suspending the introduction into the Community of specimens of certain species of wild fauna and flora

98/145/EC: Council Decision of 12 February 1998 on the approval, on behalf of the European Community, of the amendments to Appendices I and II to the Bonn Convention on the conservation of migratory species of wild animals as decided by the fifth meeting of the Conference of the parties to the Convention

15.10.30.30—Waste management and clean technology

Secondary legislation

Council Directive 75/439/EEC of 16 June 1975 on the disposal of waste oils

Council Directive 75/442/EEC of 15 July 1975 on waste

76/431/EEC: Commission Decision of 21 April 1976 setting up a Committee on Waste Management

Council Directive 78/176/EEC of 20 February 1978 on waste from the titanium dioxide industry

81/972/EEC: Council recommendation of 3 December 1981 concerning the re-use of waste paper and the use of recycled paper

Council Directive 86/278/EEC of 12 June 1986 on the protection of the environment, and in particular of the soil, when sewage sludge is used in agriculture

Council Resolution of 21 December 1988 concerning transfrontier movements of hazardous waste to third countries

Council resolution of 7 May 1990 on waste policy

Council Directive 91/157/EEC of 18 March 1991 on batteries and accumulators containing certain dangerous substances

Council Directive 91/689/EEC of 12 December 1991 on hazardous waste

Council Directive 91/692/EEC of 23 December 1991 standardizing and rationalizing reports on the implementation of certain Directives relating to the environment

Commission Directive 93/86/EEC of 4 October 1993 adapting to technical progress Council Directive 91/157/EEC on batteries and accumulators containing certain dangerous substances

Council Regulation (EEC) No. 259/93 of 1 February 1993 on the supervision and control of shipments of waste within, into and out of the European Community

94/3/EC: Commission Decision of 20 December 1993 establishing a list of wastes pursuant to Article 1a of Council Directive 75/442/EEC on waste

94/575/EC: Commission Decision of 20 July 1994 determining the control procedure under Council Regulation (EEC) No. 259/93 as regards certain shipments of waste to certain non-OECD countries (text with EEA relevance)

94/741/EC: Commission Decision of 24 October 1994 concerning questionnaires for Member States reports on the implementation of certain Directives in the waste sector (implementation of Council Directive 91/692/EEC)

94/774/EC: Commission Decision of 24 November 1994 concerning the standard consignment note referred to in Council Regulation (EEC) No. 259/93 on the supervision and control of shipments of waste within, into and out of the European Community

94/904/EC: Council Decision of 22 December 1994 establishing a list of hazardous wastes pursuant to Article 1 (4) of Council Directive 91/689/EEC on hazardous waste

European Parliament and Council Directive 94/62/EC of 20 December 1994 on packaging and packaging waste

Council Directive 94/67/EC of 16 December 1994 on the incineration of hazardous waste

96/302/EC: Commission Decision of 17 April 1996 establishing a format in which information is to be provided pursuant to Article 8 (3) of Council Directive 91/689/EEC on hazardous waste (text with EEA relevance)

Council Directive 96/59/EC of 16 September 1996 on the disposal of polychlorinated biphenyls and polychlorinated terphenyls (PCB/PCT)

Commission Decision of 28 January 1997 establishing the identification system for packaging materials pursuant to European Parliament and Council Directive 94/62/EC on packaging and packaging waste (text with EEA relevance)

Commission Decision of 3 February 1997 establishing the formats relating to the database system pursuant to European Parliament and Council Directive 94/62/EC on packaging and packaging waste (text with EEA relevance)

97/283/EC: Commission Decision of 21 April 1997 on harmonized measurement methods to determine the mass concentration of dioxins and furans in atmospheric emissions in accordance with Article 7 (2) of Directive 94/67/EC on the incineration of hazardous waste

97/622/EC: Commission Decision of 27 May 1997 concerning questionnaires for Member States reports on the implementation of certain Directives in the waste sector (implementation of Council Directive 91/692/EEC)

97/640/EC: Council Decision of 22 September 1997 on the approval, on behalf of the Community, of the amendment to the Convention on the control of transboundary movements of hazardous wastes and their disposal (Basle Convention), as laid down in Decision III/1 of the Conference of the Parties

Council Resolution of 24 February 1997 on a Community strategy for waste management

Resolution of the ECSC Consultative Committee on the classification of scrap (adopted unanimously with two abstentions during the 337th session of 10 October 1997)

98/184/EC: Commission Decision of 25 February 1998 concerning a questionnaire for Member States' reports on the implementation of Council Directive 94/67/EC on the incineration of hazardous waste (implementation of Council Directive 91/692/EEC) (text with EEA relevance)

15.10.40—International cooperation

External relations

Convention concerning the International Commission for the Rhine (Berne Convention)

Convention on international trade in endangered species of wild fauna and flora

Convention on the protection of the Mediterranean Sea against pollution (Barcelona Convention)—Protocol for the prevention of pollution of the Mediterranean Sea by dumping from ships and aircraft

Protocol concerning cooperation in combating pollution of the Mediterranean Sea by oil and other harmful substances in cases of emergency

Convention for the protection of the Rhine against chemical pollution

Additional Agreement to the Agreement, signed in Berne on 29 April 1963, concerning the International Commission for the Protection of the Rhine against Pollution

Convention on the conservation of European wildlife and natural habitats

Convention on long-range transboundary air pollution—Resolution on long-range transboundary air pollution

Convention for the Conservation of Salmon in the North Atlantic Ocean

Recommendation from the International Commission for the Protection of the Rhine against Pollution to supplement Annex IV to the Convention on the Protection of the Rhine against Chemical Pollution, signed in Bonn on 3 December 1976

Protocol for the protection of the Mediterranean Sea against pollution from land-based sources

Convention on fishing and conservation of the living resources in the Baltic Sea and the Belts

Protocol to the Conference of the representatives of the States Parties to the Convention on fishing and conservation of living resources in the Baltic Sea and the Belts (Warsaw, 9 to 11 November 1982)

Exchange of letters between the Commission and the United Nations Environment Programme on the strengthening of cooperation between the two institutions

Agreement for cooperation in dealing with pollution of the North Sea by oil and other harmful substances (Bonn Agreement)

Proposal from the International Commission for the Protection of the Rhine against Pollution to supplement Annex IV to the Convention on the protection of the Rhine against chemical pollution, signed in Bonn on 3 December 1976

International Convention for the Conservation of Atlantic Tunas

Final Act of the Conference of Plenipotentiaries of the States Parties to the International Convention for the Conservation of Atlantic Tunas

Protocol attached to the Final Act of the Conference of Plenipotentiaries of the States Parties to the International Convention for the Conservation of Atlantic Tunas

Proposal by the International Commission for the Protection of the Rhine against Pollution intended to supplement Annex IV to the Convention for the Protection of the Rhine against Chemical Pollution, signed in Bonn on 3 December 1976

Vienna Convention for the protection of the ozone layer

Montreal Protocol on substances that deplete the ozone layer—Declaration by the European Economic Community

Agreement between the Federal Republic of Germany and the European Economic Community, on the one hand, and the Republic of Austria, on the other, of cooperation on management of water resources in the Danube Basin—Statute of the Standing Committee on Management of Water Resources—Final Protocol—Declaration

Amendment to the Montreal protocol on substances that deplete the ozone layer

Basel Convention on the control of transboundary movements of hazardous wastes and their disposal

Protocol to the Convention of 8 October 1990 between the Governments of the Federal Republic of Germany and the Czech and Slovak Federal Republic and the European Economic Community on the International Commission for the Protection of the Elbe

Protocol to the 1979 Convention on long-range transboundary air pollution concerning the control of emissions of nitrogen oxides or their transboundary fluxes

Cooperation Agreement for the protection of the coasts and waters of the north-east Atlantic against pollution

Convention on biological diversity—Declarations

Amendment to the Montreal Protocol on substances that deplete the ozone layer

United Nations Framework Convention on Climate Change—Declarations

Convention on the Protection of the Marine Environment of the Baltic Sea Area, 1974 (Helsinki Convention)

Convention on the Protection of the Marine Environment of the Baltic Sea Area, 1992 (Helsinki Convention revised in 1992)

Amendment to Article VII of the Convention on fishing and conservation of the living resources in the Baltic Sea and the Belts

Convention on cooperation for the protection and sustainable use of the river Danube (Convention for the protection of the Danube)—Final act

United Nations Convention to combat desertification in those countries experiencing serious drought and/or desertification, particularly in Africa—Declaration made by the European Community in accordance with Article 34(2) and (3) of the United Nations Convention to combat desertification in countries seriously affected by drought and/or desertification, particularly in Africa

PARCOM Decision 96/1 on the phasing-out of the use of hexachloroethane in the non-ferrous metal industry

Convention for the protection of the marine environment of the north-east Atlantic

Secondary legislation

75/438/EEC: Council Decision of 3 March 1975 concerning Community participation in the Interim Commission established on the basis of resolution No. III of the convention for the prevention of marine pollution from land-based sources

Council resolution of 3 March 1975 on the convention for the prevention of marine pollution from land-based sources

85/336/EEC: Council Decision of 27 June 1985 concerning a supplement in respect to cadmium to Annex IV to the Convention for the protection of the Rhine against chemical pollution

85/613/EEC: Council Decision of 20 December 1985 concerning the adoption, on behalf of the Community, of programmes and measures relating to mercury and cadmium discharges under the convention for the prevention of marine pollution from land-based sources

88/540/EEC: Council Decision of 14 October 1988 concerning the conclusion of the Vienna Convention for the protection of the ozone layer and the Montreal Protocol on substances that deplete the ozone layer

Council Regulation (EEC) No. 3943/90 of 19 December 1990 on the application of the system of observation and inspection established under Article XXIV of the Convention on the Conservation of Antarctic Marine Living Resources

Council Regulation (EC) No. 2978/94 of 21 November 1994 on the implementation of IMO Resolution A.747(18) on the application of tonnage measurement of ballast spaces in segregated ballast oil tankers

Council Regulation (EC) No. 3062/95 of 20 December 1995 on operations to promote tropical forests

REFERENCES

The European Union and the Environment, 1997.

Trade and the Environment, Law, Economics and Policy, edited by Durwood Zaelke, Paul Orbuch, Robert F. Housman, Center for International Environmental Law, 1993.

Direct Effect of European Law and the Regulation of Dangerous Substances, Christopher J. M. Smith, Gordon and Breach Publishers, 1995.

Environmental Change and International Law, edited by Edith Brown Weiss, United Nations University Press, 1992.

International Environmental Law and Policy, David Hunter, James Salzman, Durwood Zaelke, Foundation Press, 1998.

Environmental Management Systems, Jay G. Martin and Gerald J. Edgley, Governments Institutes, 1998.

International Environmental Auditing, David D. Nelson, Government Institutes, 1998.

Precautionary Legal Duties and Principles of Modern International Environmental Law, Harold Hohmann, Graham & Trotman/Martinus Nijhoff, 1998.

Environmental Management in European Companies, Success Stories and Evaluation, edited by Jobst Conrad, Gordon and Breach Science Publishers, 1998.

Environmental Strategies Handbook, A Guide to Effective Policies & Practices, Rao V. Kolluru, McGraw-Hill, Inc. 1994.

The New Approach, Legislation and Standards on the Free Movement of Goods in Europe, CEN, 1994.

Part IV

Environmental Protection in the United States

11 U.S. Laws and Regulations

CONTENTS

NATIONAL ENVIRONMENTAL POLICY ACT (NEPA) OF 1969

NEPA is the world's first environmental impact statute and was enacted January 1, 1970. The law was passed in response to the need for the federal government to evaluate the environmental effects of its actions and establish a national policy for the environment. Congress recognized the effects of population growth, high-density urbanization, industrial expansion, resource exploitation, and new and expanding technological advances. NEPA is a broad mandate for federal agencies to create and maintain "conditions under which man and nature can exist in productive harmony and fulfill the social, economic, and other requirements of present and future generations." NEPA's purpose was

- "to declare a national policy which will encourage productive and enjoyable harmony between man and his environment;
- to promote efforts which will prevent or eliminate damage to the environment and biosphere and stimulate the health and welfare of man;
- to enrich the understanding of the ecological systems and natural resources important to the Nation;
- and to establish a Council on Environmental Quality (CEQ)."

NEPA applies to all agencies of the federal government and every major action taken by the agencies that significantly affects the quality of the human environment. The application of NEPA to federal actions is not limited to actions occurring, or having effects in, the United States. NEPA is designed to control the decision making process, not the substance of the agency decision. Because NEPA was so broadly written, federal agencies had a difficult time complying with the requirements. The newly created CEQ (USC Title II, CFR Title V) was charged with overseeing implementation of the act. The CEQ was also responsible for the analysis and development of national and international environmental policy; interagency coordination of environmental quality programs; acquisition and assessment of environmental data, including environmental quality reports; and environmental conditions and trends. Since NEPA's passage, a few thousand judicial decisions have defined nearly every word of the law and provided guidance for compliance. Additionally, in 1978, the CEQ promulgated regulations implementing the procedural provisions of the act (40 CFR 1500-1508). These regulations define the human environment (both the natural and physical environment) and the relationship of the people with their environment.

Section 101 sets forth the nation's environmental goals and a broad national policy to achieve these goals and serves as a blueprint for considering a wide range of environmental effects of proposed agency actions. Section 102 provides the process to ensure that the federal agency decision makers are aware of the policies and procedures required in NEPA. Section 102 contains the requirement that federal agencies prepare "detailed statements" for actions "significantly affecting the quality of the human environment." These "detailed statements" are now known as Environmental Impact Statements (EISs). An EIS is required to include

- Environmental impacts of the proposed action,
- Any adverse environmental effects that cannot be avoided should the proposal be implemented,
- Alternatives to the proposed action,
- The relationship between local short-term uses of man's environment and the maintenance and enhancement of long-term productivity, and
- Any irreversible and irretrievable commitments of resources that would be involved in the proposed action should it be implemented.

Federal agencies filed 513 draft, final, and supplemental EISs in 1992 with the U.S. Departments of Transportation, Agriculture, Interior, and the Army Corps of Engineers accounting for 393 of them. Subjects for which the EISs were filed include forestry and range management; natural gas and oil drilling and exploration; watershed protection and flood control; parks, recreation, and wilderness areas and national seashores; mining; power facilities; transmissions; road construction; airport improvements; and buildings for federal use.

As a means to evaluate a proposed action and provide sufficient evidence to determine the level of significance of the environmental impacts, an Environmental Assessment (EA) may be conducted. Results of an EA may be a Finding Of No Significant Impact (FONSI), which explains why an action will not have a significant impact on the quality of the human environment and, therefore, will not require preparation of an EIS or will indicate that, indeed, an EIS is required. The Record of Decision (ROD) states the decision, alternatives considered, the environmentally preferable alternative(s), factors considered in the agency's decision, mitigation measures that will be implemented, and a description of applicable enforcement and monitoring programs. Some agency actions do not have a significant effect on the human environment (individually or cumulatively); therefore, an EA or and EIS is not required. These actions are called Categorical Exclusions (CEs). Upon submission of a draft EIS, the U.S. Environmental Protection Agency (EPA) evaluates the report for environmental effect and adequacy. The "grade" for environmental effect of the action will be one of the following:

- Lack of Objections
- Environmental Concerns
- Environmental Objections
- Environmentally Unsatisfactory

The adequacy of the draft EIS is divided into categories. Category 1 is adequate, which means the draft report does not require further revision other than minor additions or clarification. Category 2 is insufficient information and requires the agency submitting the draft EIS to provide additional information, data, or analysis. A Category 3 rating for adequacy means the environmental effects were not adequately analyzed or alternatives considered. A draft EIS receiving a Category 3 rating may be referred to the CEQ.

The CEQ participates in two important activities concerning international environmental impact assessment (EIA). First, in 1992, the United States signed the Convention on Environmental Impact Assessment in a Transboundary Context negotiated under the auspices of the Economic Commission for Europe (ECE). In 1993, the EPA together with the CEQ and the State Department developed implementing procedures for the convention. Second, the CEQ provides technical assistance to other countries such as Japan, China, Australia, the former Soviet Union, African nations, and the Republic of Turkey. These countries have requested information and assistance from the United States in developing an EIA process for their own countries. The CEQ also supports the International Association of Impact Assessment.

HISTORIC, SCIENTIFIC, AND CULTURAL RESOURCES

The environment protected by NEPA includes historic and cultural resources in addition to natural resources. If an agency's proposed action affects historic or cultural resources, the action is subject to NEPA. In addition to NEPA, the following federal and state laws have been enacted to preserve important historic and cultural resources.

HISTORIC SITES ACT (1935), 16 USC

Preservation for public use—historic sites, buildings, and objects of national significance.

NATIONAL HISTORIC PRESERVATION ACT (1966), 16 USC

36 CFR 60-79 National Park Service, Dept. of Interior

"that the historical and cultural foundations of the Nation should be preserved as a living part of our community life and development in order to give a sense of orientation to the American people;"

National register of historic (and significant for architecture, archeology, and culture) districts, sites, buildings, structures, and other objects.

PRESERVATION OF HISTORICAL AND ARCHEOLOGICAL DATA (1960), 16 USC

Works together with Historic Sites Act to specifically provide for the preservation of historical and archeological data that might be destroyed by flooding, building roads, and other projects involving changing of the terrain.

PROTECTION AND ENHANCEMENT OF THE CULTURAL ENVIRONMENT (1971)

Executive Order 11593, 36 FR 8921
    Policy: "The Federal Government shall provide leadership in preserving, restoring, and maintaining the historic and cultural environment of the Nation. Agencies of the executive branch of the Government (hereinafter referred to as "federal agencies") shall

1. administer the cultural properties under their control in a spirit of stewardship and trusteeship for future generations,
2. initiate measures necessary to direct their policies, plans, and programs in such a way that federally owned sites, structures, and objects of historical, architectural, or archeological significance are preserved, restored, and maintained for the inspiration and benefit of the people, and
3. in consultation with the Advisory Council on Historic Preservation (16 USC 470i), institute procedures to assure that federal plans and programs contribute to the pres-

ervation and enhancement of nonfederally owned sites, structures, and objects of historical, architectural, or archeological significance."

ADVISORY COUNCIL ON HISTORIC PRESERVATION

Title 36, Parks, Forests, and Public Property
Chapter VIII, Parts 800-899
36 CFR 800, Protection of historic and cultural properties
36 CFR 805, Procedures for implementation of NEPA

TOXIC SUBSTANCE CONTROL ACT (TSCA)

TSCA, codified by 40 CFR, Subchapter R parts 700-799, became effective January 1, 1977, to close loopholes in existing regulations that did not allow for the regulation of chemicals that pose an unnecessary risk to health and the environment prior to their distribution. In addition, TSCA allows for the regulation of substances that are discovered to cause unnecessary risk to health and the environment.

TSCA covers manufacturers, importers, and processors and regulates materials such as chemical substances and mixtures identified in TSCA by listing or definition.

BASIC GOALS OF TSCA

- Screen new chemicals and review for potential health and environmental risks.
- Require testing of chemicals identified as presenting possible risks.
- Gather information on existing chemicals.
- Control chemicals already in distribution and use that have proven to present a health risk.

HOW EPA CAN REGULATE CHEMICAL SUBSTANCES UNDER TSCA

- Ban the manufacture, use, and distribution of the chemical
- Require special warnings and labels
- Require controls during manufacture
- Court action

ITEMS NOT COVERED BY TSCA

- Food, drugs, and cosmetics
- Tobacco, firearms, ammunition
- Nuclear materials
- Pesticides

BASIC REQUIREMENTS

Reporting and Recordkeeping (Part 704)

- How much was manufactured
- Description of the manufacturing process
- Description of the worker activities
- Potential for worker exposure
- Monitoring data
- List of personal protective equipment used to prevent worker exposure

- Inventory of the quantity of the chemical released
- Method of disposal for waste materials

Chemical Imports and Exports (Part 707)

- Must notify EPA of intent to export chemicals
- Must comply with the provisions of TSCA to import

Inventory Reporting (Part 710)

- EPA must develop an inventory of existing chemicals
- Inventory based on information from manufacturers

Chemical Information (Part 712)

- Requires information regarding production, use, and exposure related information on listed substances (ITC)
- Preliminary Assessment Information
- Must manufacture or import > 1,100 lbs

Health and Safety Data Reporting (Part 716)

- Must submit health and safety study information
- Information is used to determine testing priority and assessment by the EPA
- Applies to listed substances (716.120)

Records and Reports Regarding Substances that Cause Significant Adverse Reactions to Health or the Environment (Part 717)

- Must keep records of adverse reactions to health and environment
- Must allow review of these records and/or submit copies
- Adverse reactions include cancer or birth defects, impairment of normal activities, or other disorders

Premanufacture Notification

- Must submit form prior to manufacturing or importing a "new" substance. A new substance is one that is not on the TSCA Inventory

Significant New Use (Part 721)

- Must report to EPA on significant new uses of chemicals
- Identifies specific chemicals and special requirements

Polychlorinated Biphenyls (PCBs) (Part 761)

- Manufacturing and production
- Processing
- Distribution

- Use
- Disposal of waste materials

Asbestos (Part 763)

- Manufacturing and production
- Processing
- Distribution
- Use
- Disposal of waste materials
- Asbestos in schools rule

The disposal of PCB and asbestos waste materials, including the permitting of the disposal site is regulated by TSCA.

POLYCHLORINATED BIPHENYLS

Produced from 1929 through 1977, PCBs were used in electrical equipment as a dielectric fluid (electrical insulator) in transformers and ballast. The molecular structure of PCBs includes chlorine atoms substituted for hydrogen atoms on a double benzene ring (biphenyl ring). The location of the chlorine atoms on the ring determine properties and characteristics of the PCB molecule. There are 209 possible PCB compounds that can be identified by a trade name such as Aroclor 1254. Aroclor 1254 is Chlorodiphenyl ($C_6H_3Cl_2C_6H_2Cl_3$) and is 54% chlorine with Chemical Abstract Service number 11097-69-1. Commercial products were usually a mixture of many types of PCB molecules and possible other materials, including solvents or mineral oil. PCBs are viscous (resist flow), lipid (fat) soluble resulting in their accumulation in fatty tissue and heavy weighing in the range of 10 to 15 pounds per gallon. They are stable, nonflammable, and resistant to degradation making them ideal for use in electrical equipment. Of particular environmental concern is the fact that PCBs are persistent and accumulate in the environment (bioaccumulation) and effects of exposure include eye irritation, chloracne, liver damage, and cancer.

Regulatory Overview

- Effective Date: May 31, 1979
- Regulation: TSCA, 40 CFR Part 761
- Prohibitions and authorizations (Subpart B)
- Marking of PCBs (Subpart C)
- Storage and disposal (Subpart D)
- PCB spill cleanup (Subpart G)

PCB Classification

Non-PCB	< 50 ppm PCB
PCB-contaminated	between 50 ppm and 500 ppm
PCB (fully regulated)	500 ppm or greater PCB

Prohibitions and Authorizations

Prohibitions (761.20)—Only totally enclosed PCB or PCB items could be used. PCBs were banned for manufacture and export. PCBs cannot be processed or distributed in commerce. Oils with any detectable amount of PCB cannot be used as a dust control agent.

Authorizations (761.30)—PCB and PCB-contaminated transformers could be used the rest of their useful life. High voltage transformers in or near a commercial building had to be removed from service as of October 1, 1990. The equipment could be reclassified or equipped with protection to prevent failure. PCB transformers located in or near a commercial building had to be registered with the building owner and the fire department. PCB electrical equipment may be reclassified to PCB-contaminated or non-PCB. Storage of combustible materials near a PCB transformer is prohibited. If a leak is detected, immediate action to mitigate the leak is required. If a PCB transformer is ruptured in a fire, the National Response Center (NRC) must be notified.

Marking of PCBs (Subpart C)

- PCB containers, transformers, and large capacitors
- Transport vehicles

PCB storage areas access modes to PCB transformer locations in or near a commercial building. **CAUTION**
CONTAINS

PCBs

(Polychlorinated Biphenyls)
A toxic environmental contaminant requiring special handling and disposal in accordance with U.S. Environmental Protection Agency Regulations 40 CFR 761-For Disposal Information contact the nearest U.S. E.P.A. Office.

In case of accident of spill, call toll free the U.S. Coast Guard National Response Center:
800-424-8802

Also contact: _____

Tel. No. _____

Figure 8 Example of PCB Label

Storage and Disposal

Disposal Requirements (761.60)—All disposal facilities must be permitted under TSCA. PCB fluid, whole transformers, and contaminated soil and debris must be incinerated at approved facilities for empty transformer bodies that held PCB fluid; drained PCB fluid from transformers and contaminated soil and debris may be disposed at a TSCA-permitted landfill. Any PCB waste that is mixed with hazardous waste according to the Resource Conservation and Recovery Act (RCRA) must be disposed of at a facility permitted to handle both types of waste.

Storage Requirements (761.64)—Storage area must have adequate roof, walls, and flooring. Area must have a minimum of 6-inch curb around floor. Floor drains are prohibited in the area. All PCB items must be disposed of within one year. The generator can store waste up to 275 days. The disposal facility must dispose of waste within 90 days.

PCB Activity Tracking

- Notification of PCB activity (effective 1990)
- EPA identification number
- Required use of the hazardous waste manifest system
- Certificate of disposal
- Recordkeeping

PCB Spill Cleanup (Subpart G)

Effective May 4, 1987, this requirement establishes criteria for the cleanup of PCB spills. "Spill" means any spilling, leaking, or uncontrolled discharge of any quantity of PCB. EPA or the state authority reserves the right to require additional cleanup other than that specified. Depending on the amount spilled and impacts of the spill, spills must be reported to the EPA and possibly the NRC. The EPA PCB regional coordinator must be notified no later than 24 hours following the discovery of the spill if

1. Spill contaminates surface water, sewers, or drinking water;
2. Spill contaminates grazing lands; or
3. Spill is greater than 10 pounds of PCBs (NRC, too).

Region VIII TSCA Contact Number: 293-1603

References For Spill Cleanup

- Verification of PCB Spill Cleanup by Sampling and Analysis
- Field Manual for Grid Sampling of PCB Sites to Verify Cleanup
  Call (202) 554-1404 for copies

Spill cleanup requirements (761.125) for low concentration spills (between 50 and 100 ppm) less than one pound:

- Must double wash/rinse spill area if on a solid surface (must contain runoff). Area must be cleaned to 10 micrograms per 100 square centimeters.
- Soil; must excavate spill area plus buffer (1 foot) and restore the area by backfilling (10 ppm PCB).

High concentration spills (500 ppm or greater) and low concentration spills of one pound or more:

- Notify appropriate agency (if necessary)
- Cordon off immediate spill area (must include warning signs)
- Document area of visible contamination and delineate spill boundries
- Cleanup all visible traces of fluid

Sampling Requirements (761.30)

Sampling of the contaminated area is required to verify that cleanup has been completed. The sampling must be conducted randomly or on a grid and be reproducible. All sample control procedures must be followed. Number of samples must be sufficient to detect contamination (minimum 3, maximum 40). Sampling area must be larger than the area cleaned.

CLEAN AIR ACT

The Clean Air Act (CAA) was enacted December 31, 1970, with eight amendments since, including major amendments in 1990. The regulations are codified in 40 CFR 50-80. The U.S. EPA is the responsible federal agency with most states having authority to implement the CAA requirements and issue permits—one of the law's key provisions. For example, in Colorado, the Colorado Department of Public Health and Environment (CDPHE, Air Pollution Control Division [APCD]) implements the program statewide.

The CAA provides the regulatory framework for the prevention and control of discharges into air and identifies sources of air pollution as stationary and mobile. The EPA sets national ambient air quality standards (NAAQS) for pollutants. These standards must be met by all regions of the country.

The CAA provides a permitting system for individual dischargers. The type of permit depends on the type and age of the plant as well as the types of chemicals being emitted. The CAA Amendments have expanded the permitting program.

State Implementation Plans (SIPs) are used to enforce the law and issue permits.

HISTORY

The original CAA was passed in 1955 due to factory emissions fouling the air with black smoke. Major revisions in 1970 and 1977 established clean air as a national goal and set the regulatory framework for today. CAA Amendments will also serve to shape more stringent air pollution regulations. These amendments are having a profound impact on business and industry and include the following air issues:

- Provisions for protecting the ozone layer
- Changes in the permitting program
- Mobile source pollution
- Accidental release of toxic materials
- Hazardous Air Pollutants (HAPs)

Figure 9 Air Monitoring at Canyonlands National Park (1998)

Criteria pollutants are those pollutants for which a NAAQS has been set. Noncriteria pollutants have no established NAAQS.

PROVISIONS OF THE LAW

Establishment of NAAQS as required by the act provides a national standard for clean air. Many states such as Colorado have set their own air quality standards. The law applies to specific pollutants from both stationary (e.g., fixed industrial facilities) and mobile sources (automobiles). Local standards may be more strict due to differing pollution problems and atmospheric conditions. Mountainous terrain along the western edge of Denver, Colorado and thermal inversions that trap pollutants exacerbate pollution problems in that area. Methods of attainment of the standards vary from area to area depending on the SIP that has been developed. The EPA has set both primary and secondary standards for pollutants. Secondary standards are more stringent, and only a few have been set. Primary standards are based on protection of human health. Secondary standards are based on other issues such as pollution impacts on crops, livestock, vegetation, buildings, and visibility.

HAZARDOUS AIR POLLUTANTS

EPA has authority to set specific National Emission Standards for Hazardous Air Pollutants (NESHAPs). NESHAPs are now covered under the air toxics provisions of the CAA Amendments. HAPs are defined as those pollutants that cause serious illness or an increase in mortality. Prior to the 1990 CAA Amendments only a few NESHAP standards had been set by EPA for such pollutants as asbestos, arsenic, and mercury. The new rule now covers 189 HAPs.

NESHAPs are listed below.

Asbestos	naturally occurring mineral
Beryllium	metal
Mercury	metal
Vinyl Chloride	colorless gas
Arsenic	metal
Radionuclides	by product of radioactive decay
Benzene	component of gasoline
Coke Oven Emissions	manufacturing

NEW SOURCE PERFORMANCE STANDARDS (NSPS)

These standards apply to new sources of air pollutants. The standards are set for individual industries and applied nationwide. About 50 standards are in place for such industries as sulfuric acid production.

OTHER PROVISIONS

EPA can regulate emission sources that may affect the stratosphere, such as vehicle emissions, fuel, and fuel additives. For example, catalytic converters are required on vehicles to reduce emissions. The 1977 CAA Amendments clarified the 1970 law and allowed the EPA to act on air pollution problems before harm occurs. However, EPA must show *risk*.

SUMMARY OF THE 1990 CAA AMENDMENTS

These amendments were designed to curb three major threats to the environment and human health: acid rain, urban air pollution, and toxic air emissions.

Other provisions of the CAA Amendments include the following:

- National permits program
- Improved enforcement program
- Phase out of ozone-depleting compounds
- Research and development provisions
- Programs to address the accidental release of toxics
- To facilitate compliance, the new law uses a market based regulatory approach that includes emissions banking and trading.

EXAMPLE OF STATE AIR REGULATION: COLORADO, U.S.

Colorado has authority to implement the provisions of the CAA and Colorado Air Quality Control Act. Eighteen air regulations in the state each cover a particular air pollution issue. For example, volatile organic compounds control is Regulation No. 7 and hazardous air pollutants control is Regulation No. 8. The automobile inspection and readjustment program (AIR) is a component of the state's mobile sources program and covers primary motor vehicles in metropolitan areas such as Denver, Ft. Collins, and Colorado Springs.

AIR QUALITY CONTROL COMMISSION (AQCC)

The primary goal of the commission is to adopt an air quality program for the state and to assure that the state's program meets requirements of the CAA. The eight commissioners are appointed by the governor and serve three-year terms. The commission also exercises a judicial-type function. The commission publishes an annual "Report to the Public." This report describes air pollution levels throughout the state, including pollutant readings from monitor locations, violations and maximum levels. Proposed regulatory action and proposed SIPs are submitted to the commission for approval and forwarding to the state legislature and the governor.

AIR POLLUTION CONTROL DIVISION (APCD)

The APCD serves as staff to the AQCC. The division drafts and enforces the regulations enacted by the AQCC, issues air permits, provides technical assistance, and collects and publishes air quality data. Members of the APCD are within the CDPHE.

STATE IMPLEMENTATION PLAN

The SIPs are plans for the state to comply with regulations or pollutant standards.
  The state needs to meet an air pollution criterion. The area that needs the plan has a local planning commission devise a strategy to bring the area into compliance or prevent the deterioration of current air quality. This can be done with the help and guidance of the state APCD. The plan is submitted to the AQCC for approval. The AQCC usually has a subcommittee study the plan. The AQCC either approves the plan or suggests changes that need to be made. The plan could then be revised until accepted. Once the AQCC accepts the plan, it goes to the legislature for the authority to implement the plan. Sometimes the authority to enact an SIP is already in place.

LOCAL PLANNING COMMISSION (LPC)

The LPC is the local agency responsible for drafting the proposed SIPs and regulations regarding air pollution problems in their area. The LPC in Denver, Colorado (Central Front Range Air Quality Region) is the Denver Regional Council of Governments. Local health departments also perform air quality activities such as inspections.

COLORADO PROGRAMS

Colorado has regulations in place regarding all compounds covered under the NAAQS and has one Welfare Standard.

Visibility

Colorado is concerned about the visibility in the state and has adopted what is known as a welfare standard for visibility. This is different from a primary or secondary standard.

> *Primary Standards*—health based standards
> *Secondary Standards*—protection of plants, buildings, etc.

Air Pollution Emission Notices (APEN)

APENs are the current reporting form used by the Colorado APCD. If the threshold reporting requirement for any criteria pollutant or a listed hazardous substance is not (air toxic or HAP), APENs must be completed for the emission sources in a facility. The information submitted on the APENs will be used to determine the state inventory of air pollution emissions. This inventory is necessary for the state to comply with the 1990 Amendments of the CAA. The fee structure for permits and air emissions also changes with the new APENs. This money will enable the APCD to establish the new permit system and provide funds to comply with the 1990 CAA Amendments.

Stationary Sources

- Regulate industry and similar type emission sources
- Responsible for the APEN program
- Asbestos Program—permits, inspection, and certification
- Wood burning controls

Mobile Sources

- AIR—Auto Inspection and Readjustment Program better known as emission testing.
- Oxygenated Fuels—Requires cars in certain areas to use oxygenated fuels to help reduce carbon monoxide emissions.
- Diesel Emission Controls—High Altitude Motor Vehicle Test Facility.

Technical Services

- Air dispersion modeling
- Compile data from mobile and stationary sources
- Work with other state agencies
- Air quality surveillance

CLEAN AIR ACT OF 1990

Summary of titles are shown below.

Title	Provision
I	Provisions for Attainment and Maintenance of NAAQS
II	Provisions Relating to Mobile Sources
III	Hazardous Air Pollutants (HAPs)
IV	Acid Deposition Control
V	Permits
VI	Stratospheric Ozone Protection
VII	Provisions Relating to Enforcement
VIII	Miscellaneous Provisions
IX	Clean Air Research
X	Disadvantaged Business Concerns
XI	Clean Air Employment Transition Assistance

Provisions for Attainment and Maintenance of NAAQS (Title I)

The problems of ozone (smog), carbon monoxide (CO), and particulate matter (PM-10) persist in urban areas. Currently, more than 100 million Americans live in cities that violate the NAAQS. These provisions give cities more time to comply with the ozone NAAQS. (Los Angeles has 20 years to come into compliance). States must make constant progress toward attainment of the standards. The EPA is responsible for developing guidance for the states to control stationary sources. Areas nonattainment for ozone are classified as marginal, moderate, serious, severe, or extreme. Classification and attainment dates for carbon monoxide nonattainment areas:

Classification	*Design Value	Deadline
Moderate	9.1 - 16.4 ppm	12/31/95
Serious	16.5 & above	12/31/2000

*Design value is an EPA-computed statistic that characterizes air quality and
determines the nature of the required pollution controls.

Mandatory SIP provisions apply to nonattainment regions. The required provisions depend on the area's classification. Denver (specified in the rule) must implement the following transportation control measures:

- Decrease vehicle miles
- Promote carpooling
- Enact enhanced I & M program
- Develop oxyfuel program
- Improve public transit
- Restrict certain roads or lanes (HOV)
- Establish trip-reduction ordinances
- Institute traffic flow improvement programs
- Develop programs to limit vehicle use downtown
- Place limits on road uses to non-motorized or pedestrian use in metro area
- Develop program to remove pre-1980 model year vehicles

Requirements for PM-10 nonattainment areas are as follows:

Classification	Deadline
Moderate	12/31/94
Serious	12/31/20

Mandatory SIP provisions include

- Construction and operation permit program
- Attainment demonstration
- RACM

Program Summary

More than 2.7 billion pounds of toxic air pollutants are emitted annually. Toxic chemicals cause short-term effects (acute) such as eye irritation and/or long-term effects (chronic) such as cancer. The approach to control emissions of HAPs begins with a technology-based strategy compared to a risk- or health-based approach.

General Provisions

- A major source for purposes of HAPs is any stationary source or group of sources that emits or has the potential to emit 10 tons per year (TPY) or more of any HAP or 25 TPY or more of any combination of HAPs.
- An area source includes a smaller one such as a dry cleaners.
- 189 listed substances were part of the original bill. Most are carcinogens, mutagens, or reproductive toxins.
- The original list must be revised and updated by EPA.
- EPA must establish categories of major and area sources of HAPs called source categories. All categories are to be controlled within 10 years of enactment of the law.
- All major and area sources in each category must be regulated.
- Each source category will have standards called Maximum Achievable Control Technology (MACT) standards requiring the maximum degree of emissions reductions instead of allowing a source to emit a certain amount.
- HAP emissions must not exceed a level adequate to protect public health with an ample margin of safety.

Maximum Achievable Control Technology (MACT)

The technology the MACTs are based on will be determined primarily by the ability to prove a given technology can achieve the desired emission reductions.

MACTs will also be determined based on cost, energy requirements, and nonair environmental and health impacts. Companies that voluntarily reduce emissions can qualify for an extension on meeting MACT requirements. Following the implementation of the MACT requirements, EPA will determine the residual risk levels and assess if additional controls are needed.

REPORTING HAPs ON AIR POLLUTION EMISSION NOTICES IN THE STATE OF COLORADO

HAP reporting is dependent on type of chemical, release point height, and distance to property line. A *HAP Addendum* may be required if HAPs are emitted. The APEN and the HAP Addendum forms are available in WordPerfect 5.1 format. Contact the APCD for filing in this manner. HAP APEN reporting has undergone significant changes since 1992. Colorado recognizes the original 189 HAPs from the CAA Amendments and has added an additional 130. The additional 130 HAPs added by the state legislature are called Colorado HAPs (CHAPs). HAP APEN reporting is point- and chemical-specific.

Two important aspects of HAP reporting include a statewide inventory of HAP emissions and ongoing reporting of HAP emissions. The APCD has used a phased-in approach for HAP reporting that began in 1992. Minimum reporting levels called *de minimus* levels have been established for each HAP, which applies to each emission point. Three categories of HAPs called **BINs** (BIN A–C) are used for reporting and are based on the chemical's toxicity. BIN A chemicals are the most toxic while BIN C the least. HAP APENs revisions are due by April 30 of the year following the change.

Revised APENs are required if a significant change occurs which is an increase of the pollutant of five TPY or 50%. Although not required, decreases of emissions should be reported to save on annual fees. Fees include the filing fee and a per-ton cost based on the calculated emissions of each pollutant per point. The filing fee is due at the time of submittal. Annual fees are billed. To determine the *de minimus* level (reporting threshold) you must determine the **Scenario** that applies to the emission point. The three scenarios are based on the release point above ground level and the distance to the property line.

STATE IMPLEMENTATION PLANS

Every state must submit a plan to the EPA to implement the provisions of the CAA Amendments. EPA has established criteria to determine the adequacy of the state plans. The SIPs must contain a permitting program that allows the state to obtain authorization to review, issue, administer, and enforce operating permits. The plans must address specific strategies to bring nonattainment areas into compliance by the specified dates. Many of these control strategies are dictated by Title I. Upon approval, the SIP is promulgated into federal regulations and becomes federally enforceable.

PERMITS (TITLE V)

Permits are designed to ensure that sources are in compliance with the CAA. Operating permits will enable EPA and the states to track emission sources.

Permits cover both new and existing sources of air pollution in attainment and nonattainment areas. Each permit must include enforceable emission limitations and standards. Proof of compliance is required. Terms for monitoring and analysis are included. Most often continuous monitoring is required. The permit must set forth terms for availability for facility inspections, entry, and monitoring.

States may issue general permits that cover numerous similar sources or temporary sources. Permit applications must include a description of the air pollutants being emitted and applicable pollution control requirements. EPA has the authority to deny any permit application that does not meet requirements of the CAA.

Other permit conditions include

- Ensure compliance with all applicable air emission standards
- Monitor actual emissions
- Submit periodic monitoring reports
- Certify compliance status of the facility
- Submit applications for permit modifications when operating status of the facility changes
- Submit timely application for permit renewals
- Pay application fee

ACID DEPOSITION CONTROL (TITLE IV)

The purpose of this title is to reduce the adverse effects of acid deposition through reductions in SO_2 and NO_x emissions. Emission reductions are gained through a market-based system of emission allowances.

Allowances are allocated (and tracked) by EPA in an amount based on the facilities past fossil-fuel consumption and required emission reductions. Affected sources are required to hold sufficient allowances to cover their emissions. Allowances can be used by the source to cover emissions, banked for future use, or sold. If a facility's emissions exceed allowances held, they must pay $2,000 per excess ton and offset the excess tons the following year.

STRATOSPHERIC OZONE PROTECTION

EPA has published a list of *ozone-depleting compounds* as well as their ozone-depleting potential. Ozone-depleting substances are divided into two classes. Class I substances such as CFCs are the most potent. Other Class I substances include methyl chloroform and carbon tetrachloride. Class II compounds include the hydrochlorofluorocarbons. Production and consumption of these compounds is to be capped. Venting of ozone-depleting substances during the servicing or disposal of refrigeration equipment is prohibited. Ozone-depleting substances used in vehicle air conditioning must be recycled. Containers must be labeled. EPA is to assign global warming potential to listed substances.

WATER RESOURCES

FEDERAL WATER POLLUTION CONTROL ACT (CLEAN WATER ACT)

No or inadequate treatment of sewage wastes presents a serious risk to human health from water-borne infectious diseases. Many rivers and waterways in the United States still have unacceptable levels of bacteria or other pollutants. A 1998 report from the Colorado Water Quality Control Commission identified 90 streams in the state of Colorado that do not meet federal EPA pollution standards. Wastewater from most of the 960 million (40% of India's population) people who live in the Gangetic basin in India goes untreated.[1] Section 510 of the Federal Water Pollution Control Act recognizes the pollution of surface waters in San Diego, California, from raw sewage emanating from the city of Tijuana, Mexico and has provided construction grants for treatment works in San Diego. Providing adequate treatment systems and trained operators are key steps in addressing water pollution from sewage.

The Federal Water Pollution Control Act

"The objective of this Act is to restore and maintain the chemical, physical, and biological integrity of the Nation's waters. In order to achieve this objective it is hereby declared that, consistent with the provisions of this Act—

1. it is the national goal that the discharge of pollutants into the navigable waters be eliminated by 1985;
2. it is the national goal that wherever attainable, an interim goal of water quality which provides for the protection and propagation of fish, shellfish, and wildlife and provides for recreation in and on the water be achieved by July 1, 1983;

---

[1] South China Morning Post, Monday, October 27, 1997

3. it is the national policy that the discharge of toxic pollutants in toxic amounts be prohibited;
4. it is the national policy that Federal financial assistance be provided to construct publicly owned waste treatment works;"[2]

Designed to control pollution of surface waters, the Clean Water Act (CWA) (40 CFR 100-140 and 400-470) was enacted October 18, 1972. Fifteen amendments have been added since 1972. Releases or discharges into waterways date back to the Refuse Act of 1899, but this law was designed to protect passage in waterways not water quality. The Water Quality Act of 1965 began the establishment of water quality standards. A limited law called the Water Quality Improvement Act of 1970 addressed oil spills and recreational boat sewage treatment. However, a comprehensive water pollution control act was still necessary to protect surface waters. Over President Nixon's veto, the U.S. Congress passed the Federal Water Pollution Control Act, Public Law 92-500. The act established the framework for water pollution control and included the following:

- National effluent limitations
- Water quality standards
- Permit program (the Refuse Act also required permits)
- Provisions for oil spills and toxic materials
- Construction grants for Publicly Owned Treatment Works

Pollution means man-made or man-induced alteration of the chemical, physical, biological, and radiological integrity of water. Sources of pollutants covered by the CWA include industrial, municipal, dredge and fill material, and stormwater runoff (nonpoint source). Types of pollutants include toxics, organic wastes (e.g., agricultural processing wastes), inorganic chemicals, sediment, acids and bases, heat, oil, and grease.

The Federal Water Pollution Program

Both the state and federal programs use permits as the key enforcement tool. The act requires industries or municipalities to obtain a permit to discharge into surface waters. Wetlands are also covered under the act, while ocean dumping is mainly covered by the Marine Protection and Research and Sanctuaries Act. The program was implemented to control discharges from point sources such as end-of-pipe discharges from industries. Study information demonstrated that water pollution from nonpoint sources was also a major problem that was addressed in later amendments.

The act requires the treatment of certain wastes prior to their discharge into the sanitary sewer system. This program is called Industrial Pretreatment and is implemented at a local level by the Publicly Owned Treatment Works (POTW). Under an industrial pretreatment program, the POTW conducts surveys, draws samples, approves discharges, or requires treatment prior to discharge and may visit the site. The act also provides the regulatory vehicle to control the discharges of pollutants into surface waters or waters of the state, including both point and nonpoint sources. EPA sets effluent standards for industries and municipal sewage treatment plants. The federal permit program is called the National Pollutant Discharge Elimination System (NPDES). NPDES permits set standards for discharge limits, define type of control equipment required to treat the waste, and outline the effluent guidelines (levels of reduction) and effluent limitations. The permit will also specify the maximum amount of flow allowed, contain requirements for sampling and recordkeeping, and require the filing of Discharge Monitoring Reports. Permit conditions consider the type and amount of pollutant to be discharged as well as how much pollution the receiving body of

---

[2] 33 USC 1251, Section 101 (a)

water can tolerate. In 1987, amendments addressed the issue of runoff from farms and mining sites (nonpoint sources) that could contain pollutants.

New Regulations

Phase I of these amendments is in place, and Phase II was published as a proposed rule called "National Pollutant Discharge Elimination System—Proposed Regulations for Revisions of the Water Pollution Control Program Addressing Storm Water Discharges in the Federal Register on January 9, 1998 (Vol. 63, No. 6). Phase II NPDES Storm Water regulations will be finalized by March 1, 1999 and include the following changes:

- Expand existing program to smaller municipalities
- Expand the existing program to construction sites between one and five acres
- Provide for certain exclusions for lack of impact on water quality
- Exclude industrial facilities that have "no exposure" of activities to storm water

This proposed rule was developed to facilitate watershed planning. Watershed planning is an approach to water discharges that includes coordinated planning and a focus on highest priority water quality problems within a hydrologically defined geographic area. The states are not required to implement the watershed approach; however, EPA believes that this approach is critical to improve water quality in the United States as well as implement these new regulations. The Water Quality Control Division in the state of Colorado has begun implementation of the watershed approach and has "watershed coordinators" for areas around the state.

State Programs

Permits are most often issued by a state agency. For example, in Colorado, CDPHE, Water Quality Control Division (WQCD), Permits and Enforcement Program issue these permits as part of the Colorado Discharge Permit System. Also, many states in the U.S. have implemented certification programs that require system operators to have minimum knowledge and skills. Colorado state law established a program that requires the certification of water and wastewater treatment operators meet minimum education and experience requirements as well as pass a state-administered examination. There are four levels of wastewater treatment operators (A–D). The "D" level is the lowest and requires the least amount of experience and education. The state offers exams in two cycles during the year. The operator certification program also covers water treatment plant operators, collection and distribution, and industrial treatment. Depending on the plant, a particular level of operator will be specified by the WQCD.

Federal Assistance

In addition to construction grants to upgrade treatment plants and other financial assistance, EPA has an outreach program designed to provide technical assistance and training for small systems typically in rural areas. This outreach program is designed to assist those plants with operational deficiencies that may result in violation of their permit. A Comprehensive Performance Evaluation is conducted by a trained and experienced operator who evaluates the system and its operation. If deficiencies are discovered, a Composite Correction Program is developed to assist the plant and improve efficiency. Most cases involve on-site training of personnel and maintenance of equipment.

WETLANDS

"For all that has been done to protect the air and water, we haven't halted the destruction of wetlands."
President Bill Clinton

The environmental benefits of wetlands include providing breeding and feeding grounds for wildlife, flood protection, and erosion and pollution control. In some areas, wetlands are constructed specifically to help with the treatment of municipal wastewater. Wetlands and coastal waters are rich in natural resources and have historically been under pressure from development. Of all the species listed as threatened or endangered in 1993, 54% were found in wetland and deepwater habitats. The diversity of wetlands is extensive and ranges across the United States from coastal marshes and inland swamps of the Southeast to bogs and shrub swamps in the North to tropical rainforests in Hawaii to permafrost wetlands in Alaska.

Wetlands are defined based on the types of plants, soils, and frequency of flooding of the land. The soils are at least periodically saturated with water or saltwater. Coastal wetlands are those covered with saltwater. Wetlands can be marshes, swamps, bogs, rocky shores, ponds, or transition areas between water and land. Wetlands losses continue to decline. However, from the mid-1970s to the mid-1980s, the average annual net loss was 290,000 acres.[3] Major causes of wetland losses include agriculture and commercial development followed by residential development, highway construction, impoundments, and mining. Causes of degraded wetlands include contamination caused by sediment, pesticides, and heavy metals; excess nutrients; water diversion; weeds; and low dissolved oxygen. Sources of these pollutants include agriculture, development, channelization, road construction, and urban runoff. Twenty-two states, including Colorado lost 50% or more of their wetlands between the 1780s and mid-1980s.[4]

Regulations, Programs, and Policies to Protect Wetlands

Wetlands are primarily protected under the Federal Water Pollution Control Act (Clean Water Act), Section 404. This section of the CWA is implemented by the U.S. Army Corps of Engineers and requires a permit for dredge and fill activities in waterways, including wetlands. The Corps of Engineers and EPA signed a Memorandum of Understanding in 1989 designed to reduce or avoid impacts to wetlands. During the permit process, the Corps attempts to require the least damaging alternative and may require replacement or mitigation of wetland areas. Civil and criminal penalties may be assessed to violators.

The largest violation of this law relating to wetlands in Colorado occurred in the 1980s. The Corps determined that Telluride Mountain Village, Inc. and the Telluride Company had violated the CWA by filling in at least 17 acres of wetlands without a permit. The case was investigated by EPA and turned over to the Department of Justice which negotiated a civil settlement.

One of the difficulties in protecting wetlands is defining a wetland. Not all federal agencies use the same guidelines. Definitions and criteria have been developed by the U.S. Army Corps of Engineers, EPA, U.S. Fish and Wildlife Service (FWS), and the Agriculture Department's Soil Conservation Service (SCS). *The Federal Manual for Identifying and Delineating Jurisdictional Wetlands* delineates wetlands based on precise on-the-ground measurement techniques and focuses only on vegetated wetlands. The FWS uses *Classification of Wetlands and Deepwater Habitats of the United States*. In 1994, The SCS became the lead agency in defining wetlands and mapping wetland areas throughout the United States. The FWS is still the principal federal agency with responsibility for protecting and managing the nation's fish wildlife and their habitats.

---

[3] *Environmental Quality,* 24th Annual Report, The Council on Environmental Quality.
[4] *Wetlands Status and Trends in the Conterminous United States Mid–1970's to Mid–1980's,* Report to Congress, U.S. Fish and Wildlife Service, 1991.

Figure 10 Wetlands, Chatfield State Park, Colorado

The Emergency Wetlands Resources Act of 1986 was designed "to promote the conservation of migratory waterfowl and to offset or prevent the serious loss of wetlands by the acquisition of wetlands and other essential habitat, and for other purposes." The act prioritized wetlands for conservation in the United States as well as help meet requirements of international treaties and conventions. The U.S. FWS was to report on wetlands status and trends every ten years. The first report was published in 1991. The FWS was also mandated to map all wetland areas of the conterminous United States (National Wetlands Inventory, 1990). The act also allowed the charging of fees in National Wildlife Refuge areas to provide additional revenues for the conservation of wetlands. Also, the act required the establishment and periodic review and revision of the National Wetlands Priority Conservation Plan.

Wetlands are also protected by Executive Order 11990 (Water Resources Council), Food Security Act (1985) and its Swampbuster provisions designed to slow the loss of wetlands to agricultural development, and the 1989 "no-net-loss" policy announced by President Bush. The policy included helping fund nonregulatory programs such as the North American Waterfowl Management Plan. Partnerships with the FWS has helped restore 300,000 acres of wetlands under this plan.

Additional regulations designed to protect aquatic ecosystems and waterways of the United States include

- Ocean Dumping and the Marine Protection, Research, and Sanctuaries Act (MPRSA) (1972). The MPRSA allows the National Oceanic and Atmospheric Administration (NOAA) to designate specific areas as marine sanctuaries and thus are protected for their recreational, ecological, historical, educational, and aesthetic values. The act governs the discharges of wastes into ocean waters.
- Coastal Zone Management Act (1972) is a planning statute allowing the federal government to match state funds so that a management plan for coastal areas can be developed. These management plans must take into consideration ecological, cultural, historical, and aesthetic issues and are reviewed by the NOAA. 1990 amendments require each state to develop a nonpoint source pollution control program.

SAFE DRINKING WATER ACT (SDWA), PUBLIC LAW 93-523

The oceans store 97.6% of the free water stored on earth. Ice caps and glaciers store 1.9%, and .5% is stored in groundwater. According to the United States Geological Survey (1985), a total of 74,000 MGD of groundwater and 325,000 MGD of surface water are withdrawn in the United States. Water is pumped from underground aquifers, stored in reservoirs, and diverted from rivers to meet the demand, sometimes resulting in serious environmental problems. For example, in Northern California, water has been diverted from Mono Lake for public use resulting in a serious drop in water level and an increase in salinity. The result of this diversion may cause serious harm to the area's wildlife. In the Republic of Kazakhstan, water diverted for irrigation has dropped the water level in the Aral Sea 40% since 1960 resulting in increased salinity, collapse of the fishing industry, and damage to surrounding cropland from windblown salt from the dry lakebed.[6] The Aral Sea was once the fourth largest freshwater lake in the world. In addition to draining an aquifer dry or salt water intrusion, sinkholes as deep as 400 feet have resulted from aquifer depletion. Water is being pumped out of aquifers in many areas of the western United States and the world faster than it is being replaced—a situation known as groundwater overdraft.

The SDWA passed in 1974 and was aimed at improving drinking water quality. The act established drinking water standards that fall into two categories called primary and secondary standards. Primary standards set limits on contaminants that may affect health called Maximum Contaminant Levels (MCLs). To be in compliance, a water system must keep the level of certain contaminants below the MCL. Secondary standards are advisory levels that include nonhealth-related standards for limits on physical characteristics such as odor, color, taste, and hardness. 1986 amendments to the SDWA:

- Required EPA to issue standards for 83 drinking water contaminants
- Required underground injection well operators to monitor groundwater
- Streamlined procedures for EPA to set drinking water standards

State of Colorado, Colorado Primary Drinking Water Regulations (CPDWR)

Regulations to protect drinking water in the state of Colorado based on the federal program were adopted in March 1991 and effective April 30, 1991. The program is implemented by the Water Quality Control Division, Drinking Water Section of the CDPHE. The regulation contains 14 articles that apply to public water systems. The regulation addresses microbiological contaminants, turbidity, inorganic and organic chemicals, radioactivity, corrosivity, reporting and recordkeeping, treatment of public water supplies, public notification, sanitary surveys, and hazardous cross-connections. Like the federal program, the CPDWR are based on MCLs that have been established and published in the rule. A list of Chemical MCL Violations can be obtained from the CDPHE. Primary standards are divided into four categories:

- Inorganic compounds
- Organic compounds
- Radioactive materials
- Pathogens

In an August 17, 1998 report from the CDPHE, 29 water systems in Colorado were in violation of various MCLs statewide, including nitrate, nitrite, arsenic, fluoride, combined radium, gross alpha, and EDB (ethylene dibromide). The detected level was 0.20 ug/l while the MCL for EDB is .05 ug/L. EDB is a carcinogenic compound that has been used as a gasoline additive and pesticide to fumigate grains. MCLs have been set for fluoride (4.0 mg/l), pesticides, arsenic (0.05mg/l), lead (0.05 mg/l), mercury (0.002mg/l), cadmium (0.010mg/l), nitrate (as N), and other organic and inorganic chemicals. States must monitor the amounts of the primary contaminants.

Underground Injection Control (UIC), 40 CFR 144-147

The UIC program established a control program for the injection of fluids into the ground in areas that may affect a public drinking water supply. The program is enforced by the states and became effective in 1980. Underground injection, also called wastewell injection or deep well injection, is the process by which wastes are injected into deep, confined subsurface areas. Federal regulations established five types of wells (class I–V) each with a specified type of waste such as industrial waste Class I and hazardous waste Class IV. No new Class IV wells are allowed. Class I – III wells must be evaluated every five years and a corresponding permit reissued. In addition to obtaining a permit and monitoring groundwater, operators are required to file quarterly reports, report any system malfunctions, and meet closure and postclosure requirements.

Sole Source Aquifers

Sources of drinking water that are the only source for the area are called sole source aquifers. Because of their importance, the SDWA allows for more stringent provisions for this type of aquifer. In some cases, water system operators require special permits. The 1986 amendments to the SDWA provided state and local governments federal aid to protect these aquifers.

RESOURCE CONSERVATION AND RECOVERY ACT (RCRA)

RCRA was enacted October 21, 1976 and is codified in 40 CFR 240-299. EPA and the states have authority to implement the provisions of RCRA. To address smoldering dump sites and health and air pollution problems caused by open dumps, the Solid Waste Disposal Act of 1965 was passed. This began soil covered sanitary landfills. In the 1970 amendments to the SWDA, Congress ordered an investigation (studies and surveys) into the nation's hazardous waste management practices. and hearings were held in 1975 to update the 1970 law. RCRA was enacted as a result of those hearings and replaced the SWDA. In 1978, EPA was issued the first regulations. In 1980, EPA issued final rules on hazardous waste management.

GOALS OF RCRA

- Protect human health and the environment
- Reduce waste and conserve energy and natural resources
- Reduce or eliminate the generation of hazardous waste

Four Main Programs Under RCRA

Subtitle C: Hazardous Waste

- Establishes a management system that regulates hazardous waste from the time it is generated until its ultimate disposal ("cradle to grave").
- Identification and listing of hazardous wastes
- Requirements for generators and transporters of hazardous waste
- Requirements for Treatment, Storage and Disposal Facilities (TSDF)
- Permit standards that apply to TSDFs
- Enforcement provisions
- State authorizations

Subtitle D: Solid Waste

- Promotes and encourages the environmentally sound management of solid waste. Includes minimum federal technical standards and guidelines for state solid waste plans.
- Establishes guidelines for the development and implementation of state solid waste management plans.
- Establishes criteria for classification of solid waste disposal facilities and practices.

Subtitle I: Underground Storage Tanks

- Regulates petroleum products and hazardous substances that are stored in underground tanks.

Subtitle J: Medical Waste

- Regulates medical waste generation, treatment, destruction, and disposal. As a result of medical waste washing up on beaches in 1988, Congress has enacted a 2-year demonstration program to track medical waste.

Definition of Solid and Hazardous Wastes (40 CFR 261)

A solid waste is any discarded material (discarded means abandoned, recycled, inherently waste-like, disposed of, burned, or incinerated). Exclusions from the solid waste definition include

- Domestic sewage
- Any mixture of domestic sewage and other waste passing through the sewer system to a POTW
- Industrial wastewater point source discharges covered by the CWA
- *In situ* mining waste

Hazardous waste is a solid waste or combination of solid wastes, which because of its quantity, concentration, or physical chemical or infectious characteristics may cause illness or pose a hazard to human health or the environment. Hazardous waste exhibits the following characteristics:

- Exhibits any of the characteristics of a hazardous waste
- Has been named as a hazardous waste and listed as such in the regulation
- Is a mixture containing a listed hazardous waste and a nonhazardous solid waste
- Is a waste derived from the treatment, storage, or disposal of a listed hazardous waste

Characteristics of Hazardous Waste

- Ignitability—It has a flash point of less than 140°F. It is a solid that can spontaneously combust. It is an ignitable compressed gas.
- Corrosivity—pH less than or equal to 2 or greater than or equal to 12.5.
- Corrodes steel at a rate of 6.35 mm or more per year.
- Reactivity—the material has the capability to explode or undergo a violent chemical change (e.g., unstable and chemicals that react with water).
- Toxicity—This category is determined by a test procedure called the Toxicity Characteristic Leaching Procedure (TCLP). The test is designed to see if the material would leach hazardous constituents.

A waste is a hazardous waste if it is on one of the three lists developed by the EPA:

1. Nonspecific Source Wastes (261.31): Generic wastes commonly produced by manufacturing and industrial processes. For example, spent halogenated solvents from degreasing.
2. Specific Source Wastes (261.32): Wastes from specifically identified industries such as wood preserving and petroleum refining. These wastes usually contain sludges, still bottoms, and wastewaters.
3. Commercial Chemical Products (261.33 (e) and (f)): "P" and "U" listed wastes of specific commercial chemical products or manufacturing intermediates such as sulfuric acid, DDT, and creosote.

The EPA developed these lists by looking at the following four criteria:

- Exhibit one of the four characteristics of hazardous waste
- Meet the statutory definition of hazardous waste
- Are acutely toxic or acutely hazardous
- Are otherwise toxic

Hazardous Waste Mixtures

A waste mixture that is a nonhazardous solid waste and a listed hazardous waste is considered a hazardous waste.
   Exceptions are

- Wastewater discharge subject to regulation by the CWA which is mixed with a low concentration of a listed waste
- Mixtures of nonhazardous and listed wastes that were listed for exhibiting a characteristic that no longer exhibits that characteristic
- Mixtures of nonhazardous waste and characteristic waste that no longer exhibit the characteristic
- Certain concentrations of spent solvents and laboratory discharges
- *De minimis* losses of discarded commercial chemical products (e.g., process leaks and incidental discharges)

Wastes Specifically Excluded from Regulation

- Oil and gas production waste
- Mining wastes

- Wastes from the combustion of coal
- Household wastes
- Discarded arsenical-treated wood from the end user
- Certain residues in empty containers (all material has been removed, no more than 1-inch residue remains)
- Samples being transported or stored by the sample collector or lab
- Other exemptions apply for recycling activities (depends on item being recycled and how it is done)

REGULATIONS APPLICABLE TO GENERATORS OF HAZARDOUS WASTE

A generator is broadly defined as anyone (facility owner or operator or person) who creates a hazardous waste. EPA has identified three categories of hazardous waste generators depending on how much waste is generated in a calendar month:

- Large quantity generators (LQG)
- Small quantity generators (SQG)
- Conditionally exempt small quantity generators (CESQG)

In the United States, LQGs overall generate 274 million metric tons of hazardous waste per year and are defined as facilities that generate more than 1,000 kilograms per month of hazardous waste or more than 1 kilogram of acutely hazardous waste per month. LQGs must submit an application called a Notification of Regulated Waste Activity to acquire an *EPA ID number* prior to offering any hazardous waste for transport. The EPA uses this number to track hazardous waste. Each number is site specific.

Pretransport Regulations

- All waste must be properly packaged
- All drums must be properly marked with the hazard and with a label stating the contents are hazardous waste as well as the accumulation start date.

Accumulation Time: How long can you store the waste on site?

An LGQ can store waste on site for 90 days or less if the waste is properly stored, a written contingency plan is in place, and the facility personnel are properly trained. If waste is stored longer than 90 days, the generator is considered a TSDF. Unlike generators, a TSDF is required to obtain a permit prior to operation.

The Uniform Hazardous Waste Manifest

The manifest is the key to the objective of "cradle to grave" management of hazardous waste and is required for shipments of hazardous waste to a disposal site. The manifest tracks movement of the hazardous waste from the generator (cradle) to the ultimate disposal (grave). The 1984 amendments (HSWA) required a certification that the generator had done everything they could to minimize their waste and have selected the best method for disposal.

Recordkeeping and Reporting

- Biennial reporting—information about the generator's activities for the previous year. Due by March of each even numbered year.

  - Generator information
  - Hazardous waste
  - Efforts made to reduce hazardous waste
  - Changes in volume or toxicity of the waste

- 3-year retention period for reports, manifests and test records
- Exception Reporting—if the generator does not receive a manifest back from the disposal site within 45 days from the date the initial transporter accepted the waste, this report must be submitted to the regional administrator

In-plant Emergency Procedures (40 CFR 265.56)

The location must have a designated emergency coordinator (EC). The EC coordinates all aspects of a response to an incident involving hazardous waste and if necessary contacts the appropriate response agencies such as the local fire department.   If there is a release of hazardous waste, a fire, or explosion, the EC must

- Identify the character of the material
- Identify the source of the material
- Identify the amount of the material released
- Assess possible health and environmental hazards

If the situation threatens human health or the environment *outside* the facility, the EC must contact local authorities. If evacuation is required, the EC must contact the National Response Center (800) 424-8802 with a report containing the following information:

- EC's name and phone number
- Name and address of facility
- Name and quantity of material released
- Extent of any injuries
- Possible health or environmental hazards outside the facility

The EC must ensure that clean up of the spilled material is complete and equipment used is properly decontaminated, repaired, or replaced as necessary before resuming operations. The EPA regional administrator and in the state of Colorado the CDPHE must be notified when this has been completed and the facility is ready to resume operations. Facility operating records must contain details of the incident. A written report is due within 15 days following the incident to the EPA regional administrator and the CDPHE, which contains the following information:

- Name, address, and phone number of the facility owner/operator
- Name, address, and phone number of the facility
- Date, time, and type of incident
- Name and quantity of material involved
- Extent of any injuries
- Assessment of health and environmental hazards
- Quantity and disposition of recovered material

Transportation Emergencies Involving Hazardous Waste

Hazardous Waste Discharges (40 CFR 263, Subpart C)

Transporter must take appropriate immediate action. If a hazardous waste discharge meets the requirements of 49 CFR 171.15, reporting must be in accordance with DOT regulations. The transporter is responsible for the cleanup of spilled material. Remember, the generator of the waste still maintains ultimate responsibility for proper disposal of the waste.

Immediate notice of certain hazardous materials releases (49 CFR 171.15) and detailed hazardous materials incident reports (49 CFR 171.16) are required if, as a result of a release of hazardous waste, a

- Person is killed
- Carrier or property damage in excess of $50,000
- Evacuation of the public for more than one hour
- One or more transportation arteries are closed for more than one hour
- In the carrier's judgment, it should be reported

A written follow-up report is required for all reporting made in accordance with above. (DOT form F5800.1).

Land Disposal Restrictions

Goal—protect groundwater from improper disposal of hazardous wastes.

"Reliance on land disposal should be minimized or eliminated, and land disposal, particularly landfill and surface impoundments, should be the least favored method for managing hazardous wastes..."
42 USC 6901 (b)(7)

The Land Disposal Restrictions ("Land Ban") prohibit the land disposal of hazardous wastes unless certain treatment standards have been met. Land disposal includes

- Landfills
- Waste piles
- Land treatment facilities
- Underground mines
- Surface impoundments
- Injection wells
- Salt domes
- Concrete vaults or bunkers

The owner or operator of the disposal facility has the ultimate responsibility that only wastes meeting the treatment standards are land disposed.

References and Information

- Land Disposal Restrictions, Summary of Requirements, USEPA, OSWER 9934.0-1A, February 1991
- 40 CFR, part 268
- Hazardous and Solid Wastes Amendments of 1984, PL No. 98-616 amending RCRA 3004, 42 USC 6924

- RCRA Hotline (800) 424-9346
- EPA Region VIII, RCRA Implementation Branch (303) 293-1524

Regulatory Summary

The Hazardous and Solid Wastes Amendments of 1984 (HSWA) required tighter restrictions for liquids in landfills and more stringent performance requirements for landfills. The original rule was published in the Federal Register (FR) January 14, 1986. (51 FR 1602) EPA took the statute ban on land disposal and turned it into a risk-based environmental quality standard. However, congressional objections to this approach resulted in EPA abandoning the risk-based approach to the regulation and published final regulations on November 7, 1986. (51 FR 40572)

The final rules provided for certain exemptions but only where a petitioner could show there would be "no migration" of land disposed wastes. The land disposal restrictions contain requirements for testing, treatment, storage, notification, certification of compliance, and recordkeeping. The regulations also contain variances from the restrictions. The statute sets up categories of wastes (restricted wastes) and treatment standards for each category of waste. For each category, a date was set when the land disposal of the waste would be prohibited.

- Solvents (1986) and Dioxins (1988)
- California List (adopted from the California Department of Health Services regulations that were already in place)
- First Third (of listed hazardous wastes)
- Second Third
- Third Third

Treatment Standards (Best Demonstrated Available Technology [BDAT])

Wastes treated with BDAT are exempt from the land disposal restrictions. In addition to the specified methods of treatment, concentration levels are also considered treatment standards. The treatment standards specified in the rule were selected because they substantially reduce the toxicity of the wastes or the likelihood the hazardous constituent(s) would migrate from the disposal site. Dilution of prohibited wastes is not an acceptable treatment method and is illegal. (40 CFR 268.3)

Wastes Disposed Prior to 1986

Wastes disposed prior to November 8, 1986, do not need to be removed for treatment; however, if restricted wastes are removed, the wastes must now meet the applicable treatment standard before subsequent land disposal.

Applicability

- Generators of hazardous waste (the provisions of part 268 do not apply to CESQG)
- Transporters of hazardous waste
- TSDF

Specific Requirements

Generators must determine if each hazardous waste generated is subject to the land disposal restrictions. A notification must accompany each shipment of hazardous waste from the generator to the receiving facility and if the waste meets the treatment standard as generated, a written certification. (Note: receiving facility may also include recyclers, reclaimers, and incinerators since residues from these facilities may ultimately require land disposal.)

The notification must include (1) hazardous waste code(s), (2) corresponding treatment standard(s), (3) manifest number, (4) waste analysis data (if available), and (5) date the waste is subject to the prohibitions (applies if waste exceeds any treatment standard).

Generators that send waste directly to land disposal must certify in writing that restricted wastes meet the applicable treatment standards. Generators must keep copies of all notifications, certifications, and waster analysis data on site for at least five years. Remember, if generator knowledge is used to determine prohibition status, detailed supporting documentation must be retained.

TREATMENT, STORAGE AND DISPOSAL FACILITIES (SUBTITLE C PARTS 264 AND 265)

All TSDFs (e.g., incinerator or landfill for hazardous waste) must be permitted by the EPA and/or the state in which it will be operating and comply with specific regulations. TSDF regulations establish design and operating criteria.

Regulatory Categories of TSDF include interim status and permit standards.

Interim Status:

- These facilities have not obtained a permit.
- Allowed time for existing facilities (prior to November 1989) to apply for a permit.
- Incinerators: final application must have been in by November 1986 (or status lost by 1989).
- All others must be submitted by November 1988 or lose status by 1992.

Permit Standards establish performance standards and design and operating criteria.

Both categories include administrative, nontechnical, technical, and unit-specific requirements.

Administrative and Nontechnical Requirements

- Facilities that reuse, recycle, or reclaim hazardous waste except fuels blending, generators of hazardous waste, and transporters storing manifested shipments of hazardous waste less than 10 days are exempt.
- Subpart B: General Facility Standards

  - EPA identification number
  - Wastes must be properly identified
  - Security
  - Inspections
  - Training
  - Special requirements for ignitable, reactive, and incompatible wastes
  - Proper siting

Contingency Planning and Emergency Procedures (Subparts C and D)

- Action plan for handling emergencies
- Must designate an EC
- Coordinate with local authorities

UNDERGROUND STORAGE TANKS (USTs)

Approximately two million UST systems are used nationwide to store hazardous substances (CERCLA) and petroleum products. The types of products commonly stored in UST systems include

- Motor fuels
- Jet fuels
- Distillate fuel oils
- Lubricants
- Used oils
- Petroleum solvents

EPA estimates as many as 25% of all UST systems may now be leaking. Those not leaking now may soon be leaking in the future. UST systems more than 10 years old have a much greater chance of leaking. Leaking USTs can cause fires or explosions and threaten human health. A major concern of leaking USTs is the contamination of groundwater. Fifty percent of the U.S. population relies on groundwater as a source of drinking water.

Why USTs Fail

- No corrosion protection
- Spills and overfills
- Installation mistakes
- Piping failures (most leaks)

Regulatory Summary

The problem of leaking USTs has been addressed by Congress by adding Subtitle I to RCRA in 1984. Technical requirements for tanks and piping have been established by the EPA. The complete regulations with preamble and discussion can be found in the Federal Register of September 23, 1988 (53 FR 37194). The implementing regulations may be found in title 40 of the Code of Federal Regulations (CFR), part 280. The state of Colorado enacted a statute covering USTs in June 1989. The laws are contained in titles 8, 24, and 25 of the Colorado Revised Statutes and 6 CCR.

The EPA has developed the following goals for the UST regulations:

- Prevent leaks and spills
- Find leaks and spills
- Correct the problems created by leaks and spills
- Make sure owners/operators of USTs pay for correcting the problems created by leaking USTs
- Make sure each state has a regulatory program for USTs

Regulatory Authority

- EPA, Office of Underground Storage Tanks
- State Inspector of Oils (Colorado) ("State Inspector") (SIO) 692-3330
- CDPHE, UST group (recently merged with the SIO)

What is a UST?

Any tank, including underground piping connected to the tank, that has at least 10% of its volume underground. The regulations apply to USTs used for the storage of regulated substances. Regulated substances are

- Substances defined in CERCLA 101(14) but not hazardous wastes
- Petroleum and petroleum-based products

New UST systems are those installed after December 22, 1988. Tank systems not covered by the regulations are either excluded from regulation or excluded from the definition of a UST or are covered by a deferral.

Tanks Not Covered by the Regulations in 40 CFR 280

- Farm and residential tanks holding 1,100 gallons or less of motor fuel used for noncommercial use
- Tanks storing heating oil used on the premises where it is stored
- Tanks on or above the floor of underground areas, such as basements or tunnels
- Septic tanks and systems for collecting stormwater and wastewater
- Flow-through process tanks
- Tanks holding 110 gallons or less
- Emergency spill and overfill tanks
- Any UST system holding hazardous waste (tanks used to store hazardous waste are covered by 40 CFR 265)

40 CFR 280, Technical Standards and Corrective Action Requirements for Owners and Operators of Underground Storage Tanks (UST)

Subpart	Requirement
A	Program Scope and Interim Prohibition
B	UST Systems; Design, Construction, and Installation and Notification
C	General Operating Requirements
D	Release Detection
E	Release Reporting, Investigation, and Confirmation
F	Release Response and Corrective Action for UST Systems Containing Petroleum or Hazardous Substances
G	Out-of-Service UST Systems and Closures
H	Financial Responsibility
Appendix I	Notification for Underground Storage Tanks (form)
Appendix II	List of Agencies Designed to Receive Notifications
Appendix III	Statement for Shipping Tickets and Invoices

UST Systems; Design, Construction, and Installation and Notification

All tanks and piping must properly designed, constructed, and installed. (See codes of practice in the reference section.) A Certification of Installation must be completed on a notification form. Portions of tanks and piping underground must be protected from corrosion by

- Fiber-glass-reinforced plastic
- Cathodic protection for steel tanks
- Steel-fiberglass-reinforced-plastic composite
- Corrosion potential in area tank to be installed will not cause a release

Spill and overfill prevention equipment must be used to prevent the release of product during transfer (e.g., catch basin, flow shut off).

A notification (form OMB No. 2050-0068) for a new facility to close a UST system is located in Appendix I to part 280 40 CFR. Tank suppliers must notify the owner/operator of the registration requirements. (See Appendix III to Part 280.)

General Operating Requirements

Must ensure releases do not occur as a result of spills or overfilling. Transfer operations must be constantly monitored. All spills must be immediately contained and cleaned up (see Section 280.53). All corrosion protection systems must be maintained, including

- Inspection and testing (test required within six months of installation and every three years thereafter)
- Recordkeeping (inspections and tests)

UST systems must be compatible with the substances stored. All repairs conducted must be in accordance with a code of practice developed by a nationally recognized association or testing laboratory or by a manufacturer's authorized representative. Repaired tanks and piping must be tightness tested within 30 days following the date of completion of the repair.

- National Association of Corrosion Engineers
- National Fire Protection Association
- American Petroleum Institute
- National Leak Prevention Association

Reporting

- Notification
- Reports of releases (suspected releases, spills, overfills, and confirmed releases)
- Corrective action plans
- Notification before permanent closure

Release detection is required for all UST systems and a release must be able to be detected from any portion of the tank and connected piping.

Figure 11 Groundwater Monitoring Well

Methods of release detection include

- Inventory control (monthly)
- Manual tank gauging (liquid level measurements—stick readings)
- Tank tightness testing
- Automatic tank gauging (tests for loss of product)
- Vapor monitoring
- Groundwater monitoring

Release Reporting, Investigation, and Confirmation

- Must report a suspected release within 24 hours to SIO.
- Must report any spill of 25 gallons or more to the SIO.
- Suspected releases must be investigated within 7 days.

SUPERFUND AMENDMENTS AND REAUTHORIZATION ACT (SARA) AND EMERGENCY PLANNING AND COMMUNITY RIGHT-TO-KNOW ACT TITLE III

The law was enacted October 17, 1986 and signed by Ronald Reagan. The regulations are codified in 40 CFR subchapter J, Parts 355-372. In Colorado, Title III implementing legislation was enacted in 1990. Agencies with authority to implement the law include the EPA state and local agencies (e.g., CDPHE, Jefferson County Emergency Planning Commission). SARA is an example of an *unfunded mandate*. That is, the federal government passes legislation that requires the states and local communities to develop a program but provides no funding. The purpose of the regulation is to provide information on site-specific hazardous chemicals to the community (thus the name

Community Right-to-Know), initiate emergency planning and notification activities at the local community level, and provide training for first responders to hazardous materials incidents.

GENERAL PROVISIONS

- Emergency Planning (Sections 301-303, 40 CFR Part 355)
- Emergency Release Notification (Section 304, 40 CFR Part 355)
- Community Right-to-Know Reporting Requirements (Sections 311, 312, 40 CFR Part 370)
- Toxic Chemical Release Reporting (Section 313, 40 CFR Part 372)

EMERGENCY PLANNING

The emergency planning provisions of SARA are designed to help states and local governments develop emergency response capabilities through better planning. The governor of each state must appoint a State Emergency Response Commission (SERC) by April 17, 1987. For Colorado, the CDPHE acts in this capacity. The Colorado ERC is responsible for

- Establishing Local Emergency Planning Committees (LEPC) for each established district in the state by August 17, 1987. There are 65 districts in Colorado. Most are divided by county with three representing Indian tribal areas.
- Supervising and coordinating the activities of the LEPC.
- Establishing procedures for receiving and processing public requests for information.
- Reviewing local emergency plans.

Facilities that must comply with SARA requirements must designate a representative by September 17, 1987. This representative is responsible for participating in the emergency response planning activities between the facility and local authorities.

LEPCs must include at a minimum:

- Elected state and local officials
- Police, fire, civil defense, and public health officials
- Environmental, hospital, and transportation officials
- Representatives of facilities subject to the emergency planning requirements
- Representatives of community groups and the media

LEPCs establish rules and give public notice of its activities as well as establish procedures for handling public requests for information and developing an *Emergency Response Plan* by October 17, 1988. In developing this plan the LEPCs consider available resources for preparing and responding to a chemical incident. The plan must

- Identify facilities and transportation routes where extremely hazardous substances are stored or transported.
- Describe emergency response procedures.
- Designate a community coordinator.
- Outline the emergency notification procedures.
- Describe the methods for determining that a release has occurred and the potentially affected area and population.
- List and describe the community and industry emergency equipment and facilities available. This section must include who is responsible for these items.
- Outline the evacuation plan.

- Describe the training program for response personnel.
- Discuss the methods to be used for practicing or exercising the plan.

The emergency planning provisions of SARA established Threshold Planning Quantities (TPQs) for extremely hazardous substances.

EMERGENCY RELEASE NOTIFICATION (SECTION 304, 40 CFR PART 355)

A facility must notify the SERC and LERC immediately if a release of a reportable quantity of an extremely hazardous substance or a CERCLA hazardous substance. The initial notification can be by telephone, radio, or in person. Emergency notification requirements for transportation incidents can be satisfied by dialing 911. This emergency notification must include

- The chemical name
- An indication of whether the substance is extremely hazardous
- An estimate of the quantity released into the environment
- Time and duration of the release
- The medium into which the release occurred
- Any known or anticipated acute or chronic health risks associated with the emergency, and where appropriate, advice regarding medical attention necessary for exposed individuals
- Proper precautions such as evacuation
- Name and telephone number of the contact person

Requires written follow up report after the release that must include the following information:

- Update information for that provided in the initial notice
- Actual response actions taken
- Any known or suspected chronic health risks associated with the release
- Advice regarding medical attention necessary for exposed individuals

COMMUNITY RIGHT-TO-KNOW REPORTING REQUIREMENTS (SECTIONS 311, 312, 40 CFR PART 370)

Facilities must submit an MSDS or a list to the SERC, LEPC, and Fire Department (October 17, 1987) for over 10,000 lbs or 500 lbs, 55 gallons, or TPQ (whichever is less) and report on their chemical materials by submitting emergency and hazardous chemical inventory form (Tier I, Tier II) for chemicals requiring an MSDS for greater than 10,000 lbs or 500 lbs, 55 gallons, or TPQ (whichever is less). The first reports were due by March 1, 1988.

TOXIC CHEMICAL RELEASE REPORTING (SECTION 313, 40 CFR PART 372)

Covered facilities are primarily manufacturing. The facility must have 10 or more full-time employees and meet the following thresholds for reporting:
Manufacturing or processing of the listed chemicals:

1987 - 75,000 lbs
1988 - 50,000 lbs
1989 and thereafter - 25,000 lbs

Other uses:

10,000 lbs

EPA has developed a specific toxic chemical listing located in 40 CFR 372.65. If a facility exceeds the threshold quantities, they must submit a Form R (toxic chemical release form) if over threshold by July 1, 1988 and annually thereafter.

Hazard Category Comparison (EPA and OSHA)

EPA Category	OSHA Category
Fire Hazard	Flammable/combustible
	Pyrophoric
	Oxidizer
Sudden release of pressure	Explosive
	Compressed gas
Reactive	Unstable reactive
	Organic peroxide
	Water reactive
Immediate (acute) health hazards	Toxic/highly toxic
	Irritant
	Sensitizer
	Corrosive
	Other hazardous chemicals
Delayed (chronic) health hazards	Other hazardous chemicals with an adverse effect with long term exposure (e.g., carcinogens)

MAKING SENSE OF SARA TITLE III REGULATORY CITATIONS

USC	Requirement	40 CFR
301	Establishment of state commissions, planning districts, and LEPC	
302	Substances and facilities covered and 355.30 notification	
303	Emergency Response Plans (developed by LEPC for district)	
304	Emergency Notification 355.40	
311	MSDS 370.21	
312	Emergency and hazardous chemical 370.25 inventory forms	
313	Toxic Chemical Release Reporting 372	
324	Public Availability of plans, data sheets, 370.30 forms, and follow-up notices 370.31	

The USC Citations are from Title III, Emergency Planning and Community Right-to Know, Subtitles A, B and C

- Subtitle A: Emergency Planning and Notification
  Sections 301–305
- Subtitle B: Reporting Requirements
  Sections 311–313
- Subtitle C: General Provisions
  Sections 321–330

The provisions of SARA that do not amend existing laws are divided into five titles. Only Title III is represented here.

FEDERAL INSECTICIDE, FUNGICIDE, RODENTICIDE ACT (FIFRA)

Pesticides have provided tremendous benefits to agriculture, such as increased crop production and control of noxious weeds. However, adverse environmental effects such as water contamination and human exposure concerns for farm workers and consumers have required tighter controls on pesticide production, use, and application. EPA is currently evaluating a particular group of pesticides called organophosphates for possible restrictions in their use. Organophosphates are organic compounds that contain phosphorus and include common insecticides such as malathion and diazinon.

The Federal Environmental Pesticide Control Act (October 21, 1972) amended the original FIFRA of 1947 and serves as the current framework for pesticide regulation. The 1972 amendments were significant changes in the law that focused on health, safety, and environmental concerns. Prior to 1972, the primary focus of pesticide regulation was protecting the consumer from ineffective products although the 1947 law included registration and labeling provisions. Initially, the regulation of pesticides was handled by the United States Department of Agriculture (USDA), Pesticide Division, and dates back to the Federal Insecticide Act (FIA) of 1910. In 1970, President Nixon signed a presidential reorganization order (Reorganization Order No. 3) that created the USEPA. Functions previously carried out by other agencies were now EPA's responsibility, including pesticide regulation. Pesticide programs are contained in Subchapter E of 40 CFR Parts 152–186 and are carried out by the Office of Prevention, Pesticides, and Toxic substances at EPA.

SUMMARY OF THE REGULATION

The law covers insecticides, rodenticides, disinfectants, biological products, traps, and products to control plants, trees, weeds, bacteria, algae, viruses, or any other item that may be considered a "pesticide." A pesticide is "any substance or mixture of substances intended for preventing or destroying, repelling, or mitigating any pest, or intended for use as a plant regulator, defoliant, or desiccant..." Some items meeting this definition, such as vaccines, are covered by other laws and therefore excluded. FIFRA regulates the registration and use of pesticides and if the chemical is used on agricultural products, a residue tolerance must be established by EPA. The Food and Drug Administration (FDA) under the authority of the Food, Drug and Cosmetics Act establishes acceptable levels of pesticides in food. Activities that may be banned, controlled, or restricted by the EPA include manufacture, use, importation, and disposal of pesticides. In addition, pesticides are regulated through:

- Evaluation of risks through a registration and reregistration system
- Classification of pesticides for specific purposes and uses to control exposure and misuse
- Suspension and cancellation authority of pesticide registrations
- Labeling requirements on product containers

The registration of pesticides is an important part of the Act (40 CFR 152, 158, 162) and specifies the crops and insects for which the product may be used. Registrations are approved by EPA if

- The product performs its function without unreasonable adverse effects on the environment

- When used in accordance with the label and commonly used practice, the pesticide will not cause unreasonable adverse effects on the environment
- The labeling for the product meets the requirements of the act

In the context of FIFRA, "unreasonable adverse effects on the environment" means any unreasonable risk to man or the environment, taking into account the economic, social, and environmental costs and benefits of the use of the pesticide. When determining if the risk is unreasonable, EPA evaluates the health effects such as carcinogenicity, reproductive toxicity, target organ effects, as well as the effects on fish and wildlife.

Enforcement of the act is carried out through inspections and training and testing of applicators. The states may provide training and testing programs for applicators of pesticides and conduct facility inspections. In Colorado, the Colorado Department of Agriculture, Division of Plant Industry, Pesticide Section implements the program. Applicators of restricted-use pesticides in Colorado require a special permit and certification. Retailers of restricted use pesticides must maintain records for two years after the date of sale. These records must include name and address of applicator, certification number, quantity of the pesticide, and date of the restriction. Like EPA, Colorado divides commercial pesticide applicators into 11 categories used for registration. Colorado requires commercial applicators to take an examination. Questions are geared toward the category of applicator such as "agricultural insect control" and cover

- Pest management
- Label requirements
- Safe usage
- Worker safety
- Disposal
- Storage requirements

The state also has requirements for equipment identification, containment, and storage. Many pesticides are also hazardous or universal wastes covered under RCRA. Local authorities may also regulate pesticide use such as the Santa Cruz County Agricultural Commission in Central California, U.S.A. Penalties for violations of the act include civil penalties of $5,000 each violation and criminal penalties of $25,000 each violation and one year in jail. On the Mississippi's Gulf Coast, improper application of a cotton insecticide inside hundreds of homes caused symptoms of pesticide exposure in residents and resulted in the families moving out of their homes. The EPA estimated the cost of cleanup, including decontamination of the homes, to be $50 million. The two exterminators were unlicensed and improperly applied the pesticides. One of the exterminators was convicted by a federal jury of 48 environmental crimes and faces up to 48 years in prison and $4.8 million in fines.

ADDITIONAL INFORMATION

The FIFRA Docket (Pesticides Docket) consists of

- The Federal Register Docket
- The Special Review and Registration Standard Docket
- Special Dockets

The Federal Register Docket houses background documents and public comments on proposed actions announced by the Office of Pesticide Programs (OPP) in the Federal Register. The Special Review and Registration Standard Docket contains position documents, registration standards, science chapters, public comments, references, letters, minutes of meetings between EPA and outside parties, and other pesticide documents. Special dockets are established when the OPP Program Office wants to place documents on public display. Docket information may be reviewed in person or requested in writing. The physical location of the Pesticide Dockets is in Arlington, VA. Also, EPA as well as other organizations maintain databases and computer supported information networks. For example, the Pesticide Information Network (PIN), previously called the Pesticide Monitoring Inventory, is sponsored by the Office of Prevention, Pesticides, and Toxic Substances, Office of Pesticide Programs.

PIN is designed to collect and disseminate pesticide information and includes

- Contacts directory
- Pesticide applicator training bibliography
- Pesticide monitoring inventory
- Pesticide environmental fate and effects data summaries
- Current regulatory information, including cancelled or suspended pesticides
- Restricted use product information

The Section Seven Tracking System (SSTS) evolved from the FIFRA and TSCA Enforcement System database. SSTS tracks the registration of all pesticide-producing establishments and tracks the types and amounts of pesticides, active ingredients, and devices that are produced, sold, or distributed yearly. This information is used by EPA for enforcement-targeting purposes.

The National Pesticide Information Retrieval System (NPIRS) is a subscription database of the Center for Environmental and Regulatory Systems and emphasizes EPA product registration information with a focus on agriculture. The National Pesticide Telecommunications Network (NPTN) is managed by Texas Tech University and is a free service providing information about pesticide products, such as emergency treatment, toxicology, health and environmental effects and safety information.

REFERENCES

A Guide to Federal Environmental Requirements for Small Governments, Office of the Administrator (H1501), EPA 270-K-93-001, September 1993.

American National Standards Institute (ANSI), "Liquid Petroleum Transportation Piping Systems," Standard B31.4.

American National Standards Institute (ANSI), "Petroleum Refinery Piping," Standard B31.3.

American Petroleum Institute, "Cathodic Protection of Underground Petroleum Storage Tanks and Piping Systems," Publication 1632.

American Petroleum Institute, "Installation of Underground Petroleum Storage Systems," Publication 1615.

American Petroleum Institute, "Recommended Practice for Bulk Liquid Stock Control at Retail Outlets," Publication 1621.

American Society of Testing and Materials (ASTM), "Standard Specification for Glass-Fiber-Reinforced Polyester Underground Petroleum Storage Tanks," Standard D4021-86.

Association for Composite Tanks, "Specification for the Fabrication of FRP Clad Underground Storage Tanks," ACT-100.

Code of Federal Regulations, parts 355, 370, and 372.

Chemicals in Your Community, A Guide to the Emergency Planning and Community Right-to-Know Act, USEPA, September 1988.

Colorado Environmental Law Handbook, Holme, Roberts, & Owen, Governments Institute.

Colorado Revised Statutes (CRS), titles 8, 24, 25, and 6 Colorado Code of Regulations (CCR).

Community Awareness and Right-to-Know, Emergency Management Institute, IG 305.6, September 1990.

Community Right-to-Know and Small Business, Understanding Sections 311 and 312 of the Emergency Planning and Community Right-to-Know Act of 1986, USEPA, Office of Solid Waste and Emergency Response, September 1988.

Environmental Planning for Small Communities, A Guide for Local Decision Makers, Regional Operations & State/Local Relations, EPA/625/R-94/009, September 1994.

Exercising Emergency Plans Under Title III, Emergency Management Institute, IG 305.4, February 1991.

Hazardous Materials, An Introduction for Public Officials, FEMA, USEPA, USDOT, SM 300, September 1990.

Hazardous Materials Information Management, Emergency Management Institute, Federal Emergency Management Agency, National Emergency Training Center, IG 305.2, September 1990.

Hazardous Materials Planning Guide, National Response Team, March 1987.

HELP! EPA Resources for Small Governments, Regional Operations & State/Local Relations (H1501), 21V-1001, September 1991.

Letter from Cheryl A. Crisler, Chief, Prevention Section USEPA Region VIII to LEPC and SERC members regarding the identification of industries that may have to comply with SARA, Letter includes an enclosure identifying by SIC code and industry the type of EHS used in that industry.

Musts for USTs, A Summary of the Regulations for Underground Storage Tank Systems, USEPA, Office of Underground Storage Tanks, July 1990.

National Association of Corrosion Engineers, "Control of External Corrosion on Metallic Buried, Partially Buried, or Submerged Liquid Storage Systems, RP-02-85.

National Fire Protection Association (NFPA), "Flammable and Combustible Liquids Code," Standard 30 and NFPA 385 (transfer procedures).

National Leak Prevention Association, "Spill Prevention, Minimum 10 Year Life Extension of Existing Steel Underground Tanks by Lining Without the Addition of Cathodic Protection," Standard 631.

Petroleum Equipment Institute, "Recommended Practices for Installation of Underground Liquid Storage Systems," publication RP100.

Steel Tank Institute, "Specification for STI-P3 System of External Corrosion Protection of Underground Steel Storage Tanks."

Superfund Amendments and Reauthorization Act of 1986, Public Law 99-499, October 17, 1986.

Title III Fact Sheet, Emergency Planning and Community Right-to-Know, USEPA.

Underwriter's Laboratory, "Standard for Glass-Fiber-Reinforced Plastic Underground Storage Tanks for Petroleum Products," Standard 1316.

Underwriter's Laboratory, "Corrosion Protection Systems for Underground Storage Tanks." Standard 1746.

Underwriter's Laboratory of Canada, "Standard for Reinforced Plastic Underground Storage Tanks for Petroleum Products," Standard CAN4-S615-M83.

When All Else Fails! Enforcement of the Emergency Planning and Community Right-to-Know Act, A Self Help Manual for Local Emergency Planning Committees, USEPA, Solid Waste and Emergency Response (OS-120), September 1989.

40 CFR 280.

42 USC 6912, 6991, 6991a, 6991b, 6991c, 6991d, 6991e, 6991f, and 6991h.

53 FR 37194, September 23, 1988.

Part V

Sources of Additional Information

12 Directory of Organizations and Agencies

CONTENTS

ENVIRONMENTAL AGENCIES OF THE WORLD

ALBANIA

Ministry of Health and Environmental Protection
Committee of Environment Preservation and Protection
Bulevardi Zhan D'Arc, Tirana
Telephone: (355) 423-2937, (355) 422-5486
FAX: (355) 423-2937

ANGOLA

Secretariat of the State of the Environment
Secretaria do Estado do Ambiente
Luanda
Telephone: (244-2) 332-611
(244-2) 334-709

ANTIGUA AND BARBUDA

Ministry of Tourism, Culture and the Environment
Queen Elizabeth Highway, St. John's, Antigua
Telephone: (268) 462-0787
FAX: (268) 462-2836

ARGENTINA

Secretaria de Recursos Naturales y Ambiente Humano de la Nacion
San Martin 459, Piso 2,
1004 Buenos Aires
Telephone: (541) 348-8200
FAX: (541) 394-3856

ARMENIA

Ministry of Environment
Moskovian St. 35, Erevan 375002
Telephone: (374-2) 533-629, (374-2) 530-741
FAX: (374-2) 538-613, (374-2) 534-902

AUSTRALIA

Department of Environment, Sport and Territories
GPO Box 787, Canberra, ACT 2601
Telephone: (616) 274-1111
FAX: (616) 274-1123

AUSTRIA

Umweltbundesamt
(Federal Environment Agency)
Spittelauer Lande 5, A-1090 Vienna, Austria
Telephone: 43-1-31304-0
FAX: 43-1-31304-5400
http://www.ubavie.gv.at/

AZERBAIJAN

Ministry of Public Health
4, Todorskovo Street, Baku 370014
9-9412-93-29-77
State Ecology Committee of Azerbaijan
31 Istiglaliyat Street
Baku 370001
Telephone: 9-9412-92-41-73
FAX: 9-9412-92-59-07

BAHAMAS

Ministry of Health and the Environment
P.O. Box N-3730
Nassau
Telephone: (809) 322-7425, (809) 322-3607
FAX: (809) 322-7788

BANGLADESH

Ministry of Environment and Forestry
Bangladesh Secretariat
Building 6, Room No. 1309
Dhaka
Telephone: (8802) 860481
FAX: (8802) 869210

BARBADOS

Ministry of Health and the Environment
Sir Frank Walcott Building, Culloden Road
St. Michael
Telephone: (809) 431-7680
FAX: (809) 437-8859
e-mail: envdivn@mail.caribsurf.com

BELARUS

Ministry of Natural Resources and Environmental Protection
ul. Kazintsa 4
Minsk 220855
FAX: (172) 260-084

BELGIUM

Brussels Region
Administration of Environment of the Region of Brussels
Guledelle 100, B-1200 Brussels
Telephone: (32 2) 775-7575
FAX: (32 2) 775-7611

Flemish Region
Administratie Milieu, Natuur en Landinrichting
(Department of the Environment and Agriculture)
14-18 rue Belliard, B-1040 Brussels
Telephone: (32 2) 507-3901
FAX: (32 2) 507-3005

Walloon Region
Administration de l'Environnement et de l'Agriculture
(Department of Environment and Agriculture)
15 avenue Prince de Liege, B-5100 Jambes
Telephone: (32 81) 32 12 11
FAX: (32 81) 32 59 84

BELIZE

Ministry of Tourism and Environment
Belmopan
Telephone: (501 8) 22816
FAX: (501 8) 22862

BENIN

Ministére de l'Environnement de l'Habitat et de l'Urbanisme (MEHU)
(Ministry of the Environment, Housing and Town Planning)
P.O. Box 01-3621
Cotonou
Telephone: (229) 314-661, (229) 314-839

BHUTAN

National Environment Commission
Thimphu
Telephone: (975) 222-056
FAX: (975) 223-385

BOLIVIA

Ministerio de Desarrollo Sostenible y Medio Ambiente
Casilla 12814, La Paz
Telephone: (591 2) 354522
FAX: (591 2) 392892

BOSNIA AND HERZEGOVINA

Ministry of Physical Planning and Environment
Sarajevo, M.Tita 7a
Telephone: (387-71) 473-124
FAX: (387-71) 663-548

BOTSWANA

National Conservation Strategy
Private Mail Bag 0068
Gaborone
Telephone: 302-050 (267) 302 051

BRAZIL

Instituto Brasileiro do Meio Ambiente e dos Recursos Naturais Renovaveis (IBAMA)
(Brazilian Institute of Environment and Renewable Natural Resources)
SAIN, Av. L4 Norte Quadra 604, Ed. Sede
Brasilia, DF 70818-000
Telephone: (55 61) 226-8221
FAX: (55 61) 322-1058
http://www.ibama.gov.br/

Companhia de Tecnologia de Saneamento Ambiental (CETESB)
(Environment Sanitation Agency)
Av. Prof. Frederico Hermann Jr., 345, CEP - 05489-900
Sao Paulo
Telephone: (011) 3030-6000
FAX: (011) 3030-6402
http://www.cetesb.br/

Fundacao Instituto Brasileiro de Geografia e Estatistica (IBGE)
(Brazilian Institute of Geography and Statistics)
Avenue Franklin Roosevelt No. 166-10th Andar, 21021-120
Rio de Janeiro
Telephone: (55-21) 220-6671

FAX: (55-21) 220-5943
http://www.IBGE.gov.br/

FEEMA, State of Rio de Janeiro
Fundacao Estadual de Engenharia do Meio Ambiente
Rua Fonseca Teles 121/15, Sao Cristovao
Rio de Janeiro 20940-200
Telephone: (55-21) 589-3724

BULGARIA

Ministry of Environment
ul. W.Gladstone 67
Sofia 1000
Telephone: (359 2) 876-156, (359 2) 874-777
FAX: (359 2) 521-634, (359 2) 521-634

BURKINA FASO

Ministre d'Etat, Ministre de l'Environnement et de l'Eau
(Ministry of Environment and Water)
03 BP 7044, Ouagadougou 03
Telephone: (226) 324-074, 307-098
FAX: (226) 314-605, (226) 316-491

BURUNDI

National Institute for Environment and Nature Conservation
Gitega
Telephone: (257) 402-071, (257) 402-073

CAMBODIA

The Ministry of Environment
Sihanouk Boulevard
Phnom Penh
Telephone: (855 23) 426 814
FAX: (855 23) 427 844

CAMEROON

Ministére de l'Environnement et des Fôrets
(Ministry of Environment and Forests)
Yaoundé
Telephone: (237) 229-483
FAX: (237) 229-489, 225-091

CANADA

Environment Canada
Inquiry Centre
351 St. Joseph Boulevard, Hull
Quebec K1A 0H3

Telephone: 819-997-2800 or 1-800-668-6767
FAX: 819-953-2225, 819-953-0966
e-mail: enviroinfo@ec.gc.ca
http://www.ec.gc.ca/

Minister of Environmental Protection
Communications Division
Telephone: (403) 427-8636
FAX: (403) 422-6339.
Information Centre (403) 422-2079
FAX: (403) 427-4407.
Public Information: (403) 944-0313

Canadian Departments of Environment and Industries'
National Environmental Training Initiative
13th Floor
351 St. Joseph Blvd.
Hull, Quebec K1A 0H3
Telephone: (819) 994-7977

CANUTEC – Transport Dangerous Goods
Transport Canada
Ottawa, Ontario
Canada, KiAON5
Telephone: (613) 992-4624 (information)

CAPE VERDE

Ministry of Agriculture, Food and Environment
Praia
Telephone: (238) 614-054

CARIBBEAN CENTRE FOR DEVELOPMENT ADMINISTRATION (CARICAD)

Ground Floor, I.C.B. Building, Roebuck Street
Saint Michael, Barbados
Telephone: (809) 427 8535/6
FAX: (809) 436 1709
e-mail: caricad@caribsurf.com

CENTRAL AFRICAN REPUBLIC

Ministry of Environment and Tourism
Bangui
Telephone: (236) 610-216

CHAD

Ministre de Tourisme et l'Environnement
(Ministry of Tourism and Environment)
N'Djemana
Telephone: (235) 523-919, (235) 522-247

CHILE

Comision Nacional del Medio Ambiente (CONAMA)
(National Environment Commission)
Obispo Donoso 6, Providencia, Santiago
Telephone: (56 2) 244-1262
http://www.conama.cl/

CHINA

National Environmental Protection Agency (NEPA)
115 Xizhimennei, Nanxiaojie
Beijing 100035
Telephone: (86) 106 - 615 17 89
FAX: (86) 106 - 615 17 68

COLOMBIA

Ministerio del Medio Ambiente
(Ministry of Environment)
Edificio Avianca, Calle 16 No. 6 - 66
Santafé de Bogotá
Telephone: (57-1) 336-1166
FAX: (57-1) 336-3984

Instituto de Hidrologia, Meteorologia y Estudios Ambientales, Colombian
Santafé de Bogotá D.C.
Columbia
www@ideam.gov.co
http://www.ideam.gov.co/

COMOROS

Ministry of Agriculture
Moroni

CONGO

Ministry of Tourism and Environment
Brazzaville

COSTA RICA

Ministerio de Recursos Naturales, Energia y Minas (MIRENEM)
(Ministry of Natural Resources, Energy and Mines)
Apartado 10104, 1000 San Jose
Telephone: (506) 233-4533
FAX: (506) 257-0697

COTE D'IVOIRE

Ministere du Logement, du Cadre de Vie et de l'Environnement
(Ministry of Housing, Quality of Life and the Environment)

B.P. V 153, Tour D - 19 eme, Etage
Telephone: (225) 219-406, (225) 214-662
FAX: (225) 214-581

CROATIA

State Directorate for Environment
Ul. grada Vukovara 78/III, Zagreb
Telephone: (385-1) 613-3444
FAX: (385-1) 537-203
e-mail: duzo@ring.net
http://www.vlada.hr/tijela/tijelae.html or http://www.vlada.hr/

CUBA

Ministerio de Ciencia, Tecnologia y Medio Ambiente
(Ministry of Science, Technology and Environment)
Ciudad de La Habana
Cuba
Telephone: (53-7) 223-594, 628-631
FAX: (53-7) 338-654

CYPRUS

Ministry of Agriculture, Natural Resources and Environment
Loukis Akritas Ave., 1411 Nicosia
Telephone: (357) 230-2247
FAX: (357) 244-5156

CZECH REPUBLIC

Czech Ministry of Environment
Vrsovicka 65, CS-100 10 Praha 10
Telephone: 420-2-67121111
FAX: 420-2-67310308
e-mail: info@env.cz
http://www.env.cz/

DEMOCRATIC REPUBLIC OF THE CONGO

Ministry of Environment and Tourism
B.P. 12348, Kinshasa-Gombe
Telephone: (243-12) 343-90

DENMARK

Danish Ministry of Environment and Energy
Højbro Plads 4
1200 København K
Telephone: +45 33 92 76 00
FAX: +45 33 32 22 27
e-mail: mem@mem.dk
http://www.mem.dk/

Djibouti

Ministry of Environment, Tourism and Art
Djibouti, Djibouti
Telephone: (253) 351-020

Dominica

Ministry of Agriculture and the Environment
Government Headquarters, Roseau, Dominica
Telephone: (767) 448-2401 ext: 3424
FAX: (767) 448-5200

Dominican Republic

Comision Nacional de Saneamiento Ecologico
Palacio de Nacional, Cesar Nicolas Pension
Santo Domingo
Telephone: (809) 562-3500

Ecuador

Consejo Nacional de Desarrollo (CONADE), Programa de Recursos Naturales y Medio
  Ambiente
(National Council of Development, Program of Natural Resources and Environment)
Arenas y Av. 10 de Agosot, Quito
Telephone: (593 2) 540-455
FAX: (593 2) 565-809

El Salvador

Ministerio de Medo Ambiente y Recursos Naturales (MARN)
(Ministry of Environment and Natural Resources)
Edificio IPSFA, Avenue Roosevelt
San Salvador
Telephone: (503) 260-8900
FAX: (503) 260-5637

Equatorial Guinea

Ministry of Health and Environmental Protection
Malabo, Bioko Norte
Telephone: (240) 92-686, (240) 92-501

Eritrea

Ministry of Land, Water and Environment
P.O. Box 976, Asmara
Telephone: (291-1) 118-021
FAX: (291-1) 123-285

ESTONIA

Ministry of Environment
Nature Protection Department/ Keskkonnaministeerium
Toompuiestee 24, Tallinn 0001
Telephone: (372 2) 443-210, (372 2) 451-864
FAX: (372 2) 626-2801
e-mail: min@ekm.envir.ee
http://www.envir.ee/

ETHIOPIA

Ministry of Natural Resource Development and Environmental Protection
P.O. Box 1034, Addis Ababa
Telephone: (251-1) 510-455
FAX: (251-1) 513-042

EUROPEAN COMMISSION, DIRECTORATE GENERAL XI: ENVIRONMENT, NUCLEAR SAFETY AND CIVIL PROTECTION

200, Rue de la loi
Brussels B-1049 Belgium
Telephone: +32 2 299.11.11
FAX: +32 2266 995

EUROPEAN ENVIRONMENT AGENCY

Kongens Nytorv 6, DK-1050
Copenhagen K, Denmark
Telephone: +45 33 36 71 00
FAX: +45 33 36 71 99.
http://www.eea.eu.int/

FIJI

Ministry for Local Government and Environment
Fiji Football Association Building, Gladstone Road
Suva
Telephone: (679) 211-310
FAX: (679) 303-515

Department of Environment
P.O. Box 2131, Government Buildings
Suva
Telephone: (679) 312-789
FAX: (679) 311-699

FINLAND

Ministry of the Environment/ Ympäristöministeriön
Ratakatu 3, FIN-00121 Helsinki
Telephone: (358 0) 19911

FAX: (358 0) 1991499
http://www.vyh.fi/

FRANCE

Ministère de l'Aménagement du Territoire et de l'Environnement
(Ministry of the Environment)
41 Avenue Georges Mandel, F-75016 Paris
Telephone: (33 1) 46 47 31 32
FAX: (33 1) 46 47 38 95
http://www.environnement.gouv.fr/

Agence de l'Environnement et de la Maîtrise de l'Énergie (ADEME)
(The French Environment and Energy Control Agency)
27, rue Louis Vicat
F-75737 PARIS Cedex 15
France
Telephone: (+33) (0)1 47 65 20 00
FAX: (+33) (0)1 46 45 52 36
http://www.ademe.fr/

Institut français de l'environnement (IFEN)
61, boulevard Alexandre Martin
F - 45058 ORLÉANS CEDEX 1
Telephone: 33 (0) 2 38 79 78 78
FAX: 33 (0) 2 38 79 78 70
e-mail: ifen@ifen.fr
http://www.ifen.fr/

Institut National de l'Environnement Industriel et des Risques
(The National Institute of Industrial Environment and Risks)
Parc Technologique ALATA B.P. N°2
60550 Verneuil-en-Halatte - France
Telephone: 03 44 55 66 77
FAX: 03 44 55 66 99
http://www.ineris.fr/

FRENCH GUYANA

Prefecture, rue Fiedmond, 97307 Cayenne Cedex
Telephone: (594) 300520
FAX: (594) 300277

GABON

Ministere du la Planification, Environnement, Tourisme
(Ministry of Planning, Environment and Tourism)
B.P. 2119, Libreville
Telephone: (241) 763-755

GAMBIA

National Environment Agency
5 Fitzgerald Street, Banjul
Telephone: (220) 224-867 224-868, 224-869;
FAX: (220) 229-701

Ministry of Agriculture and Natural Resources
The Quadrangle, Banjul
Telephone: (220) 228-291, (220) 228-292

GEORGIA

Ministry of Environmental Protection
Kostava str. 68a, Tbilisi 380015
Telephone: (995-32) 230-664
FAX: (995-32) 955-006
http://www.parliament.ge/SOEGEO/english/institut/moe/moe.htm

GERMANY

German Federal Environment Ministry
Umweltbundesamt, Bismarckplatz 1
14193 Berlin, Germany
Telephone: 49 30 89 03-0
FAX: 49 30 89 03-2285
http://www.bmu.de/

GHANA

Ministry of Environment, Science and Technology
P.O. Box M 232
Ministries Post Office
Accra
Telephone: 233-21-662013
FAX: 233-21-666828

GREECE

Ministry of Environment, Planning and Public Works
147 Patision Street, GR-112 51
Athens
Telephone: (30 1) 865-033
FAX: (30 1) 865-0476

GRENADA

Ministry of Health and the Environment
Carenage, St. Georges
Telephone: (473) 440-2649
FAX: (473) 440-4127

GUATEMALA

Comision Nacional del Medio Ambiente
(National Environment Commission)
Presidencia de la Republica, 5a. Avenida 8-07, Zone 10
Guatemala
Telephone: (502 2) 312723
FAX: (502 2) 341708
http://www.concyt.gob.gt/concyt/sistema/comision/medamb/medio.htm

GUINEA

Ministry of Environment
Conakry
Telephone: (224) 414-012, (224) 412-513
FAX: (224) 414-001, (224) 414-913

GUINEA BISSAU

Ministerio do Desenvolvimentio Rural, Recursos Naturais e do Ambiente
(Ministry of Rural Development, Natural Resources and Environment)
Bissau
Telephone: (245) 201-038, (245) 211-756

GUYANA

Office of the Advisor on Science, Technology, and Environment
New Marice Street, Georgetown
Telephone: (592 2) 66543, (592 2) 73849

HAITI

Ministry of the Environment
Haut Turgeau
Port-au-Prince
Telephone: (509) 45-7585

Ministry of Agriculture, Natural Resources and Rural Development
Rte Nationale 1
Damien
Telephone: (509) 22-3595

HONDURAS

Ministerio de Recursos Naturales
(Ministry of Natural Resources)
Tegucigalpa
Telephone: (504) 32-8817

HONG KONG

Environmental Protection Department
24-28th Floors, Southern Centre

130 Hennesy Road
Wanchai
Telephone: (852) 2835-1018
FAX: (852) 2838-2155

HUNGARY

Ministry for Environment and Regional Policy
H-1011 Budapest Fõ utca 44-50
Telephone: (36-1) 457-3369
FAX: (36-1) 201-4361
e-mail: gridbp@kik.ktm.hu
http://www.gridbp.meh.hu

ICELAND

Ministry for the Environment
Vonarstraeti 4, IS-150
Reykjavik
Telephone: (354) 560-9600
FAX: (354) 562-4566

INDIA

Ministry of Environment and Forests
Paryavaran Bhavan, CGO Complex
Lodhi Road, New Delhi, 110 003
India
Telephone: 91-11-4361896
e-mail: secy@envfor.delhi.nic.in
http://www.nic.in/envfor/

Central Pollution Control Board
Parivesh Bhawan, CBD-cum-Office Complex,
East Arjun Nagar, New Delhi, 110 032
India
e-mail: cpcb@envfor.delhi.nic.in
http://www.nic.in/envfor/cpcb/

INDONESIA

Environmental Control Agency (Bapedal)
11/F Arhaloka Building, Jl. Jendral Sudirman No. 2
Jakarta Pusat
Telephone: (62) 21 - 58 39 16
FAX: (62) 21 - 58 37 93

IRELAND

Department of Environment
Custom House
Dublin 1

Telephone: (353 1) 679-3377
FAX: (353 1) 742-710

Department of the Environment for Northern Ireland
Clarence Court, 10-18 Adelaide Street
Belfast BT2 8GB
Telephone: (01232) 540540
FAX: (01232) 540782
e-mail: csrb.doe@nics.gov.uk
http://www.doeni.gov.uk/doe.htm

ITALY

Ministero dell'Ambiente
(Ministry of the Environment)
Via Volturno 58,
I-00185 Rome
Telephone: (39 6) 445-1317
FAX: (39 6) 446-9112

National Agency for New Technologies, Energy and Environment (ENEA)
http://www.enea.it/ or http://wwwamb.casaccia.enea.it/

JAMAICA

Ministry of Public Service and the Environment
Ministry of Environment and Mining
Citibank Building, 63-67 Knutsford Blvd., Kingston 5
Telephone: (876) 926-3235
FAX: (876) 929-6616

Jamaica Conservation and Development Trust
95 Dumbarton Avenue, Kingston 10
P.O. Box 1225, Kingston 8
Jamaica
Telephone: 1 (809) 960 2848/9
FAX: 1 (809) 960 2850
e-mail: jamcondt@uwimona.edu.jm

National Environmental Societies Trust
Natural Resources Conservation Authority
National Environmental Education Action Plan
531/2 Molynes Road, Kingston 10
Jamaica, West Indies
Telephone: (876)923-5125
FAX: (876)923-5070

JAPAN

Environment Agency
1-2-2 Kasumigaseki, Chiyoda-ku, 100
Tokyo

Telephone: (81) 3 - 3580 13 75
FAX: (81) 3 - 3504 16 34

KAZAKSTAN

Ministry of Ecology and Bioresources
Panfilov str. 106, Almaty 480091
Telephone: (73-272) 636-862
FAX: (73-272) 636-973
e-mail: neapkz@online.ru

KENYA

Ministry of Environment and Natural Resources
Kencom House, Moi Avenue
P.O. Box 30126
Nairobi
Telephone: (254-2) 229-261
FAX: (254-2) 338-272
http://www.kenyaweb.nolo
www.com/kenyagov/environm/environm.html

KIRIBATI

Ministry of Environment and Social Development
P.O. Box 234, Bikenibeu
Tarawa
Telephone: (686) 280-00
FAX: (686) 282-02

Ministry of Natural Resource Development
P.O. Box 64, Bairiki
Tarawa
Telephone: (686) 210-99
FAX: (686) 211-20

KYRGYZSTAN

Ministry of Environmental Protection
Isanov Str. 131, Bishkek 720 033
Telephone: 9963312 - 26 42 44
FAX: 9963312 - 26 23 21

LAOS

Department of Forestry and Environment
Ministry of Agriculture and Forestry
Lane Xang Avenue, Vientiane
Telephone: (856-21) 412-340
FAX: (856-21) 412-341

LATVIA

Ministry of Environmental Protection and Regional Development
Peldu Street 25, Riga LV-1494
Telephone: (371) 702-6503, (371) 702-6560
FAX: (371) 782-0402, (371) 782-0442
e-mail: ralfs@varam.gov.lv
http://www.varam.gov.lv/

LESOTHO

Ministry of Natural Resources
P.O. Box 772, Maseru, 100
Telephone: (266) 323-163
FAX: (266) 310-520

LIBERIA

Ministry of Lands, Mines and Energy
Capitol Hill, P.O. Box 10-9024
1000 Monrovia 10
Telephone: (231) 226-281, (231) 221-580

LIECHTENSTEIN

Ministry of Agriculture, Forestry, and Environment
FL-9490 Vaduz
Telephone/FAX: (41 75) 66411

LITHUANIA

Ministry of Environmental Protection
A. Juozapaviciaus 9, 2600 Vilnius
Telephone: (370 2) 725-868
FAX: (370 2) 728-020/722-029
e-mail: info@nt.gamta.lt
http://neris.mii.lt/aa/index.html
http://www.gamta.lt/

LUXEMBOURG

Ministere de l'Environnement
(Ministry of Environment)
18 Montee de la Petrusse
L-2918 Luxembourg-Ville
Telephone: (352) 478 6824
FAX: (352) 400 410
http://www.mev.etat.lu/home.html

MACEDONIA

Ministry of Urban Planning, Construction and Environment
14, Dame Gruev Street, Skopje, 91 000

Telephone: (389 91) 116-141
FAX: (389 91) 117-163
e-mail: gradba@unet.com.mk
http://www.republic-of-macedonia.unet.com.mk/

MADAGASCAR

L'Office National pour l'Environnement
(National Environment Office)
B.P. 822, Antananarivo 101
Telephone: (261-2) 641-06, 641-07, (261-2) 259-99
FAX: (261-2) 306-93
one@dts.mg
http://www.madonline.com/one/

MALAWI

Ministry of Forestry, Fisheries and Environmental Affairs
P.O. Box 30048, Lilongwe 3
Telephone: (265) 781-000, (265) 781-111
FAX: (265) 781-487, (265) 784-26833

MALAYSIA

Ministry of Science, Technology and the Environment
Department of Environment
14th Floor, Wisma Sime Darby, Jalan Raja Laut
50662 Kuala Lumpur
Telephone: (60-3) 293-8955
FAX: (60-3) 293-6006
http://pop.jaring.my/jas/start.html

MALDIVES

Ministry of Planning Human Resources and Environment
Ghazee Building, Male' 20-50
Telephone: (960) 323-852, (960) 324-861
FAX: (960) 327-351

MALI

Ministry of Environment
Bamako

MALTA

Environment Protection Department
Education Buildings
Floriana CMR02
Telephone: +356 232022
FAX: +356 241378
http://www.environment.gov.mt

MARSHALL ISLANDS

Ministry of Health and Environment
Majuro, MH 96960
Telephone: (692) 625-3632
FAX: (692) 625-3432

MAURITANIA

Ministry of Rural Development and Environment
Nouakchott
Telephone: (222-2) 51-500

MAURITIUS

Ministre de l'Administration Regional et l'Environnement
(Ministry of Local Government and Environment)
9th Floor, Kenlee Tower, Corner of Georges and Barracks Streets
Port Louis
Telephone: (230) 208-1944
FAX: (230) 212-9407
http://ncb.intnet.mu/environ.htm

MEXICO

Instituto Nacional de Ecologia (INE)
(National Institute of Ecology)
Rio Elba 20, 16, Col Cuauhtemoc
06500 Mexico, DF
Telephone: (5) 553-9968
FAX: (5) 286-6872
http://www.ine.gob.mx/

Secretaria de Medio Ambiente, Recursos Naturales y Pesca (SEMARNAP)
(Secretariat of the Environment, Natural Resources and Fish)
http://www.semarnap.gob.mx/

Procuraduría Federal de Protección al Ambiente (PROFEPA)
Asesor del C. Procurador, Periférico Sur 5000, 5° Piso Col. Insurgentes - Cuicuilco México,
   D.F. 04530
Telephone: (528 54 02) 528 53 73
FAX: (528 54 02) 528 53 91
http://www.semarnap.gob.mx/profepa

Secretaria de Comunicaciones y Transportes
Direccion General de Autotransporte Federal
Direccion de Transporte de Materiales y Residuos Peligrosos
Calzada de las Bombas No. 411, 7o. Piso
Couoacan 04800, D. F.

MICRONESIA

Department of Health Services
Environmental Protection Agency
P.O. Box PS70
Palikir, Pohnpei, FM 96941
Telephone: (691) 320-2872
FAX: (691) 320-5263

MOLDOVIA

State Department of Environmental Protection and Natural Resources
Chisinau 73, Stefan cel Mare Blvd.
Telephone: (373-2) 226-161
FAX: (373-2) 233-806

MONGOLIA

Ministry for Nature and Environment
Khudaldaany gudamji 5, Ulaanbaatar 11
Telephone: (976 1) 329-619
FAX: (976 1) 321-401

MOZAMBIQUE

Ministry of Environmental Action Coordination
Av. Acordos de Lusaka 2115, Caixa Postal 2020
Maputo
Telephone: (258-1) 466-245
FAX: (258-1) 465-849

MYNAMAR (BURMA)

National Committee for Environmental Affairs
No. 37 Thantaman Road
Yangon
Telephone: (95 1) 21689
FAX: (95 1) 21546

NAMIBIA

Ministry of Environment and Tourism
4th Floor, Old Swabou Building, Post Street Mall,
Private Bag 13346
Windhoek
Telephone: (264-61) 284-2111
FAX: (264-61) 229-936

NEPAL

Ministry of Environment and Population
Singh Durbar

Kathmandu
Telephone: (977 1) 228-555

THE NETHERLANDS

Ministerie van Volkshuisvesting, Ruimtelijke Ordening en Milieubeheer
(Ministry of Housing, Spatial Planning and the Environment)
Rijnstraat 8, 2515 XP Den Haag
Telephone: 31(0)70 339 39 39
http://www.minvrom.nl/

NEW ZEALAND

Ministry for the Environment
P.O. Box 10-362, Wellington
Telephone: (64 4) 917 7400
FAX: (64 4) 917 7523
http://www.mfe.govt.nz/

NICARAGUA

Ministerio del Ambiente y Recursos Naturales
(Ministry of Environment and Natural Resources)
Apartado 5123
Managua
Telephone: (505 2) 31110
FAX: (505 2) 31274

NIGER

Ministre de Hydroulique et l'Environnement
(Ministry of Hydraulics and Environment)
Niamey
Telephone: (227) 734-722
FAX: (227) 734-646

NIGERIA

Federal Environmental Protection Agency (FEPA)
Independence Way South, Private Mail Bag 265
Garki, Abuja
Telephone: (234-9) 523-3379

NORTH KOREA

State Environmental Committee
Pyongyang

NORWAY

Miljøverndepartementet
(Ministry of the Environment)
Myntgata 2, Postboks 8013 Dep, 0030

Oslo
Telephone: 22 24 90 90
FAX: 22 24 95 62
http://www.odin.dep.no/md/

Norwegian Pollution Control Authority
Box 8100 Dep, N-0032 Oslo; Strømsveien 96
Telephone: (+47) 22 57 34 00
FAX: (+47) 22 67 67 06,
e-mail: postmottak.sft@sftospost.md.dep.telemax.no
http://www.sft.no

PAKISTAN

Ministry of Environment
5/F, Shaheed-i-Millat Secretariat
Islamabad
Telephone: (92) 51 - 920 22 11
FAX: (92) 51 - 920 27 25

PANAMA

Instituto de Recursos Naturales Renovables (INRENARE)
(Institute of Renewable Natural Resources)
Apartado 2016, Paraiso
Ancon
Telephone: (507) 32-6601
FAX: (507) 32-6612
http://www.ccad.org.gt/miempana.htm

PAPUA NEW GUINEA

Department of Environment and Conservation
Central Government Buildings, P.O. Box 6601
Boroko
Telephone: (675) 271-788
FAX: (675) 271-900

PARAGUAY

Ministerio de Agricultura y Ganaderia, Subsecretaria de Recursos Naturales y Medio Ambiente
(Ministry of Agriculture and Livestock, Subsecretariat of Natural Resources and Environment)
Calle Presidente France 473, Asuncion
Telephone: (595 21) 443-971
FAX: (595 21) 449-951

PERU

Consejo Nacional Del Ambiente (CONAM)
Av. San Borja Norte 226
San Borja
Lima, Peru

Telephone: (51 1) 225.5370
FAX: (51 1) 225.5369
e-mail: Postmaster@Conam.gob.pe

PHILIPPINES

Department of Environment and Natural Resources
Visayas Avenue, Diliman
1100 Quezon City
Philippines
Telephone: (63-2) 929-6626 to 29
http://www.denr.gov.ph/
webmaster@denr.gov.ph

POLAND

Ministerstwo Ochrony Srodowiska
(Ministry of Environmental Protection, Natural Resources and Forestry)
Zasobow Naturalnych i Lesnictwa (MOSZNIL)
ul. Wawelska 52/54
Warsaw 00-922
Telephone: (48 22) 251-111, (48 22) 250-001
FAX: (48 22) 253-355, (48 22) 254-693
http://www.mos.gov.pl/

PORTUGAL

Ministry of Environment
Rua do Século, no. 63, 1200 Lisbon
Telephone: (351-1) 321-1360
FAX: (351-1) 343-2777
e-mail: ipamb@mail.telepac.pt

Direcção Geral do Ambiente
R. Murgueira, Alfragide
Telephone: (351-1) 472-8200
FAX: (351-1) 471-9074
e-mail: web.coord@dga.min-amb.pt
http://www.dga.min-amb.pt

ROMANIA

Ministry of Waters, Forests and Environment Protection
Bd. Líbertatíí 12, Sector 5
Bucharest
Telephone: (401) 631-6146, (401) 410-0243
FAX: (401) 312-4227

RUSSIAN FEDERATION

Ministry of Environmental Protection and Natural Resources
Bol.Gruzinskaya ul. 4/6, Moscow 123812

Telephone: (7-095) 254-6733, (7-095) 254-7683
FAX: (7-095) 254-8283, (7-095) 230-2792
http://www.fcgs.rssi.ru/eng/mepnr/index.htm

Environmental Federal Information Agency
http://russia.refia.msu.ru/

RWANDA

Ministry of Tourism and Environment
B.P. 2378
Kigali
Telephone: (250) 720-59, (250) 720-93
FAX: (250) 779-58

SAMOA

Department of Lands Survey and Environment
Private Mail Bag, Apia
Telephone: (685) 22-481
FAX: (685) 23-176, 21-504

SAO TOME AND PRINCIPE

Ministry of Social Equipment and Environment
Sao Tome, Principe, West Africa
Sao Tome
Telephone: (239-12) 23375

SENEGAL

Ministère de l'Environnement et de la Protection de la Nature
(Ministry of Environment and Natural Protection)
Building Administratif
Dakar
Telephone: (221) 823-1088

SEYCHELLES

Ministry of Agriculture and Marine Resources
P.O. Box 166, Victoria
Mahe
Telephone: (248) 225-333
FAX: (248) 225-245

SINGAPORE

Ministry of Environment
40 Scotts Road, 0922
Telephone: (65) 732-773
FAX: (65) 738-4468
http://www.gov.sg/env/

SLOVAK REPUBLIC

Ministerstvo zivotneho prostredia
(Ministry of Environment)
Hlboka 2, SK-812 35 Bratislava
Telephone: (421-7) 392-451
FAX: (421-7) 391-201

Slovenska Agentura Zivotneho Prostredia
(Slovak Environmental Agency [SEA])
Tajovskeho 28, 975 90 Banska Bystrica
Telephone: 42 (0)88 732157
FAX: 42 (0)88 732160
http://sun.sazp.sk/

SLOVENIA

Ministry of Environment and Urban Planning
Zupanciceva ulica 6, 1000 Ljubljana
Telephone: (386-61) 178-5211
FAX: (386-61) 224-548
http://www.sigov.si/cgi-bin/wpl/mop/index.htm
http://www.sigov.si/cgi-bin/spl/upp/uk_index.htm/language=csz

SOLOMON ISLANDS

Ministry of Forests, Environment and Conservation
P.O. Box G24, Honiara
Telephone: (677) 25-848, (677) 22-944

SOUTH AFRICA

Department of Environmental Affairs and Tourism
240 Vermeulen Street, 7th Floor, Pretoria 0002
Telephone: (27-12) 321-9587
FAX: (27-12) 323-5181
http://www.gov.za/envweb/
med_js@ozone.pwv.gov.za

Ministry of Water Affairs and Forestry
15th Floor, 120 Plein Street, Cape Town
Telephone: (27-0) 21-45-5585
FAX: (27-0) 21-45-3362
http://www-dwaf.pwv.gov.za/idwaf/web-pages/minister/minister.htm

SOUTH KOREA

Ministry of Environment
1 Chungang-dong, Gwacheon-shi
Kyonggi-do 427 760
Telephone: (822) 504-9284

FAX: (822) 504-9284
http://www.moenv.go.kr/

SPAIN

Ministerio de Medio Ambiente
(Environment Ministry)
Pza. San Juan de la Cruz s/n
28017 Madrid
Telephone: (34) 1 597 6577
http://www.mma.es/

SRI LANKA

Ministry of Transport, Environment and Women's Affairs
Central Environmental Authority
6th Floor, Unity Plaza Building, Colombo 4
Telephone: (941) 549-455

ST. KITTS AND NEVIS

Ministry of Tourism, Culture and Environment
Pelican Mall, Bay Road, Basseterre
Telephone: (869) 465-4040
FAX: (869) 465-8794

ST. LUCIA

Ministry of Agriculture, Fisheries and the Environment
Government Building, The Waterfront, Castries
Telephone: (758) 452-2611

ST. VINCENT AND THE GRENADINES

Ministry of Health and the Environment
Kingstown
St. Vincent and the Grenadines
Telephone: (809) 457-2586
FAX: (809) 457-2684

SUDAN

Ministry of Tourism and Environment
P.O. Box 300
Khartoum
Telephone: (249-11) 779-700

SURINAME

Ministry of Natural Resources and Energy
Dr. J.C. de Miranda Street 13-15, Paramaribo
Telephone: (597) 4733420

SWAZILAND

Ministry of Natural Resources and Energy
P.O. Box 57, Mbabane
Telephone: (268) 46-244
FAX: (268) 42-436

SWEDEN

Naturvårdsverket
(Swedish Environmental Protection Agency)
S-106 48 Stockholm, Besöksadress: Blekholmsterrassen 36
Telephone: 08-698 10 00
FAX: 08-20 29 25
e-mail: natur@environ.se
http://www.environ.se/

SWITZERLAND

Office Federal de l'Environnement, des Forets et du Paysage
(Federal Office for Environment, Forests, and Landscape)
Hallwylstrasse 4, CH-3003 Bern
Telephone: (41 31) 322-9311
FAX: (41 31) 322-9981

TAIWAN

Environmental Protection Administration
41 Chunghwa Road, Section 1
Taipei
Telephone: (886) 311-7722
FAX: (886) 311-6071
http://www.epa.gov.tw/english.now/index1.htm

Industrial Pollution Control Center
16/F, 77 Tun Hwa South Road, Sec. 2
Taipei
Telephone: (886) 2 - 325 54 86
FAX: (886) 2 - 700 04 64

TAJIKISTAN

Ministry of Environmental Protection
12 Bokhtar Street, Dushanbe
Telephone: (2-66) 232-878

TANZANIA

Prime Minister's Office, Environment Section
P. O. Box 3021, Dar es Salaam, 255-51
Telephone: (255-51) 112-849, (255-51) 110-365
FAX: (255-51) 112-608

THAILAND

Pollution Control Department
404 Phaholyothin Center Building
Phaholyothin Road, Sam Sen Nai Phayathai
Bangkok 10400
Telephone: (662) 619-2286, ext. 8
FAX: (662) 619-2285
http://www.pcd.go.th/

Ministry of Science, Technology and Environment
Rama VI Road, Ratchathewi
10400 Bangkok
Telephone: (66) 2 - 246 00 64
FAX: (66) 2 - 246 81 06
http://www.nectec.or.th/bureaux/moste/moste.html

TOGO

Ministry of Rural Development, Environment and Tourism
Avenue de Sarawaka
Lomé
Telephone: (228) 210-305
FAX: (228) 218-927, 210-333

TONGA

Ministry of Lands, Survey and Natural Resources
P.O. Box 5, Nuku'alofa
Telephone: (676) 22-655

Ministry of Agriculture, Forestry, Fisheries, and Marine Affairs
P.O. Box 14, Nuku'alofa
Telephone: (676) 21-300
FAX: (676) 23-888

TRINIDAD AND TOBAGO

Ministry of Food Production, Marine Exploration, Forestry and Environment
St. Clair Circle, St. Clair
Port of Spain
Telephone: (868) 622-1221

Institute of Marine Affairs
P.O. Box 3160
Carenage Post Office
Trinidad and Tobago
Telephone: (809) 634-4291/4
FAX: (809) 634-4433

TURKEY

Ministry of Environment
Eskisehir Yolu 8.Km, Ankara 06100
Telephone: (90-312) 285-3283
FAX: (90-312) 286-2271, 342-4001

TURKMENISTAN

Ministry of Natural Resource and Environmental Protection
28 Azady Street, Ashgabat 744000
Telephone: (99-312) 254-317
FAX: (99-312) 252-505

UGANDA

Ministry of Natural Resources
Amber House, 33 Kampala Road,
P. O. Box 7270
Kampala
Telephone: (256-41) 233-331
FAX: (256-41) 230-220

National Environment Management Authority
Ministry of Natural Resources
P.O. Box 22255
Kampala
Telephone: (256-41) 236-817, (256-41) 251-064
FAX: (256-41) 257-521
e-mail: neic@starcom.co.ug
http://www.uganda.co.ug/nema/

UKRAINE

Ministry of Environmental Protection and Nuclear Safety
Khreschatyk str. 5, Kiev 252001
Telephone: (380-44) 229-4292, (380-44) 228-7798
FAX: (380-44) 228-2922, (380-44) 228-7798
e-mail: mep@mep.freenet.kiev.ua

UNITED KINGDOM

Department of Environment, Transport and the Regions
Eland House, Bressenden Place
London, SW1E 5DU
Telephone: 0171-890 3000
http://www.detr.gov.uk/

Environment Agency
Hampton House, 20 Albert Embankment
London SE1 7TJ
Telephone: 0171 587 3000

FAX: 0171 587 5258
http://www.environment-agency.gov.uk/

United States Environmental Protection Agency (U.S. EPA)

Environmental Protection Agency Headquarters
401 M Street, SW
Washington, DC 20460
(202) 260-2090

Region 1 (CT, MA, ME, NH, RI, VT)
Environmental Protection Agency
One Congress Street
John F. Kennedy Building
Boston, MA 02203-0001
Telephone: (617) 565-3420
FAX: (617) 565-3660

Region 2 (NJ, NY, PR, VI)
Environmental Protection Agency
290 Broadway
New York, NY 10007-1866
Telephone: (212) 637-3000
FAX: (212) 637-3526

Region 3 (DC, DE, MD, PA, VA, WV)
Environmental Protection Agency
1650 Arch Street
Philadelphia, PA 19103-2029
Telephone: (215) 814-2900
Customer Service Center: (215) 814-5000 or (800) 438-2474
Emergency Response Hotline: (215) 814-9016

Region 4 (AL, FL, GA, KY, MS, NC, SC, TN)
Environmental Protection Agency
100 Alabama Street, SW
Atlanta, GA 30303
Telephone: (404) 562-9900
FAX: (404) 562-8174
Toll free: (800) 421-1754

Region 5 (IL, IN, MI, MN, OH, WI)
Environmental Protection Agency
77 West Jackson Boulevard
Chicago, IL 60604-3507
Telephone: (312) 353-2000
Toll free: (800) 621-8431
FAX: (312) 353-4135

Region 6 (AR, LA, NM, OK, TX)
Environmental Protection Agency

Fountain Place 12th Floor, Suite 1200
1445 Ross Avenue
Dallas, TX 75202-2733
Telephone: (214) 665-6444
Toll free: (800) 887-6063
FAX: (214) 665-7113

Region 7 (IA, KS, MO, NE)
Environmental Protection Agency
726 Minnesota Avenue
Kansas City, KS 66101
Telephone: (913) 551-7000
Toll free: (800) 848-4568
FAX: (913) 551-7467

Region 8 (CO, MT, ND, SD, UT, WY)
Environmental Protection Agency
999 18th Street Suite 500
Denver, CO 80202-2466
Telephone: (303) 312-6312
Toll free: (800) 227-8917
FAX: (303) 312-6339

Region 9 (AZ, CA, HI, NV)
Environmental Protection Agency
75 Hawthorne Street
San Francisco, CA 94105
Telephone: (415) 744-1305
FAX: (415) 744-2499

Region 10 (AK, ID, OR, WA)
Environmental Protection Agency
1200 Sixth Avenue
Seattle, WA 98101
Telephone: (206) 553-1200
Toll free: (800) 424-4372
FAX: (206) 553-0149

Other U.S. EPA Locations

U.S. EPA
26 Martin Luther King Drive
Cincinnati, Ohio 45268

U.S. EPA
Research Triangle Park, NC 27711

U.S. EPA Laboratories

U.S. EPA
National Air and Radiation Environmental Laboratory (NAREL)

540 South Morris Avenue
Montgomery, AL 36115-2601
Telephone: (334) 270-3400
FAX: (334) 270-3454

U.S. EPA
National Enforcement Investigations Center Laboratory
Box 25277, Bldg. 53
Denver Federal Center
Denver, CO 80225
Telephone: (303) 236-5132

U.S. EPA
National Exposure Research Laboratory (NERL)
Ecosystems Research Division
960 College Station Road
Athens, GA 30605-2700
Telephone: (706) 355-8005

U.S. EPA
National Exposure Research Laboratory (NERL)
Environmental Sciences Division
P.O. Box 93478
Las Vegas, NV 89193-3478
Telephone: (702) 798-2100
FAX: (702) 798-2637

U.S. EPA
National Exposure Research Laboratory (NERL)
Microbiological and Chemical Exposure Assessment Research Division (MCEARD)
26 West Martin Luther King Drive
Cincinnati, OH 45268

U.S. EPA
National Health and Environmental Effects Research Laboratory (NHEERL)
Atlantic Ecology Division
27 Tarzwell Drive
Naragansett, RI 02882
Telephone: (401) 782-3001
FAX: (401) 782-3030

U.S. EPA
National Health and Environmental Effects Research Laboratory (NHEERL)
Gulf Ecology Division
Sabine Island Drive
Gulf Breeze, FL 32561
Telephone: (850) 934-9200
FAX: (850) 934-9201

U.S. EPA
National Health and Environmental Effects Research Laboratory (NHEERL)

Mid-Continent Ecology Division
6201 Congden Boulevard
Duluth, MN 55804
FAX: (218) 720-5703

U.S. EPA
National Health and Environmental Effects Research Laboratory (NHEERL)
Western Ecology Division
200 SW 35th Street
Corvallis, OR 97333
Voice: 541-754-4600
FAX: 541-754-4799

National Risk Management Research Laboratory (NRMRL)
Subsurface Protection and Remediation Division
P.O. Box 1198
Ada, OK 74820
Telephone: (580) 436-8500

National Risk Management Research Laboratory (NRMRL)
Water Supply and Resources Division
Urban Watershed Management Branch
2890 Woodbridge Avenue (MS-104)
Edison, NJ 08837

U.S. EPA
National Vehicle and Fuel Emissions Laboratory (NVFEL)
2565 Plymouth Road
Ann Arbor, MI 48105
Telephone: (734) 668-4333

U.S. EPA
Radiation and Indoor Environments National Laboratory
P.O. Box 98517
Las Vegas, NV 89193-8517
Telephone: (702) 798-2476

URUGUAY

Ministerio de Vivienda, Ordenamiento Territorial y Medio Ambiente
(Ministry of Housing, Land-use Planning and Environment)
Zabala 1427, 11000 Montevideo
Telephone: (598 2) 961925
FAX: (598 2) 962914

UZBEKISTAN

State Committee for Environmental Protection
A. Kadiry Str. 5a, Tashkent 700128
Telephone: 73712 - 41 48 12
FAX: 73712 - 413 990

VANUATU

Ministry of Land and Natural Resources
Private Mail Bag 007, Port Vila
Telephone: (678) 23-105

VENEZUELA

Ministerio del Ambiente y de los Recursos Naturales Renovables (MARNR)
(Ministry of Environment and Renewable Natural Resources)
Apartado Postal 167, Caracas 1010-A
Telephone: (58 2) 408-1001
FAX: (58 2) 483-1148
http://www.marnr.gov.ve/

VIETNAM

Ministry of Forestry
123 Lo Duc, Hanoi
Telephone: (844) 53236

Ministry of Water Resources
164 Tran Quang Khai, Hanoi
Telephone: (844) 68141

ZAMBIA

Ministry of Environment
P.O. Box 30055, Lusaka
Telephone: (260-1) 253-040, (260-1) 253-046
FAX: (260-1) 252-852
e-mail: menr@zamnet.zm

ZIMBABWE

Ministry of Environment and Tourism
14th Floor Karigamombe Centre
53 Samora Machel Avenue
Private Mail Bag 7753, Causeway
Harare
Telephone: (263-4) 757-881, (263-4) 704-701
FAX: (263-4) 794-450

INSTITUTIONS OF THE EU

European Parliament
Rue Belliard, 97-113
B-1040 Brussels
Telephone: 284 28 86
FAX 284 90 64

Council of the European Union
Rue de la Loi, 175
B-1048 Brussels
Telephone: 285 61 11

European Commission
Rue de la Loi, 200
B-1049 Brussels
Telephone: 299 11 11

Court of Justice of the European Communities
Boulevard Konrad Adenauer
L-2925 Luxembourg
Telephone: 43 03-1

European Court of Auditors
12, rue Alcide De Gasperi
L-1615 Luxembourg
Telephone: 43 98-1
FAX: 43 98-45630

Economic and Social Committee
Rue Ravenstein, 2
B-1000 Brussels
Telephone: 546 90 11

Committee of the Regions
Rue Ravenstein, 2
B-1000 Brussels
Telephone: 546 21 55
FAX: 546 20 85

European Investment Bank
L-2950 Luxembourg
Telephone: 43 79-1

U.S. STATE ENVIRONMENTAL AGENCIES

Alabama

Forestry Commission
513 Madison Avenue
Montgomery, AL 36130

Alabama Office of Water Resources
401 Adams Ave. Suite 360
Montgomery, AL 36103
Telephone: (334) 242-5506
FAX: (334) 242-0776
e-mail: alaowr@usenet.com

Alabama State FSA Office
P.O. Box 235013
Montgomery, AL 36123-5013
Telephone: (334) 279-3500
FAX: (334) 279-3550

Auburn University Extension Service
116 Extension Hall
Auburn, AL 36849
Telephone: (334) 844-5533
FAX: (334) 844-4586

Department of Environmental Management
1751 Cong WL Dickinson Dr.
Montgomery, AL 36130
Telephone: (334) 213-4310
FAX: (334) 213-4399

Department of Environmental Management
P.O. Box 301463
Montgomery, AL 36130-1463
Telephone: (334) 270-5655

Soil and Water Conservation Commission
P.O. Box 304800
Montgomery, AL 36130-4800
Telephone: (334) 242-2620
FAX: (334) 242-0551

ALASKA

Alaska Department of Environmental Conservation
555 Cordova
Anchorage, AK 99501
Telephone: (907) 269-7554

Alaska Department of Fish and Game
1300 College Rd.
Fairbanks, AK 99701
Telephone: (907) 459-7287

Alaska Division of Forestry
3601 C St. Suite 1034
Anchorage, AK 99503-5925
Telephone: (907) 465-3379
FAX: (907) 586-3113

Alaska Department of Natural Resources
400 Willoughby Dr. 5th Floor
Juneau, AK 99801

Telephone: (907) 465-2401
FAX: (907) 465-3886

Department of Environmental Conservation
410 Willoughby Ave. Suite 105
Juneau, AK 99811
Telephone: (907) 465-5158
FAX: (907) 465-5274
e-mail: swilling@envircon.state.ak.us

ARIZONA

Arizona Department of Environmental Quality
3033 N. Central
Phoenix, AZ 85012
Telephone: (602) 207-4518
FAX: (602) 207-4467

Environmental Service Division
1688 W. Adavis
Phoenix, AZ 85007
Telephone: (602) 542-3578
FAX: (602) 542-0466

Natural Resource Conservation Division
1616 W. Adams
Phoenix, AZ 85007
Telephone: (602) 542-2699
FAX: (602) 542-4668
e-mail: wwarskow@ind.state.az.us

State Land Department
1616 W. Adams
Phoenix, AZ 85007

ARKANSAS

Arkansas Forestry Commission
3821 West Roosevelt Rd.
Little Rock, AR 72204-6396

Arkansas Soil and Water Conservation Commission
101 E. Capitol Suite 350
Little Rock, AR 72201
Telephone: (501) 682-3954
FAX: (501) 682-3991

Department of Health
4815 W. Markham Slot 37
Little Rock, AR 72205
Telephone: (501) 661-2623

CALIFORNIA

California Coastal Commission NPS
921 11th St. Rm. 1200
Sacramento, CA 95814
Telephone: (916) 445-6096
FAX: (916) 324-6832
e-mail: cccnps@cwo.com

Department of Conservation
801 K St. MS2020
Sacramento, CA 95814
Telephone: (916) 323-1777

Department of Forestry and Fire Protection
P.O. Box 944246 1416 9th St.
Sacramento, CA 94244-2460
Telephone: (916) 653-7772
FAX: (916) 653-4171

Department of Health Services
2151 Berkeley Way
Berkeley, CA 94704
Telephone: (510) 540-2192
FAX: (510) 540-2181

Office of Land Conservation
801 K St. MS 13-71
Sacramento, CA 95814
Telephone: (916) 324-0850

Resources Agency
1416 9th St. Rm. 1311
Sacramento, CA 95814
FAX: (916) 653-8102

Water Resources Control Board
901 P St. 4th Floor
Sacramento, CA 95814
Telephone: (916) 657-0682
FAX: (916) 657-2388

COLORADO

Colorado Department of Public Health and Environment
Cherry Creek Drive South
Denver, CO 80222-1530
Telephone: (303) 692-2000

Colorado Forestry Association
12820 Evans
Elbert, CO 80106

Colorado State Forest Service
203 Forestry Bldg.
Fort Collins, CO 80523
Telephone: (970) 491-6303
FAX: (970) 491-7736

Department of Agriculture
700 Kipling Suite 4000
Lakewood, CO 80215
Telephone: (303) 239-4100

Department of Natural Resources
1313 Sherman St. Rm. 718
Denver, CO 80203
Telephone: (303) 866-3311

Soil Conservation Board
1313 Sherman St. Rm. 219
Denver, CO 80203
Telephone: (303) 866-3351
FAX: (303) 832-8106

State Conservation Board
1313 Sherman St. Rm. 219
Denver, CO 80203-2243
Telephone: (303) 866-3351
FAX: (303) 832-8106

CONNECTICUT

Coastal Zone Management Agency, Office of Long Island
79 Elm Street
Hartford, CT 06106-5127
Telephone: (860) 424-3034
FAX: (860) 424-4054

Department of Environmental Protection
79 Elm Street
Hartford, CT 06106
Telephone: (860) 424-3730
FAX: (860) 424-4055
e-mail: stanley.zaremba@po.state.ct.us

Division of Forestry
79 Elm Street
Hartford, CT 06106

Telephone: (860) 424-3630
FAX: (860) 566-6024

DELAWARE

Department of Agriculture
2320 S. DuPont Highway
Dover, DE 19901

Division of Water Resources
89 Kings Highway Box 1401
Dover, DE 19903
Telephone: (302) 739-4590
FAX: (302) 739-6140

DNREC, Division of Soil and Water Conservation
P.O. Box 1401
Dover, DE 19903
Telephone: (302) 739-4793

89 Kings Highway P.O. Box 1401
Dover, DE 19903
Telephone: (302) 739-4411
FAX: (302) 739-6724

DISTRICT OF COLUMBIA

Department of Conservation and Regulatory Affairs
2100 Martin Luther King Ave. SE
Washington, D.C. 20020
Telephone: (202) 645-6601
FAX: (202) 645-6622

Trees and Land, District Government
2750 South Capitol St. SE
Washington, D.C. 20032

FLORIDA

Department of Environmental Planning
2600 Blair Stone Rd.
Tallahassee, FL 32399
Telephone: (904) 488-3601

Department of Agriculture and Consumer Services
3125 Conner Blvd. Admin. Bldg. Rm. B-
Tallahassee, FL 32399-1650
Telephone: (904) 488-6249
FAX: (607) 488-2164

Division of Forestry
3125 Conner Blvd.

Tallahassee, FL 32399-1650
Telephone: (904) 488-4274

GEORGIA

Department of Natural Resources
Environmental Protection Division
7 MLK Jr. Dr. Suite 643
Atlanta, GA 30334

Georgia Forestry Commission
P.O. Box 819
Macon, GA 31298-4599
Telephone: (912) 751-3480
FAX: (912) 751-3465

Georgia Soil and Water Conservation
P.O. Box 8024
Athens, GA 30603
Telephone: (706) 542-3065
FAX: (706) 542-4242

Soil Water Conservation Commission
P.O. Box 8024
Athens, GA 30603
Telephone: (706) 542-3065
FAX: (706) 542-4242

Water Quality Management Program
7 Martin Luther King Dr. AG Annex
Atlanta, GA 30334
Telephone: (404) 656-4905
FAX: (404) 651-9425

HAWAII

Coastal Zone Management Program
P.O. Box 3540
Honolulu, HI 96811-3540
Telephone: (808) 587-2879
FAX: (808) 587-2899
e-mail: caroly@aloha.net

Department of Health
P.O. Box 3378
Honolulu, HI 96801
Telephone: (808) 586-4348
FAX: (808) 586-4370

Department of Lands and Natural Resources
P.O. Box 373

Honolulu, HI 96802
Telephone: (808) 587-0230
FAX: (808) 587-0283

Division of Forestry and Wildlife
1151 Punchbowl Street
Honolulu, HI 96813
Telephone: (808) 587-0160

Division of Water Resource Management
P.O. Box 373
Honolulu, HI 96809
Telephone: (808) 587-0248

IDAHO

Department of Agriculture
P.O. Box 790
Boise, ID 83701

Department of Water Resources
1301 N. Orchard
Boise, ID 83706
Telephone: (208) 327-7887

Idaho Department of Environmental Quality
1118 F St.
Lewiston, ID 83501
Telephone: (208) 799-4378

Department of Environmental Quality
Community Services
1410 N. Hilton
Boise, ID 83706
Telephone: (208) 373-0115

Idaho Department of Lands
1910 Northwest Blvd. Ste. 201
Coeur d'Alene, ID 83814
Telephone: (208) 769-1535

Soil Conservation Commission
P.O. Box 83720
Boise, ID 83720-0083
Telephone: (208) 334-0217
FAX: (208) 334-2339

Water Quality Bureau
1410 N. Hilton Street
Boise, ID 83706

Telephone: (208) 334-5860
FAX: (208) 334-0576

ILLINOIS

Committee-River and Stream Protection
509 W. Washington Street
Urbana, IL 61801
Telephone: (217) 333-7734
e-mail: bullard@uiuc.edu

Department of Agriculture
P.O. Box 19281
Springfield, IL 62794-9281
Telephone: (217) 782-6297

Division of Forest Resources
P.O. Box 19255
Springfield, IL 62794-9225

Department of Natural Resources
524 S. 2nd
Springfield, IL 62701
Telephone: (217) 785-8577

Illinois Environmental Protection Agency (IEPA)
1701 S. First Ave. Ste. 600
Maywood, IL 60153
Telephone: (708) 338-7900
FAX: (708) 338-7930

Illinois EPA
1001 N. Grand Ave. East
Springfield, IL 62794-9276
Telephone: (217) 782-3362
FAX: (217) 785-1225
e-mail: epa1184@epa.state.il.us

INDIANA

Department of Environmental Management
P.O. Box 6015
Indianapolis, IN 46206-6015
Telephone: (317) 243-5145
FAX: (317) 243-5092

Department of Environmental Management
504 N. Broadway Ste. 418
Gary, IN 46402
Telephone: (219) 881-6712
FAX: (219) 881-6745

Indiana Department of Natural Resources
402 W. Washington St. Rm. W265
Indianapolis, IN 46204
Telephone: (317) 233-3870
FAX: (317) 233-3882

IDNR Coastal Zone Management, Division of Water
402 W. Washington St. Rm. W264
Indianapolis, IN 46204-2748
Telephone: (317) 232-1106
FAX: (317) 233-4579

IDNR Division of Soil Conservation
402 W. Washington Street
Indianapolis, IN 46204
Telephone: (317) 233-3870
FAX: (317) 233-3882

Iowa

Department of Natural Resources
Wallace Office Building East 9th
Des Moines, IA 50319

Iowa Environmental Council
7031 Douglas Ave.
Des Moines, IA 50322
Telephone: (515) 237-5532
FAX: (515) 237-5385
e-mail: johnson_iec@commonlink.com

Fish and Wildlife Private Lands Coordinator
P.O. Box 399
Prairie City, IA 50228
Telephone: (515) 994-2415
FAX: (515) 994-2104

Iowa Department of Agriculture, Soil Conservation Division
Wallace State Office Bldg.
Des Moines, IA 50319
Telephone: (515) 281-6146
FAX: (515) 281-6170

Iowa Natural Heritage Foundation
444 Ins. Exchange Bldg.
Des Moines, IA 50309
Telephone: (515) 288-1846
FAX: (515) 288-0137

Iowa Watershed Coalition
RR 2 Box 144

Mt Ayr, IA 50854
Telephone: (515) 464-2669

Kansas

Conservation Commission
109 SW 9th St. Ste. 500
Topeka, KS 66612-1299
Telephone: (913) 296-3600
FAX: (913) 296-6172

Department of Health and Environment
Forbes Fld. Bldg. 283
Topeka, KS 66620
Telephone: (913) 296-5560

Department of Health and Environment
800 W. 24th Street
Lawrence, KS 66046
Telephone: (913) 842-4600

Kentucky

Department of Fish & Wildlife Resources
1 Game Farm Rd.
Frankfort, KY 40601
Telephone: (502) 564-4762

Kentucky Division of Conservation
691 Teton Trail
Frankfort, KY 40601
Telephone: (502) 564-3080
FAX: (502) 564-9195
e-mail: coleman@mail.nr.state.ky.us

Kentucky Division of Forestry
627 Comanche Trail
Frankfort, KY 40601

Louisiana

Department of Environmental Quality
P.O. Box 82215
Baton Rouge, LA 70884-2215
Telephone: (504) 765-0546
FAX: (504) 765-0635
e-mail: jan_b@eq.state.la.us

Department of Natural Resources
Coastal Management Division
P.O. Box 44487
Baton Rouge, LA 70804-4487

Telephone: (504) 342-7591
FAX: (504) 342-9439

LA DEQ
Office of Water Resources
Jeany Anderson-LaBar
P.O. Box 82215
Baton Rouge, LA 70884-2215

Louisiana Department of Agriculture and Forestry
P.O. Box 3554
Baton Rouge, LA 70821-3554
Telephone: (504) 922-1270
FAX: (504) 922-2577

MAINE

Department of Environmental Protection
312 Canco Rd.
Portland, ME 04103
Telephone: (207) 822-6300
FAX: (207) 822-6303

Department of Environmental Protection
Bureau of Land
State House #17
Augusta, ME 04333
Telephone: (207) 287-7727
FAX: (207) 287-7191

Maine Department of Agriculture
State House Station 28
Augusta, ME 04333
Telephone: (207) 287-3117
FAX: (207) 287-7548

Maine Forest Service
22 State House Station Harlow Building
Augusta, ME 04333

Soil and Water Conservation Commission
State House Station 28
Augusta, ME 04333
Telephone: (207) 287-2666
FAX: (207) 287-7548

State of Maine Drinking Water Program
10 State House Station
Augusta, ME 04333-0010
Telephone: (207) 287-2070

FAX: (207) 287-4172
e-mail: brian.tarbuck@state.me.us

MARYLAND

Department of Natural Resources
580 Taylor Ave.
Annapolis, MD 21401
Telephone: (410) 974-5780
FAX: (410) 974-2833

Department of Natural Resources
Coastal Zone Management Division
5800 Taylor Ave. E-2
Baltimore, MD 21401
Telephone: (410) 974-2784
FAX: (410) 974-2833

Department of Environment
2500 Broening Hwy.
Baltimore, MD
Telephone: (410) 631-3323

Department of Natural Resources
Forest Service
Tawes State Office Building 580
Annapolis, MD 21401

Maryland Department of Agriculture
50 Harry S. Truman Pkwy.
Annapolis, MD 21401
Telephone: (410) 841-5863
FAX: (410) 841-5914
e-mail: lawrenl@mda.state.md.us

MASSACHUSETTS

Coastal Zone Management
100 Cambridge Street
Boston, MA 02202
Telephone: (617) 727-9530
FAX: (617) 727-2754
e-mail: jsmith_eoea@state.ma.us

Department of Environmental Protection
1 Winter St. 9th Floor
Boston, MA 02108
Telephone: (617) 292-5529

Division of Forest and Parks
100 Cambridge Street

Boston, MA 02202
Telephone: (617) 727-3180
FAX: (617) 727-9402

Executive Office of Environmental Affairs
100 Cambridge St. Rm. 2000
Boston, MA 02202
FAX: (617) 727-2754

Massachusetts Watershed Coalition
10 Monument Sq. P.O. Box 577
Leominster, MA 01453-0577
Telephone: (508) 534-0379
FAX: (508) 534-1329
e-mail: mwc@ultranet.com

Office of Environmental Affairs
Paula Jewell
100 Cambridge Street
Boston, MA 02202
Telephone: (617) 727-9800
FAX: (617) 727-2754

MICHIGAN

Coastal Management Program
Stevens T. Mason Bldg.
P.O. Box 30458
Lansing, MI 48909-7958
Telephone: (517) 373-1950
FAX: (517) 335-3451
e-mail: heinm@deq.state.mi.us

Department of Agriculture
P.O. Box 30017
Lansing, MI 48909
Telephone: (517) 373-2620
FAX: (517) 335-3329

Department of Public Health
301 E. Michigan Ave.
Paw Paw, MI 49079
Telephone: (616) 657-6113

Department of Natural Resources
Surface Water Quality Division
P.O. Box 128
Roscommon, MI 48653
Telephone: (517) 275-5151
FAX: (517) 275-5167

MI DNR
350 Ottawa Ave. NW
Grand Rapids, MI 49503
Telephone: (616) 456-5071
FAX: (616) 456-1239

MI DNR
1990 U.S. 41 S.
Marquette, MI 49855
Telephone: (906) 228-6561

Radiological Protection Division
P.O. Box 30630
Lansing, MI 48909-8130
Telephone: (577) 335-8326

Forest Management Division
Mason Bldg. 8th Floor Box 30452
Lansing, MI 48909-7952

Michigan Department of Environmental Quality
P.O. Box 30473
Lansing, MI 48909-7973
Telephone: (517) 335-6928

MI DEQ Land and Water Management
P.O. Box 30458
Lansing, MI 48909
Telephone: (517) 373-8804

MINNESOTA

Board of Water and Soil Resources
1 W. Water St. Suite 200
St. Paul, MN 55107

Department of Natural Resources
1201 E. Highway 2
Grand Rapids, MN 55744
Telephone: (218) 327-4416
FAX: (218) 327-4263
e-mail: dan.retka@dnr.state.mn.us

Division of Forestry
500 Lafayette Road
St. Paul, MN 55155-4044

Department of Natural Resources
500 Lafayette Rd.
St. Paul, MN 55155-4032
Telephone: (612) 296-0431

Minnesota Department of Health
Doug Mandy
P.O. Box 64975
St. Paul, MN 55164-0975
Telephone: (612) 215-0757

Minnesota Pollution Control Agency
520 Lafayette Road N.
St. Paul, MN 55155-4194
Telephone: (612) 296-0550

MISSISSIPPI

Department of Marine Resources
Coastal Programs
2620 W. Beach Blvd.
Biloxi, MS 39531
Telephone: (601) 385-5860
FAX: (601) 385-5917

Department of Environmental Quality
P.O. Box 10385
Jackson, MS 39289
Telephone: (601) 961-5354

Mississippi Soil & Water Conservation
P.O. Box 23005
Jackson, MS 39225-3005
Telephone: (601) 359-1281
FAX: (601) 354-6628

Mississippi Forestry Commission
Suite 300 301 Building
Jackson, MS 39201

MISSOURI

Department of Natural Resources
P.O. Box 176
Jefferson City, MO 65102
Telephone: (314) 751-1820
FAX: (314) 751-7376

Department of Natural Resources
P.O. Box 250
Rolla, MO 65401
Telephone: (314) 368-2130
FAX: (314) 368-2111

Missouri Watershed Association
Route 8 Box 300

Poplar Bluff, MO 63901
Telephone: (573) 686-4482

Missouri Department of Conservation
2901 West Truman Blvd. P.O. Box 18
Jefferson City, MO 65102

MONTANA

Bureau of Reclamation
P.O. Box 36900
Billings, MT 59107
Telephone: (406) 247-7749

Montana Department of Environmental Quality
1520 East Sixth Avenue
Helena, MT 59620-0901
Telephone: (406) 444-5319
FAX: (406) 444-1374
e-mail: stlehman@mt.gov

Department of Natural Resources and Conservation
Forestry Division
2705 Spurgin Road
Missoula, MT 59801

Department of Natural Resources and Conservation
1520 E. 6th Ave.
Helena, MT 59620-2301
Telephone: (406) 444-6664
FAX: (406) 444-0533

DNRC Water Resources Division
48 N. Last Chance Gulch
P.O. Box 201601
Helena, MT 59620-1601
Telephone: (406) 444-6628

Montana NRCS
10 East Babcock Street, Suite 443
Bozeman, MT 59715-4704
Telephone: (402) 437-4103
FAX: (402) 437-5327

NEBRASKA

Department of Environmental Quality
P.O. Box 98922
Lincoln, NE 68509-8922
Telephone: (402) 471-4264
FAX: (402) 471-2909

Department of Forestry, Fish and Wildlife
Rm. 101 Plant Ind. Bldg.
Lincoln, NE 68583-0814

Department of Health
301 Centennial Mall S.
Lincoln, NE 68509
Telephone: (402) 471-2541
FAX: (402) 471-6436

Natural Resource Commission
P.O. Box 94876
Lincoln, NE 68509-4876
Telephone: (402) 471-2081
FAX: (402) 471-3132

Water Control and Environmental Protection
101 Natural Resources Hall
Lincoln, NE 68583
Telephone: (402) 472-1632
FAX: (402) 472-3574

NEVADA

Nevada Division of Environmental Protection
333 W. Nye Lane
Carson City, NV 89710
Telephone: (702) 687-4670
FAX: (702) 687-6396

Division of Forestry
123 W. Nye Lane Suite 142
Carson City, NV 89710
Telephone: (702) 687-4353
FAX: (702) 687-4244

Environmental Protection Agency
P.O. Box 11
Carson City, NV 89702
Telephone: (702) 887-7528

NEW HAMPSHIRE

Department of Environmental Services
P.O. Box 95
Concord, NH 03301
Telephone: (603) 271-2961
FAX: (603) 271-2867

Division of Forests and Lands
Box 1856 172 Pembroke Road
Concord, NH 03302-1856

New Hampshire Coastal Program Office
2 1/2 Beacon Street
Concord, NH 03301
Telephone: (603) 271-2155
FAX: (603) 271-1728

NEW JERSEY

Department of Environmental Protection
CN 418 401 E. State Street
Trenton, NJ 08625
Telephone: (609) 633-1166
FAX: (609) 292-4608

New Jersey DEP Office of Environmental Planning
401 E. State St. CN418
Trenton, NJ 08625-0418
Telephone: (609) 633-0536
FAX: (609) 292-0687

State Forestry Service
CN404
Trenton, NJ 08625

Department of Agriculture
State Soil Conservation Committee
CN 330
Trenton, NJ 08625
Telephone: (609) 292-5540
FAX: (609) 633-7229

NEW MEXICO

State Office
Ecological Services
2105 Osuna NE
Albuquerque, NM 87107
Telephone: (505) 761-4525

State of New Mexico
Environment Department
P.O. Box 26110
Santa Fe, NM 87502
Telephone: (505) 827-0152
FAX: (505) 827-0160

Forestry and Resources Conservation
P.O. Box 1948
Santa Fe, NM 87504-1948

Soil and Water Conservation Bureau
P.O. Box 1865
Lovington, NM 88260
Telephone: (505) 396-2715
FAX: (505) 396-2715

NEW YORK

Coastal Zone Management Program
162 Washington Avenue
Albany, NY 12231
Telephone: (518) 474-6000

DEC Bureau of Water Quality
50 Wolf Rd.
Albany, NY 12233-3508
Telephone: (518) 457-3656
FAX: (518) 485-7786

Division of Coastal Resources and Waterfront
162 Washington Ave.
Albany, NY 12231
Telephone: (518) 474-6000
FAX: (518) 473-2464

Division of Lands and Forests
50 Wolf Rd.
Albany, NY 12233-4250

Division of Water Resources
50 Wolf Rd. Rm 308
Albany, NY 12233-3504
Telephone: (518) 457-1626

New York City Department of Environmental Protection
Box D
Downsville, NY 13755
Telephone: (607) 363-7008

New York State Department of Environmental Conservation
205 Belle Mead Road
E. Setauket, NY 11733
Telephone: (516) 444-0468
FAX: (516) 444-0474

New York State Department of Environmental Conservation
50 Wolf Rd.
Albany, NY 12333-3508
Telephone: (518) 457-0635
FAX: (518) 485-7786

New York State Soil and Water Conservation
1 Winners Circle Capitol Plaza
Albany, NY 12235
Telephone: (518) 457-3738
FAX: (518) 457-3412

New York State Office of Parks, Recreation and History
80 Blue Spruce Ln.
Ballston Lake, NY 12019-1321
Telephone: (518) 474-3714
FAX: (518) 474-7013

New York State Department of Environmental Conservation
205S Belle Meade Rd.
East Setauket, NY 11733
Telephone: (516) 444-0467
FAX: (516) 444-0474

Suffolk County Department of Health Services
Suffolk County Center
Riverhead, NY 11901-3397
Telephone: (516) 852-2077
FAX: (516) 852-2743

NORTH CAROLINA

Department of Environmental Health and Natural Resources
P.O. Box 27687
Raleigh, NC 27611-7687
Telephone: (919) 733-2302
FAX: (919) 715-3559

Department of Environmental Health and Natural Resources
Division of Environmental Management
P.O. Box 29535
Raleigh, NC 27626-0535
Telephone: (919) 733-5083
FAX: (919) 715-5637
e-mail: david@clem.ehnr.state.ne.us

Department of Environmental Health and Natural Resources
Soil and Water Conservation
512 N. Salisbury
Raleigh, NC 27604
Telephone: (919) 715-6106

Division of Coastal Management
P.O. Box 27687
Raleigh, NC 27611

Division of Forest Resources
P.O. Box 29581
Raleigh, NC 27626-0581
Telephone: (919) 733-2162
FAX: (919) 715-4350

Division of Water Quality
Environmental Science Branch
4401 Reedy Creek Road
Raleigh, NC 27607
Telephone: (919) 733-6946

Division of Water Resources
George Norris
P.O. Box 27687
Raleigh, NC 27611

North Carolina State University
Box 8107
Raleigh, NC 27695
Telephone: (919) 515-1676
FAX: (919) 515-2610

North Dakota

North Dakota Department of Health
Division of Water Quality
Rm. 203 1200 Missouri Ave. Box 5520
Bismarck, ND 58506-5520
Telephone: (701) 328-5210

Groundwater Program
Environmental Engineer
1200 Missouri Avenue
Bismarck, ND 58502
Telephone: (701) 221-5233
FAX: (701) 221-5200
e-mail: ccmail.sradig@ranch.state.nd.us

North Dakota Forest Service
First and Brander
Bottineau, ND 58318
Telephone: (701) 228-5422
FAX: (701) -22-8-54

North Dakota State Water Commission
900 E. Boulevard

Bismarck, ND 58505
Telephone: (701) 328-4989
FAX: (701) 328-3696
e-mail: lklap@water.swc.state.nd.us

OHIO

Ohio Department of Natural Resources
4435 Fountain Square B3
Columbus, OH 43224
Telephone: (614) 265-6926

Ohio Department of Natural Resources
Coastal Management
1952 Belcher Dr. Bldg. C-4
Columbus, OH 43224
Telephone: (614) 265-6413
FAX: (614) 267-2981
e-mail: mike.colvin@dnr.ohio.gov

Division of Forestry
1855 Fountain Square Ct. H-1
Columbus, OH 43224

DNR Soil and Water Conservation
1939 Fountain Sq. Ct. Bldg. E2
Columbus, OH 43224-1336
Telephone: (614) 265-6610
FAX: (614) 262-2064

Ohio EPA
Environmental Planning Section
P.O. Box 1049
Columbus, OH 43215-1049
Telephone: (614) 644-3660
FAX: (614) 644-3689

Ohio EPA
1800 Watermark Drive
Columbus, OH 43216-1049
Telephone: (614) 644-2877
FAX: (614) 644-2329
e-mail: larry_antosch@central.epa.ohio.gov

Pollution Abatement/Land Treatment
1939 Fountain Square Ct. E-2
Columbus, OH 43224-1336
Telephone: (614) 265-6637
FAX: (614) 262-2064
e-mail: jerry.wager@dnr.ohio.gov

OKLAHOMA

Oklahoma Conservation Commission
2800 North Lincoln Blvd. Suite 160
Oklahoma City, OK 73105
Telephone: (405) 521-2384
FAX: (405) 521-6686

Oklahoma Department of Environmental Quality (DEQ)
1000 Northeast 10th Street
Oklahoma City, OK 73117-1212

Oklahoma Water Resources Board
3800 N. Classen
Oklahoma City, OK 73118
Telephone: (405) 530-8800
FAX: (405) 530-8900
e-mail: drsmithee@owrb.state.ok.us

Secretary of the Environment
P.O. Box 1075
Oklahoma City, OK 73101
Telephone: (405) 231-2500
FAX: (405) 231-2690

OREGON

Department of Environmental Quality
811 SW 6th Avenue
Portland, OR 97204
Telephone: (503) 229-6893
FAX: (503) 229-6124
e-mail: roger.s.wood@state.or.us

Department of Fish and Wildlife
P.O. Box 59
Portland, OR 97207

Oregon Department of Forestry
2600 State Street
Salem, OR 97310

Oregon Coastal Management Program
800 NE Oregon St. No. 18
Portland, OR 97232
Telephone: (503) 731-4065
FAX: (503) 731-4068
e-mail: jeff.weber@state.or.us

Oregon Department of Agriculture
635 Capitol St. NE

Salem, OR 97310-0110
Telephone: (503) 986-4705
FAX: (503) 986-4730

PENNSYLVANIA

Bureau of Forestry
P.O. Box 8552
Harrisburg, PA 17105-8552
Telephone: (717) 787-2703
FAX: (717) 783-5109

Department of Environmental Protection
P.O. Box 8555
Harrisburg, PA 17105-8555
Telephone: (717) 787-2529
FAX: (717) 787-9549
e-mail: walsh.james@a1.dep.state.pa.us

Department of Environmental Regulation
555 N. Ln. Ste. 6010
Conshohocken, PA 19428
Telephone: (610) 832-6170

Department of Environmental Resources
P.O. Box 2063
Harrisburg, PA 17105-2063
Telephone: (717) 783-8727

Department of Environmental Resources
Land and Water
P.O. Box 8555
Harrisburg, PA 17105-8555
Telephone: (717) 772-5651
FAX: (717) 787-9549

Division of Coastal Zone Management
P.O. Box 8555 Rachel Carson Bldg.
Harrisburg, PA 17105-8555
Telephone: (717) 787-2529
FAX: (717) 787-9549
e-mail: tabor.james@a1.dep.state.pa.us

Environmental Council
239 Fourth Ave. Ste. 1808
Pittsburgh, PA 15222

State Conservation Committee
2301 Cameron Street
Harrisburg, PA 17110-9408

Telephone: (717) 787-8821
FAX: (717) 787-2271

PUERTO RICO

Bureau of Coasts
P.O. Box 5887-Puerta de Tierra Statio
San Juan, PR 00906
Telephone: (787) 721-7593
FAX: (787) 721-7591

Department of Agriculture
P.O. Box 10163
Santurce, PR 00908
Telephone: (809) 721-2120
FAX: (809) 723-9747

Forest Service Bureau
P.O. Box 5887 Puerta de Tierra
San Juan, PR 00906
Telephone: (809) 724-3647

Puerto Rico Environmental Quality Board
P.O. Box 11488
Santruce, PR 00910
Telephone: (809) 767-8056
FAX: (809) 766-2483

Soil Conservation Committee
P.O. Box 10163
Santurce, PR 00908-1163
Telephone: (809) 725-5982
FAX: (809) 723-9747

RHODE ISLAND

Coastal Resources Management Council
4808 Tower Hill Road
Wakefield, RI 02879
Telephone: (401) 277-2476
FAX: (401) 277-3922

Department of Environmental Management
83 Park Street
Providence, RI 02903
Telephone: (401) 277-3434
FAX: (401) 277-2591

Department of Environmental Management
Groundwater Section
291 Promenade Street

Providence, RI 02908
Telephone: (401) 277-2234
FAX: (401) 521-4230

Division of Forest Environment
1037 Hartford Pike
North Scituate, RI 02857

SOUTH CAROLINA

Bureau of Water
2600 Bull Street
Columbia, SC 29205
Telephone: (803) 734-8910
e-mail: rungeaw@columb32.dhec.state.sc.us

Department of Health and Environmental Control
2600 Bull Street
Columbia, SC 29201
Telephone: (803) 734-5228
FAX: (803) 734-5355

Department Natural Resources
2221 Devine St. Ste. 222
Columbia, SC 29205
Telephone: (803) 734-9100
FAX: (803) 734-9200

Department of Natural Resources
Land Resource Division
2221 Devine St. Ste. 222
Columbia, SC 29205
Telephone: (803) 734-9330
FAX: (803) 734-9200

Land Resource and Conservation
2221 Devine St. Ste. 222
Columbia, SC 29205

South Carolina Forestry Commission
P.O. Box 21707
Columbia, SC 29221
Telephone: (803) 896-8800
FAX: (803) 798-8097

SCDHEC
Bureau of Water
2600 Bull Street
Columbia, SC 29201
Telephone: (803) 734-5402
FAX: (803) 734-4435

SOUTH DAKOTA

Department of Environment and Natural Resources
523 E. Capitol
Pierre, SD 57501
Telephone: (605) 773-4216
FAX: (605) 773-4068

Department of Agriculture
445 E. Capitol
Pierre, SD 57501-3185
Telephone: (605) 773-5276
FAX: (605) 773-5926

Division of Resource Conservation and Forestry
Foss Building 523 E. Capitol Avenue
Pierre, SD 57501

TENNESSEE

Tennessee Department of Environment and Conservation
CGL 8th Fl. L & C Tower 401 Church
Nashville, TN 37243-1533

Tennessee Department of Environment and Conservation
Division of National Heritage
8th Fl. L & C Tower 401 Church St.
Nashville, TN 37243-0447

Tennessee Department of Environment and Conservation
Division Water Pollution Center
425 N 15th Street
Nashville, TN 37206

Office of Surface Mining
530 Gay Street, Suite 400
Knoxville, TN 37902
Telephone: (423) 545-4103

Tennessee Department of Agriculture
Ellington Agriculture Ctr. Hollman B
Nashville, TN 37204
Telephone: (615) 360-0690
FAX: (615) 360-0335

TEXAS

Texas Natural Resource Conservation Commission
P.O. Box 13087
Austin, TX 78711
Telephone: (512) 239-4813

Soil and Water Conservation Board
P.O. Box 658
Temple, TX 76503
Telephone: (817) 773-2250
FAX: (817) 773-3311

Texas Forest Service
College Station, TX 77843-2136

Texas Soil and Water Conservation Board
P.O. Box 658
Temple, TX 76503
Telephone: (817) 773-2250
FAX: (817) 773-3311

UTAH

Utah Department of Agriculture
P.O. Box 146500
Salt Lake City, UT 84114-6500
Telephone: (801) 538-7171
FAX: (801) 538-9436

Department of Environmental Quality
Division of Water Quality
288 N. 1460 W.
Salt Lake City, UT 84114
Telephone: (801) 538-6065
FAX: (801) 538-6016

Department of Natural Resources
Triad Center Suite 425
Salt Lake City, UT 84180-1204
Telephone: (801) 538-5555
FAX: (801) 533-4111

VERMONT

Department of Environmental Conservation
103 South Main St. Bldg. 10 North
Waterbury, VT 05671-0408
Telephone: (802) 241-3770
FAX: (802) 241-3287
e-mail: rickh@waterq.anr.state.vt.us

Department of Forest Parks and Recreation
103 S. Main Street
Waterbury, VT 05671-0601
Telephone: (802) 241-3670
FAX: (802) 244-1481

VIRGINIA

Department of Conservation
203 Governor St. Ste. 206
Richmond, VA 23219
Telephone: (804) 786-6523
FAX: (804) 786-1798

Department of Environmental Quality
P.O. Box 10009 9th Floor
Richmond, VA 23240-0009
Telephone: (804) 762-4488

Virginia Department of Forestry
P.O. Box 3758
Charlottesville, VA 22903
Telephone: (804) 977-6555
FAX: (804) 977-7749

WASHINGTON

Coastal Zone Management
MS 7600
Olympia, WA 98504-7600
Telephone: (360) 407-7297
FAX: (360) 407-6902

Department of Ecology
N. 4601 Monroe Ste. 202
Spokane, WA 99205-1295
Telephone: (360) 407-7060

Department of Ecology
P.O. Box 47775
Olympia, WA 98504-7775
Telephone: (360) 407-7060

Department of Ecology
300 Desmond Drive
Lacey, WA 98504
Telephone: (360) 407-6412
FAX: (360) 407-6426

Department of Natural Resources
Box 47001 1111 Washington Street
Olympia, WA 98504-7001

WEST VIRGINIA

Department of Environmental Protection
1201 Greenbrier Street

Charleston, WV 25311-1088
Telephone: (304) 558-3615

Division of Environmental Protection
Office of Water Resources
1201 Greenbrier Street
Charleston, WV 25311
Telephone: (304) 558-2108
FAX: (304) 558-5905

Division of Environmental Protection
10 McJunkin Road
Nitro, WV 25143-2506
Telephone: (304) 759-0512
FAX: (304) 759-0528

Division of Forestry
329C Percival Hall WV University
Morgantown, WV 26505-6125
Telephone: (304) 293-2441

West Virginia State Soil Conservation
1900 Kanawha Blvd. E.
Charleston, WV 25305-0193
Telephone: (304) 558-2204
FAX: (304) 558-1635

WISCONSIN

Department of Natural Resources
P.O. Box 12436
Milwaukee, WI 53212
Telephone: (414) 263-8685

Wisconsin DNR Forestry
101 S. Webster Street
Madison, WI 53707-7921
Telephone: (608) 264-9247

Wildlife Assistance Office
700 Rayovac Dr. Ste. 207
Madison, WI 53711
Telephone: (608) 264-5469
FAX: (608) 264-5472
e-mail: Kitchen, Art

Wisconsin Coastal Management Program
Department of Administration, Box 7868
Madison, WI 53707-7868
Telephone: (608) 267-7982

FAX: (608) 267-6931
e-mail: gylung@mail.state.wi.us

Wisconsin Land Conservation Association
559 D'Onofrio Drive, Suite 12
Madison, WI 53719
Telephone: (608) 833-1833

WYOMING

Department of Environmental Quality
122 W. 25th 4 W.
Cheyenne, WY 82002
Telephone: (307) 777-7079
FAX: (307) 777-5973

State Engineer
Herschler Bldg.
Cheyenne, WY 82002
Telephone: (307) 777-6150
FAX: (307) 777-5451

Wyoming State Forestry Division
1100 West 22nd Street
Cheyenne, WY 82002
e-mail: wyjo36@wydsprod.state.wy.us

Wyoming Department of Agriculture
2219 Carey Avenue
Cheyenne, WY 82002-0100
Telephone (307) 777-6579
FAX: (307) 777-6593
e-mail: gstumb1@missc.state.wy.us

ENVIRONMENTAL STANDARDS AND MANAGEMENT ORGANIZATIONS

American National Standards Institute
11 West 42nd Street
13th floor
New York, NY 10036
Telephone: (212) 642-4900
Telefax: (212) 398-0023
Telex: 42 42 96 ansi ui

ANSI Registrar Accreditation Board (RAB)
National Accreditation Program for ISO 14000 Environmental Management Standards
7315 Wisconsin Ave. Suite 250-E
Bethesda, MD 20814
Telephone: (301) 469-3363

American Society for Testing and Materials (ASTM)
100 Barr Harbor Drive
West Conshohocken, PA 19428-2959
Telephone: (610) 832-9585

American Society for Quality Control (ASQC)
611 East Wisconsin Avenue
P.O. Box 3005
Milwaukee, WI 53201
Telephone: (800) 248-1946

Standards Association of Australia
P.O. Box 458
North Sydney NSW 2059
Australia
Telephone: (02) 963-4111

Austria, EU Environmental Management and Audit Scheme (EMAS)

Johannes Mayer, Director
Department Information-Documentation-Library
Federal Environment Agency
Spittelauer Laende 5, A
1090 Vienna, Austria
Telephone: +43-1-31304-3240
e-mail: Mayer@uba.ubavie.gv.at

British Standards Institution
389 Chiswick High Road
GB-London W4 4AL
Telephone: + 44 181 996 90 00
Telefax: + 44 181 996 74 00

United Kingdom Accreditation Service (UKAS)

Brussels Institute for Environmental Management

Canadian Standards Association
178 Rexdale Blvd.
Etobicoke, Ontario M9W-1R3
Canada
Telephone: (416) 747-4044
FAX: (416) 747-2475

Commission for Environmental Cooperation (CEC)

Linda F. Duncan
Law and Enforcement Cooperation Program
393 Rue Street Jacques Bureau 200
Montreal, Quebec, H2Y 1N9

Telephone: 514-350-4334
e-mail: Lduncan@ccemtl.org

Environmental Management Secretariat for Latin America and the Caribbean (LAC)

Alexis Ferrand
Environmental Management Secretariat
c/o CIID/IDRC
Casilla de Correo 6379
Montevideo, Uruguay
Telephone: +598-2-922031/4
e-mail: Aferrand@indrc.ca

Japanese Standards Association
1-24, Akasaka 4
Minato-Ku
Tokyo 107
Japan
Telephone: (33) 585-8003

NSF International
2100 Commonwealth Blvd.
Ann Arbor, MI 48105
Telephone: (313) 332-7333

Organization for Economic Cooperation and Development (OECD)

Carlo Pesso
OECD
2 Rue Andre Pascal
75775 Paris, France
Telephone: +33-1-45-24-16-82
e-mail: Carlo.pesso@oecd.org

Quality Management Systems/Environmental Management Systems (QMS/EMS)

Reto Felix
University of St. Gall
Institute for Management of Technology
(ITEM-HSG)
Unterstrasse 22
CH-9000 St. Gall
Switzerland
Telephone: +41-71-228-24-14
e-mail: Reto.felix@item.unisg.ch

Sustainable Development Networking Programme (SDNP)

Sustainable Development and Environment Ministry
Juan Pablo Arce, National Coordinator
RDS/UNDP
P.O. Box 12814
La Paz, Bolivia
Telephone: +591-2-317320
sdnp@coord.rds.org.bo

World Bank

David Hanrahan
Environment Department, Room S3069
World Bank
1818 H Street NW
Washington, D.C. 20433
Telephone: (202) 458-5686
e-mail: Dhanrahan@worldbank.org

U.S. EPA VOLUNTARY STANDARDS NETWORK COORDINATION OF ISO 14001 ACTIVITIES

USEPA

401 M Street, SW
Washington, D.C. 20460

Mary McKiel, Director, (202) 260-3584
Eric Wilkinson, Coordinator, (202) 260-3575
Voluntary Standards Network

Pep Fuller, EPA Standards Executive
Office of Pollution Prevention, Pesticides and Toxic Substances
Telephone: (202) 260-3584

Jim Horne, Standards Coordinator
Environmental Management System (EMS), Sub-TAG Lead
EPA EMS Workgroup Chair
Office of Water, 4201
Telephone: (202) 260-5802

Cheryl Wasserman, Auditing Sub-TAG Lead
Office of Enforcement and Compliance Assurance, 2251A
Telephone: (202) 260-564-7219

David Schaller, Standards Coordinator, Region 8
999 18th Street, Suite 500
Denver, CO 80202-2466
Telephone: (303) 312-6146

Specific USEPA Initiatives

Tom Link, OAQPS Webmaster
USEPA (MD-12)
Research Triangle Park, NC 27711
Telephone: (919) 541-5456

Tom Link, OAR Superstructure Manager
USEPA (MD-12)
Research Triangle Park, NC 27711
Telephone: (919) 541-5456

Natural Gas STAR Program
Rhone Resch, Program Manager
Telephone: (202) 233-9793

Tai-ming Chang, Director
Environmental Leadership Program (ELP)
Office of Enforcement and Compliance
Office of Compliance (2223-A)
401 M Street, SW
Washington, D.C. 20460
Telephone: (202) 564-5081

Cheryl Wasserman
EMS Audit Procedural Guidelines
Associate Director for Policy Analysis
Office of Federal Activities

Julie Lynch
Consumer Labeling Initiative
Division of Pollution Prevention (MC-7409)
Office of Pollution Prevention and Toxics
USEPA (see address above)
Telephone: (202) 260-4000

Susan McLaughlin
Environmental Accounting Project
Division of Pollution Prevention (MC-7409)
Office of Pollution Prevention and Toxics
USEPA
Telephone: (202) 260-3844

Eun-Sook Goidel
Environmentally-Preferable Public Purchasing
Division of Pollution Prevention (MC-7409)
Office of Pollution Prevention and Toxics
USEPA
Telephone: (202) 260-3296

Ed Weiler
Expanding the Use of Environmental Information by the Banking Industry through ISO 14000
Division of Pollution Prevention
Telephone: (202) 260-2996

Penelope Hansen
USEPA, ORD (8301)
Environmental Technology Verification Program
Telephone: (202) 260-2600

Greg Ondich
USEPA, ORD (8301)
Implementing EMS in the Metal Finishing Industry
Telephone: (202) 260-5753

Jim Horne
EMS Demonstration Project
Using EMS to Meet Watershed Protection Goals
OW EMS Implementation Workgroup
USEPA, Office of Water
Telephone: (202) 260-5802

Jim Edwards, Deputy Director
Code of Environmental Management Principles for Federal Agencies (CEMP)
USEPA
Office of Planning, Prevention and Compliance
Telephone: (202) 564-2462

Andrew Cherry
EMS Primer for Federal Facilities
USEPA
Federal Facilities Enforcement Office (2261-A)
Telephone: (202) 586-2417

Steve Sisk
Compliance-Focused EMS
USEPA – NEIC
Box 25227, Building 53
Denver Federal Center
Denver, CO 80225
Telephone: (303) 236-3636 ext. 540

Jon Kessler
Project XL
Telephone: (202) 260-3761

Organizations for Environmental Auditing

Environmental Auditing Round Table
P.O. Box 23798, Washington, D.C. 20026

The Institute for Environmental Auditing
P.O. Box 23686, Washington, D.C. 20026-3686

The Environmental Auditing Forum
Telephone: (916) 723-3710

13 List of Relevant Documents and Publications

CONTENTS

PUBLICATIONS FOR U.S. ENVIRONMENTAL REGULATIONS

RESOURCE CONSERVATION AND RECOVERY ACT (RCRA)

A Catalog of Hazardous and Solid Waste Publications, EPA/530-SW-91-013, 5th edition, May 1991

RCRA Orientation Manual, 1990 Edition, EPA/530-SW-90-36

The Law of Hazardous Waste, Management, Cleanup, Liability, and Litigation, Susan M. Cooke, Matthew Bender and Co., Inc.

Guide to the Generator Requirements of the Colorado Hazardous Waste Regulations, CDPHE, 692-3300

Colorado Hazardous Waste Regulations, 6 CCR 1007-3

Hazardous Waste Guidance, Personnel Training for Large Quantity Generators of Hazardous Waste, CDPHE, March 1997

Hazardous Waste Guidance, Preparedness and Prevention, Contingency Plan Emergency Procedures for Large Quantity Generators of Hazardous Waste, CDPHE, March 1997

Law of Chemical Regulation and Hazardous Waste, Donald W. Stever, Environmental Law Series

The Generator's Guide to Hazardous Materials/Waste Management, Leo H. Traverse

Solid Waste Disposal Act, 42 USC 6901 - 6991i

Code of Federal Regulations, Title 40, Parts 260 - 299

Multi-media Investigation Manual, EPA 330989003R, USEPA, Office of Enforcement

RCRA Enforcement Policy Compendium, PB92-963624 OWPE 92 RE001C, Office of Waste Programs Enforcement, Sept. 1992

The Complete Guide to the Hazardous Waste Regulations, Travis P. Wagner, Van Nostrand Reinhold

Waste Management Guide, Laws, Issues and Solutions, Deborah Hitchcock Jessup, BNA Books, (800) 372-1033, 1992

RCRA Permit Policy Compendium, EPA/530-SW-91-062K, Sept. 1994

Land Disposal Restrictions, Summary of Requirements, USEPA, OSWER 9934.0-1A, February 1991

40 Code of Federal Regulations, Part 268

Hazardous and Solid Wastes Amendments of 1984, Public Law No. 98-616 amending RCRA 3004, 42 USC 6924

RCRA Hotline (800) 424-9346

EPA Region VIII, RCRA Implementation Branch (303) 293-1524

UNDERGROUND STORAGE TANKS

National Association of Corrosion Engineers, "Control of External Corrosion on Metallic Buried, Partially Buried, or Submerged Liquid Storage Systems," RP-02-85

Underwriter's Laboratory, "Standard for Glass-Fiber-Reinforced Plastic Underground Storage Tanks for Petroleum Products," Standard 1316

Underwriter's Laboratory, "Corrosion Protection Systems for Underground Storage Tanks," Standard 1746

Underwriter's Laboratory of Canada, "Standard for Reinforced Plastic Underground Storage Tanks for Petroleum Products," Standard CAN4-S615-M83

American Society of Testing and Materials (ASTM), "Standard Specification for Glass-Fiber-Reinforced Polyester Underground Petroleum Storage Tanks," Standard D4021-86

Association for Composite Tanks, "Specification for the Fabrication of FRP Clad Underground Storage Tanks," ACT-100

American Petroleum Institute, "Installation of Underground Petroleum Storage Systems," Publication 1615

American Petroleum Institute, "Recommended Practice for Bulk Liquid Stock Control at Retail Outlets," Publication 1621

Petroleum Equipment Institute, "Recommended Practices for Installation of Underground Liquid Storage Systems," publication RP100

American National Standards Institute (ANSI), "Petroleum Refinery Piping," Standard B31.3

American National Standards Institute (ANSI), "Liquid Petroleum Transportation Piping Systems," Standard B31.4

American Petroleum Institute, "Cathodic Protection of Underground Petroleum Storage Tanks and Piping Systems," Publication 1632

Steel Tank Institute, "Specification for STI-P3 System of External Corrosion Protection of Underground Steel Storage Tanks"

National Fire Protection Association (NFPA), "Flammable and Combustible Liquids Code," Standard 30 and NFPA 385 (transfer procedures)

National Leak Prevention Association, "Spill Prevention, Minimum 10 Year Life Extension of Existing Steel Underground Tanks by Lining Without the Addition of Cathodic Protection," Standard 631

40 CFR 280

42 USC 6912, 6991, 6991a, 6991b, 6991c, 6991d, 6991e, 6991f, and 6991h

53 FR 37194, September 23, 1988

Colorado Revised Statutes (CRS), titles 8, 24, 25 and 6 Colorado Code of Regulations (CCR)

Colorado Environmental Law Handbook, Holme, Roberts, & Owen, Governments Institute

Musts for USTs, "A Summary of the Regulations for Underground Storage Tank Systems," USEPA, Office of Underground Storage Tanks, July 1990

SUPERFUND AMENDMENTS AND REAUTHORIZATION ACT (SARA)

Superfund Amendments and Reauthorization Act of 1986, Public Law 99-499, October 17, 1986

Code of Federal Regulations, Parts 355, 370, and 372

Title III Fact Sheet, Emergency Planning and Community Right-to-Know, USEPA

Community Right-to-Know and Small Business, "Understanding Sections 311 and 312 of the Emergency Planning and Community Right-to-Know Act of 1986," USEPA, Office of Solid Waste and Emergency Response, September 1988

Letter from Cheryl A. Crisler, Chief, Prevention Section USEPA Region VIII to LEPC and SERC members regarding the identification of industries that may have to comply with SARA; letter includes an enclosure identifying by SIC code and industry the type of EHS used in that industry.

When All Else Fails! Enforcement of the Emergency Planning and Community Right-to-Know Act, "A Self Help Manual for Local Emergency Planning Committees," USEPA, Solid Waste and Emergency Response (OS-120), September 1989

Chemicals in Your Community, "A Guide to the Emergency Planning and Community Right-to-Know Act," USEPA, September 1988

Hazardous Materials, "An Introduction for Public Officials," FEMA, USEPA, USDOT, SM 300, September 1990

Hazardous Materials Information Management, Emergency Management Institute, Federal Emergency Management Agency, National Emergency Training Center, IG 305.2, September 1990

Exercising Emergency Plans Under Title III, Emergency Management Institute, IG 305.4, February 1991

Community Awareness and Right-to-Know, Emergency Management Institute, IG 305.6, September 1990

Hazardous Materials Planning Guide, National Response Team, March 1987

HELP! EPA Resources for Small Governments, Regional Operations and State/Local Relations (H1501), 21V-1001, September 1991

Environmental Planning for Small Communities, "A Guide for Local Decision Makers," Regional Operations & State/Local Relations, EPA/625/R-94/009, September 1994

A Guide to Federal Environmental Requirements for Small Governments, Office of the Administrator (H1501), EPA 270-K-93-001, September 1993

WETLANDS

Classification of Wetlands and Deepwater Habitats of the United States, U.S. Department of the Interior, Fish and Wildlife Service, FWS/OBS-79/31, reprinted 1992, December 1979

Wetlands Status and Trends in the Conterminous United States Mid-1970's to Mid 1980's, Report to Congress, U.S. Department of the Interior, Fish and Wildlife Service

America's Wetlands: Our Vital Link Between Land and Water, EPA 170OPA87016, Government Printing Office, National Technical Information Service: PB93-206399

Design Manual: Constructed Wetlands and Aquatic Plant Systems for Municipal Wastewater Treatment, EPA 625188022, National Center for Environmental Publications and Information

Handbook for Constructed Wetlands Receiving Acid Mine Drainage, EPA 540SR93523, National Center for Environmental Publications and Information

American Wetlands: Protecting America's Wetlands: A Fair, Flexible, and Effective Approach (President Clinton's Press Package), EPA 843E93001, August 24, 1993

Environmental Fact Sheet: Controlling the Impacts of Remediation Activities in or around Wetlands, EPA 530F93020, National Center for Environmental Publications and Information

Interagency Statement of Principles Concerning Federal Wetlands Programs on Agricultural Lands, EPA 843K93002

Report on the Use of Wetlands for Municipal Wastewater Treatment and Disposal, EPA 430988005, National Center for Environmental Publications and Information, Government Printing Office, National Technical Information Service: PB88-233481

Office of Wetlands, Oceans, and Watersheds, 1993 Publications List, EPA 840B93002, National Center for Environmental Publications and Information

Wetlands Protection Hotline, (800) 832-7828, this hotline responds to requests for information on wetlands and options for their protection. It is a central point of contact for the Wetlands Division of the Office of Wetlands, Oceans, and Watersheds. The hotline has an extensive contact list for additional information. Wetlands Division publications are also distributed directly from the hotline.

ENVIRONMENTAL AUDITING

Environmental Audits, 6th edition, Lawrence B. Cahill and Raymond W. Kane, Governments Institute, Inc. 1989

Professional Environmental Management and Auditing, 3rd printing, Dr. Valcar A. Bowman, Cahners Publishing Company

A Practical Guide to Plant Environmental Audits, Blakeslee and Grabowski, Van Nostrand Reinhold, 1985

Environmental Auditing: Fundamentals and Techniques, Greeno, John Wiley & Sons, 1985

Environmental Auditing Handbook, Harrison, McGraw-Hill Book Company, 1983

Environmental Auditing Handbook: Basic Principles of Environmental Compliance Auditing, second edition, Truitt, 1983

Annotated Bibliography on Environmental Auditing, USEPA, Office of Policy, Planning, and Evaluation, 7th edition, March 1988

Current Practices in Environmental Auditing, USEPA Report Number 230-09-83-006, February 1984

Recordkeeping Requirements, Donald S. Skupsky, Information Requirements Clearinghouse (IRCH), 1989

Chemical Safety Audit Program, Fact Sheets, EPA550F91100

Site Auditing: Environmental Assessment of Property, Specialty Technical Publishers

Federal Register, Vol. 51, No. 131, Environmental Auditing Policy Statement, Notice, Wednesday, July 9, 1986

Compendium of Audit Standards, Walter Willburn, American Society for Quality Control, 1983

Standards for the Professional Practice of Internal Auditing, The Institute of Internal Auditors, Inc., 1981

FREE PUBLICATIONS AVAILABLE FROM THE EU

The European Union and the Environment, (a 36-page booklet briefly describing the Fifth Environmental Action Programme "Towards Sustainability," EU environmental policies and their sustainable development actions

The ABC of Community Law (booklet)

The European Commission 1995-2000 (booklet)

Working together—the institutions of the European Community (booklet)

How is the European Union protecting our environment (booklet)

How does the European Union work? (booklet)

The European Union: Key figures (booklet)

How does the European Union manage agriculture and fisheries? (booklet)

The common agricultural policy in transition (leaflet)

Environmental protection: a shared responsibility, An 8-page fold-out leaflet that outlines policy and programs being implemented in the EU

Towards the 5th Framework Programme, Scientific and Technological Objectives, a 52-page booklet that describes the programs' content and objectives, implementation, legal aspects and management, published by the European Commission, ISBN 92-827-9259-5, 1997

SELECTED TITLES IN INTERNATIONAL ENVIRONMENTAL LAW AND MANAGEMENT

Environmental Law, Nancy K. Kubasek, Gary S. Silverman, Prentice Hall, 294 p. with index, $31.25, 1994

Final chapter presents 27 pages on international law. Authors are an attorney/teacher and an environmental program department head at Bowling Green State University. [6x9 pbk]

---

[1] Available from: USEPA, Regulatory Reform Staff, PM-223, 401 M Street SW, Washington, D.C. 20460

ISO 14000 Certification: Environmental Management Systems, W. Lee Kuhre, Prentice Hall, 365 p. with appendix and index, $60.00, 1995

> Initial book in Prentice Hall International Series on ISO 14000. Primer on ISO 14000. Appendix includes templates of ISO documentation, which are replicated on enclosed disk. [7x9½ hdc]

Environmental Strategies Handbook: A Guide to Effective Policies and Practices, Rao V. Kollurn, Ed., McGraw-Hill, 1,030 p. with index, $79.50, 1994

> A collection of writings (personal experience and case studies) on environmental strategies by leading environmental professionals, attorneys, and scientists. Topics address the issue of incorporating environmental, health, and safety priorities into management thinking and business operations. [6x9½ hdc]

ISO 14000: A Guide to the New Environmental Management Standards, Tom Tibor, Ira Feldman, Irwin Professional Publishing, 237 p. with apdx. and index, $35.00, 1996

> Presents history, elements, and how-to of ISO 14000. Chapter on European Union's Eco-Management and Audit Scheme, similar to ISO. Authors are a technical writer and ISO Technical Advisory Group member, and an environmental consultant. [6x9½ pbk]

Corporate Environmental Strategy: The Avalanche of Change Since Bhopal, Bruce W. Piasecki, John Wiley & Sons, 180 p. with index, $25.95, 1995

> Analyzes, through case studies, the adoption of environmental strategy alongside other business concerns. Environmental strategies of Union Carbide, ARCO, AT&T, and Warner-Lambert are discussed and compared. Author is owner of American Hazard Control Group. [6x9½ hdc]

Environmental Profiles: A Guide to Projects and People, Linda Sobel Katz, Sarah Orrick, Robert Honig, Garland Publishing, 1,083 p. with index, $125.00, 1993

> Volume describes projects and entities supporting environmental projects and agreements worldwide. Presented country by country. [8½x11]

What is ISO 14000? Questions and Answers, Caroline G. Hemenway, CEEM Information Services, ASQC Quality Press, 2nd Edition, 44 p. with appendix, $24.95, 1996

> Introductory book on ISO 14000, includes history, status, and applications. [7x9½ pbk]

The ISO 14000 Handbook, Joe Cascio, Editor, CEEM Information Services, ASQC Quality Press, 450 p. with glossary, $75.00, 1996

> Explains ISO 14000, its elements, status, and other initiatives. Includes copies of the draft ISO 14000 standards. [7x9½ pbk]

Trade and the Environment: Law, Economics, and Policy, Durwood Zaelke, Paul Orbuch, Robert F. Housman, Island Press, 312 p. with index, $24.95, 1993

> A collection of writings by economists, attorneys, and environmental professionals on the growing debate over trade and environmental issues. [6x9½ pbk]

Costing the Earth: The Challenge for Governments, The Opportunities for Business, Frances Caircross, Harvard Business School Press, 341 p. with index, $14.95, 1993

> While topics of sustainability and pollution prevention are discussed, significant space is given to global environmental issues and agreements, which hinge on self-restraint. Author is the environmental editor of *The Economist*. [6x9½ pbk]

World Resources: 1992-93, A Report by The World Resources Institute in collaboration with The United Nations Environment Programme and The United Nations Development Programme, Oxford University Press, 1992

> A premier to the United Nations Conference on Environment and Development in Rio de Janeiro, in 1992, this volume discusses sustainable development, regional issues, and conditions and trends worldwide. Dedicated chapter to participation in conventions and agreements. [8½x11 hdc]

OTHER TITLES

Colorado Journal of International Environmental Law and Policy, Allen, David, Editor in Chief, Niwot, Colorado, The University Press of Colorado

Biodiversity and International Law: The Effectiveness of International Environmental Law, Bilderbeek, Simone, Editor, Report of the Global Consultation on the development and Enforcement of International Environmental Law, organized by the Netherlands National Committee for the International Union for Conservation of Nature and Natural Resources] Amsterdam, IOS Press, 1992

Codification of Environmental Law: Proceedings of the International Conference, Bocken, Hubert and Donatienne Ryckbost, Editors, In Ghent, February 21 and 22, 1995, International Environmental Law & Policy Series, The Hague, Kluwer Law International, 1996

Environmental Guide to the Internet, Briggs-Erickson, Carol and Toni Murphy, Government Institutes, Inc., Rockville, MD, 2nd edition, 1996

Vital Signs 1995: The Trends That Are Shaping Our Future, Brown, Lester R., et al. New York, W.W. Norton & Company, 1995

State of the World 1995: A Worldwatch Institute Report on Progress Toward a Sustainable Society, Brown, Lester R., et al. New York, W.W. Norton & Company, 1995

The Environment After Rio: International Law and Economics, Campiglio, Luigi, Laura Pineschi, Dominico Siniscalco, and Tullio Treves, Editors, International Environmental Law and Policy Series, London, Graham & Trotman/Martinus Nijhoff, 1994

The European Environmental Law Guide, Clifford Chance, [International Law Firm], January 1, 1995

Trends in International Environmental Law, Harvard Law Review, Editors, American Bar Association, Section of International Law and Practice, 1992

International Law Environment and Development: Dossier UNCED No. 1, International Juridical Organization for Environment and Development, United Nations Conference on Environment and Development, International Juridical Organization for Environment and Development, 1992

Environmental Profiles: A Guide to Projects and People, Katz, Linda Sobel, Sarah Orrick, and Robert Honig, New York, Garland Publishing, Inc., 1993

Reforming U.S. Trade Policy to Protect Global Environment: A Multilateral Approach, Kennedy, Kevin C., The Harvard Environmental Law Review, Volume 18, Number 1, 1994

Environmental Law, Kubasek, Nancy K. Englewood Cliffs, New Jersey, Prentice Hall, 1994

Environmental Law: A Guide for Corporations, Lintz, Valarie, Editor, London: Euromoney Publications, 1992

International Environmental Law & Policy, Nanda, Ved P., Irvington-on-Hudson, New York, Transnational Publishers, Inc., 1995

International and European Trade and Environmental Law after the Uruguay Round, Petersmann, Ernst-Ulrich, London, Kluwer Law International, 1995

Corporate Environmental Strategy: The Avalanche of Change Since Bhopal Piasecki, Bruce W. New York, John Wiley & Sons, Inc., 1995

International Environmental Law: Emerging Trends and Implications for Transnational Corporations, United Nations, New York, United Nations Department of Economic and Social Development, Transnational Corporations and Management Division, 1993

Multilateral Treaties: Index and Current Status, Bowman and Harris, 1984

SUPPLIERS OF ENVIRONMENTAL, HEALTH, AND SAFETY RESOURCE MATERIALS (BOOKS, MANUALS, STANDARDS, PERIODICALS, TRAINING MATERIALS, SOFTWARE, VIDEOS)

ENVIRONMENT, GENERAL

Executive Enterprises Publications Company, Inc.
22 West 21st Street
New York, New York 10010-6904
Telephone: (800) 332-1105

> Remediation, pesticides, chemical hazard communication, waste reduction, environmental compliance, hazardous waste, underground storage tanks, environmental audits, environmental law.

U.S. Government Books
U.S. Government Printing Office
Superintendent of Documents
Washington, D.C. 20402
Telephone: (202) 512-2250

> Testing and managing solid waste, hazardous chemicals, toxic substances, water quality, air quality, health and the environment, environmental protection, resources. The Clearinghouse for Inventories and Emission Factors (CHIEF) is available on CD ROM.

Van Nostrand Reinhold
Academic Sales and Marketing
115 Fifth Ave.
New York, NY 10003
Telephone: (800) 497-4VNR

> Publications in safety management, emergency planning, energy and environment, air/water science and technology, pollution prevention, hazardous waste, environmental compliance, occupational medicine, public health, toxicology, environmental policy, professional reference, environmental auditing, and environmental decision making.

ENVIRONMENTAL, INTERNATIONAL

Bureau of National Affairs (BNA)
1231 25th Street N.W.
Washington, D.C. 20037
International Sales Telephone: (301) 948-0540

Environmental and safety publications and subscription services for managers and legal professionals, including

- Environment Reporter (U.S.)
- International Environment Reporter

Many titles are available in various languages.

Business & Legal Reports, Inc.
39 Academy Street
Madison, CT 06443-1513
Telephone: (800) 727-5257

Environmental Manager's Guide to ISO 14000. Updates are sent out every two months. Other publications on ISO include

- ISO 14000 Understanding the Environmental Standards
- ISO 14001 An Executive Report

Other Selected Titles

- South Korea Environmental Report: A Resource for Business
- China Environmental Report: A Resource for Business
- Mexico Environmental Report: A Resource for Business
- 1994 International symposium on Environmental Contamination in Central and Eastern Europe
- The Greening of World Trade
- Environmental Impact of NAFTA

Oxford University Press
198 Madison Avenue
New York, NY 10016
Telephone: (800) 451-7556

International environmental science, natural history, ecology, and management. Selected titles include

- Basic Documents on International Environmental Law
- International Management of Hazardous Wastes, The Basel Convention and Related Legal Rules
- Resource and Environmental Management in Canada
- Canadian Environmental Policy
- Green Globe Yearbook of International Co-Operation on Environment and Development 1995
- Australian Environmental History
- Environmental Policy in New Zealand
- Environmental Diplomacy
- International Law and the Environment
- Yearbook of International Environmental Law

World Data Center A (WDC-A) for Human Interactions in the Environment
2250 Pierce Road
University Center, MI 48710-0001
Telephone: (517) 797-2727

Clearinghouse for international environmental agreements

ENVIRONMENTAL LAW/COMPLIANCE

Bureau of National Affairs (BNA)
1231 25th Street N.W.
Washington, D.C. 20037
Telephone: (800) 372-1033
International Sales: (301) 948-0540

Environmental and safety publications and subscription services for managers and legal professionals, including

- Air & Water Pollution Control
- Environment Reporter
- International Environment Reporter
- Hazardous Materials Transportation
- Occupational Safety & Health Reporter
- World Pharmaceutical Standards Review

Many titles are available in various languages.

Business & Legal Reports, Inc.
39 Academy Street
Madison, CT 06443-1513
Telephone: (800) 727-5257

OSHA compliance and safety, environmental management, including The Environmental Manager's Compliance Advisor, hazard communication, and human resources. Publications include compliance reports that are published twice monthly.

Clark Boardman Callaghan
155 Pfingsten Road
Deerfield, IL 60015
Telephone: (800) 323-1336

Environmental law series includes

- Baxter's Environmental Compliance Manual
- Clean Air Act Handbook
- Law of Environmental Protection
- Law of Solid Waste: Pollution Prevention and Recycling
- Law of Wetlands Regulation
- Environmental Spill Reporting Handbook
- NEPA Law and Litigation
- Environmental Law & Practice: Compliance/Litigation/Forms

Governments Institute, Inc.
4 Research Place
Rockville, MD 20850
Telephone: (301) 921-2355

Environmental regulations courses and books and publications. Videotape distribution service also available. Topics include environmental laws, hazard communication, radon, emergency planning and community-right-to-know, waste minimization, groundwater monitoring wells, PCBs, hazardous waste. The Code of Federal Regulations, CD-ROMs, and other electronic EH & S information is also available.

STP Special Technical Publishers, Inc.
267 West Esplanade, Suite 306
North Vancouver, B.C., Canada V7M1A5
Telephone: (604) 983-3434

Environmental compliance (U.S.) publications

Thompson Publishing Group
1725 K Street, N.W.
Washington, D.C. 20006
Telephone: (202) 872-4000

Offices also located in New York, Brussels, and Moscow. Publications include more than 50 handbooks, newsletters, and reference services. Some topics include energy, environment, food and drug law, and health care and safety. Environmental publications include

- Environmental Compliance Tool Kit (U.S.)
- Environmental Packaging – U.S. Guide to Green Labeling, Packaging, and Recycling
- EPA Enforcement Manual
- Risk Management Program Handbook: Accidental Release Prevention Under the 1990 Clean Air Act

ENVIRONMENTAL MANAGEMENT

American Society for Testing and Materials (ASTM)
100 Barr Harbor Drive
West Conshohocken, PA 19428-2959
Telephone: (610) 832-9585
service@astm.org

The ISO standards series and the ISO Standards Handbook

CEEM, Inc.
10521 Braddock Road
Fairfax, VA 22032
Telephone: (703) 250-5900, (800) 745-5565
inquiry@ceem.com

Publications include International Environmental Systems Update, which includes 12 issues covering ISO 14000 series standards, certification, special reports, strategies for EMS development, and calendars of training and conferences. Other titles include

- The ISO 14000 Handbook
- The ISO 14000 Resource Directory

- ISO 14000 Questions and Answers
- ISO 14000 Case Studies
- ISO 14000 In Focus, a Business Perspective for Sound Environmental Management, Video Training Workbook

CRC Press/Lewis Publishers
2000 Corporate Blvd. NW
Boca Raton, FL 33431
Telephone: (800) 272-7737, (407) 994-0555

Selected Titles:

- Environmental Law and Enforcement
- Environmental Management Handbook
- Pollution Prevention Handbook
- Environmental Tools on the Internet
- The ISO 14000 EMS Audit Handbook
- Inside ISO 14000: The Competitive Advantage of Environmental Management
- Environmental and Safety Auditing: Program Strategies for Legal, International, and Financial Issues
- Environmental Life Cycle Analysis
- International Environmental Risk Management
- Moving Beyond Environmental Compliance: A Handbook for Integrating Pollution Prevention with ISO 14000

Global Engineering Documents
15 Inverness Way East
P.O. Box 1154
Englewood, CO 80150
Telephone: (800) 854-7179

A variety of environmental publications and regulations from around the world. Standards from ISO, ANSI, ASTM, BSI, and many others. A CD ROM product locator is available.

Governments Institute, Inc.
4 Research Place
Rockville, MD 20850
Telephone: (301) 921-2355

Topics include environmental audits, industrial environmental management, pollution prevention, and ISO 14000. A video series on compliance auditing is also available.

ENVIRONMENTAL SCIENCE/ENGINEERING/CONSERVATION

Battelle Press
505 King Avenue
Columbus, OH 43201-2693
Telephone: (800) 451-3543

Books on management and leadership, health physics, toxicology nuclear energy, and radioactive waste and waste management.

CRC Press/Lewis Publishers
2000 Corporate Blvd. NW
Boca Raton, FL 33431
Telephone: (800) 272-7737, (407) 994-0555

Many titles in environmental science and technology, pollution prevention, remediation, hazardous waste planning, hydrology, environmental compliance, health and safety, pesticides, sampling, and wetlands. Some titles available on CD ROM. Selected titles:

- Environmental Law and Enforcement
- Environmental Management Handbook
- Pollution Prevention Handbook

Gordon and Breach
Harwood Academic Publishers
PTT
P.O. Box 566
Williston, VT 05495-0080
Telephone: (800) 326-8917

Book titles in environmental science include

- Encyclopedia of Environmental Science and Engineering
- Environmental Toxicology
- Environmental Management in European Companies: Success Stories and Evaluation
- Direct Effect of European Law and the Regulation of Dangerous Substances
- Pollution Prevention
- Air Toxics

Noyes Publications
Noyes Data Corporation
Mill Road at Grand Ave.
Park Ridge, NJ 07656
Telephone: (201) 391-8484

Pollution control technologies, environmental engineering books, including topics in hazardous chemicals, managing hazards, air pollution control, water quality management, pesticides, land/soils/leachate, incineration, corrosion control, resource recovery, spills/cleanup/disposal, drum handling, decontamination, storage tanks, remedial action, waste reduction, and treatment technologies.

Springer-Verlag New York, Inc.
J. Keller, Dept. B400
175 Fifth Ave.
New York, NY 10010
Telephone: (800) 777-4643

Water chemistry, acid deposition effects, environmental toxicology and environmental toxins series, chemical water and wastewater treatment, environmental science, chemistry of plant protection series, *Acidification in Finland, The Silent Countdown* (essays in European Environmental History).

National Wildlife Federation
1400 Sixteenth Street N.W.
Washington, D.C. 20036-2266

> *The Conservation Directory*

HAZARDOUS MATERIALS

CRC Press/Lewis Publishers
2000 Corporate Blvd. NW
Boca Raton, FL 33431
Telephone: (407) 994-0555

> Water/wastewater treatment, environmental science, chemistry, industrial hygiene, air pollution, solid and hazardous waste, groundwater, environmental and occupational health, environmental management and compliance, and general reference. Certified Hazardous Materials Manager (CHMM) preparation database is also available.

National Fire Protection Association (NFPA)
1 Batterymarch Park
P.O. Box 9101
Quincy, MA 02269-9101
Telephone: (800) 344-3555

> NEC products, life safety code and standards, fire safety education, fire service, hazardous materials, and special topics. Videos and CD ROM products also available.

INTERNATIONAL (ENVIRONMENTAL STANDARDS AND TECHNICAL PUBLICATIONS)

Africa

Book and journal orders for Gordon and Breach:

Marston Book Services Ltd.
P.O. Box 269
Abingdon, Oxon OX14 4YN, U.K.
Telephone: 44(0) 1235 465500
direct.order@marston.co.uk

Asia

Book and journal orders for Gordon and Breach:

International Publishers Distributor
Kent Ridge
P.O. Box 1180
Singapore 911106
Telephone: 65 741 6933
jpdmktg@sg.gphap.com

Australia

ACEL Information
Pty. Ltd.
58 Atchison Street

P.O. Box 471
St. Leonards NSW 2065
Australia
Telephone: (02) 906-5566

ACGIH publications:

Australian Institute of Occupational Hygiene (AIOH)
P.O. Box 1205
Unit 3 34 Carrick Drive
Tullamarine, VIC 3043, Australia
Telephone: +61-3-9-335-2577
aioh@ibm.net

DA Books and Journals
648 Whitehorse Road
Mitcham 3132
Victoria, Australia
Telephone: (03) 873-4411

Global Info Centre
HIS Australia Pty. Ltd.
Building A, Ground Floor
244 Beecroft Road
Epping NSW 2121
Australia
Telephone: 61-2-9876-5333

Book and journal orders for Gordon and Breach:

Fine Arts Press
Level 1, Tower A
112 Talavera Road
North Ryde, NSW 2113
Australia
Telephone: 61 2 9878 8222
infor@gbpub.com.au

Standards Association of Australia
P.O. Box 458
North Sydney NSW 2059
Australia
Telephone: (02) 963-4111

Standards and Technical Publications
P.O. Box 1019
Unley, South Australia 5061
Telephone: 8373-1540

Brazil

ACGIH publications:

> Associacao Brasileira de Higienistas Ocupacionais (ABHO)
> c/o Trikem S.A.
> Av. Assis Chateaubriand, 5260
> 57010-009 Maceio – AL.
> Brazil
> Telephone: +55-82-218-2411
> saeed@trikem.com.br

Publicacoes Tecnicas Internacionais Ldtda.
R. peixoto Gomide, 209
01409 Sao Paulo – SP
Brazil
Telephone: (11) 259-664

SAE Brazil
Av. Paulista, 2.073
Horsa II – CJ.2.001
01311-940 Sao Paulo
Brazil
Telephone: (11) 287-2627

China

China National Sci-Tech Information
9/F Real Estate Bldg.
Renminnan Road
Shenzhen 518001
P.R. China
Telephone: 23929

CNPIEC
No. 15 Xueyuan Lu
Beijing 10083
P.R. China
Telephone: 2010821

Europe

ASTM European Office:

27-29 Knowl Piece
Wilbury Way
Hitchin, Herts SG4 0SX
England

Book and journal orders for Gordon and Breach:

> Marston Book Services Ltd.
> P.O. Box 269
> Abingdon, Oxon OX14 4YN, U.K.
> Telephone: 44(0) 1235 465500

Hong Kong/Pacific Rim

Global Info Centre Hong Kong
Unit 1 11/F
Multifield Plaza
3-7A Prat Avenue
Tsim Sha Tsui
Kowloon, Hong Kong
Telephone: 852-2368-5733

India

Allied Publishers Private Inc.
ASTM Department
751 Mount Road
Madras 600-002
India
Telephone: (44) 863938

Indonesia

C.V. Djakarta Raya
32, Jalan Raya
Jatinegara Timur
Jakarta Timur 13310
Indonesia
Telephone: (21) 819-3487

Book Supply Bureau, A-68
South Extension – 1
New Delhi – 110049
India
Telephone: 461-1991

Japan

Book and journal orders for Gordon and Breach:

> YOHAN (Western Publications Distribution Agency), 3-14-9
> Okubo, Shinjuku-ku
> Toyko 169-0072 Japan
> Telephone: 81 03 3208 0186

Japanese Standards Association
1-24, Akasaka 4
Minato-Ku
Tokyo 107
Japan
Telephone: (33) 585-8003

Kinokuniya Co. Ltd. Book Division
P.O. Box 5050
Tokyo Intl. 100-31

Japan
Telephone: (33) 439-0124

Maruzen Company, Ltd. Book Division
P.O. Box 5050
Tokyo 156
Telephone: (33) 275-8595

Korea

International Professional Associates
C.P.O. Box 246
Seoul 100-602
Korea
Telephone: (02) 704-1733

Korean Standards Association
13-31 Yoido-dong
Youngdungpo-gu
Seoul, Korea 150-010
Telephone: (02) 369-8114

Korea Stock Book Centre
P.O. Box 34, Yeoeido
Seoul, Korea
Telephone: (02) 785-1631

Kumi Trading Company
P.O. Box 3553
Seoul 100
Korea
Telephone: (02) 588-6667

Latin America

Global Info Centre
3909 NE 163rd Street, Suite 110
North Miami Beach, FL 33160 U.S.A.
Telephone: (305) 944-1099

International Library Service
2722 N. 650 East
P.O. Box 735
Provo, UT 84610
Telephone: (801) 374-6214

Mexico

ACGIH publications:

Analisis Ambiental
Jose antonio Torres N0 691
Col. Ampliacion asturias
06890 Mexico D.F., Mexico
Telephone: +915-740-3073

His De Mexico
AZ. Rio Churubusco
No. 364
Col. El Carmen
Coyoacan D.F. 04001
Mexico
Telephone: (5) 659-5889

Middle East

Book and journal orders for Gordon and Breach
IPD Marketing Services Ltd.
P.O. Box 310
St. Helier, Jersey JE4 0TH
Channel Islands
Telephone: 44 (0) 118 956 0080

New Zealand

Standards Association of New Zealand
Private Mail Bag 2439
Wellington, New Zealand
Telephone: (04) 384-2108

Singapore

ACGIH publications:
Lee Hung Technical Co.
Blk 2022 Bukit Batok
St. 23 #03-134/136/138
Bukit Batok Industrial Park A
Singapore 659527
Telephone: +65-5606900

Singapore Institute of Standards and Industrial Research
1 Science Park Drive
Singapore 0511
Telephone: 722-9686

South America

ACGIH publications:

Siafa S.R.L.
Avda. De Mayo 1370
Piso 7 Of. 160
1362 Buenos Aires, Argentina
Telephone: +54-1-383-0517
siafa@interprov.com

High Tec Higiene Industrial LTDA
Calle 50 No. 79-54 Int. 11
Santa Fe de Bogota, D.C.

Columbia
Telephone: +57-1-4161751

American Books

Tucuman 994
1049 Buenos Aires
Argentina
Telephone: 396-3704

Book and journal orders for Gordon and Breach

PTT
P.O. Box 566
Williston, VT 05495-0080
Telephone: (800) 326-8917
info@gbhap.com

Journal orders

IPD P.O. Box 32160
Newark, NJ 07102
Telephone: (800) 545-8398
info@gbhap.com

Tradinco SRL
Tucuman 423 Piso 1
1049 Buenos Aires
Argentina
Telephone: (01) 311-2908

Surninistros Asociados
S.A.
Belgrano 333
1642-San Isidro B.A.
Republica, Argentina
Telephone: 732-2653

Taiwan

ACGIH publications

Asia Leadtech Corp.
10F-1, No. 150, Chiu Ru 2nd Road
Kaohsiung, Taiwan, R.O.C.
Telephone: +886-7-311-9049
cfi@ksts.seed.net.tw

Tao's Publisher's & Book Co., Ltd.
Room 502, 5th Floor No. 121
Chung Ching S. Road
Sec, 1 Taipei, Taiwan
R. O. C.
Telephone: (02) 331-5773

United Kingdom

ACGIH publications

H & H Scientific Consultants Ltd.
P.O. Box MT27
Leeds LS17 8QP
United Kingdom
Telephone: +44-113-268-7189
hhsc@dial.pipex.com/hhsc/

SAFETY

ACGIH Publications
Kemper Woods Center
1330 Kemper Meadow Drive
Cincinnati, OH 45240-1634
Telephone: (513) 742-2020
pubs@acgih.org

> Publications in industrial hygiene, environment, safety and health, toxicology, hazardous materials and waste, workplace controls, physical agents, ergonomics, laboratory, and computer resources.

National Safety Council
444 N. Michigan Ave.
Chicago, IL 60611-3991
Telephone: (800) 621-7619

> Safety professional's library, ergonomics, hazard control, occupational health, motor transportation safety, health care facility safety, safety management and administration, industrial hygiene, noise control, wellness.

Safety Quarterly – Safety Training Videos for Professionals
The Idea Bank
P.O. Box 4115
Santa Barbara, CA 93140-9916
Telephone: (800) 621-1136

> Safety training videos, including safety shorts, office safety, equipment safety, and a four-volume safety series on working with pesticides; some videos available in English or Spanish.

National Fire Protection Association (NFPA)
1 Batterymarch Park
P.O. Box 9101
Quincy, MA 02269-9101
Telephone: (800) 344-3555

> NEC products, life safety, codes and standards, fire safety education, fire service, hazardous materials, and special topics. Videos and CD ROM products also available.

STANDARDS

American Society for Testing and Materials (ASTM)
100 Barr Harbor Drive

West Conshohocken, PA 19428-2959
Telephone: (610) 832-9585
service@astm.org

All 10,000 ASTM technical standards are available (CD ROM editions also). Periodicals, training courses and testing program information, manuals, ISO documents, data series, directories, adjuncts such as photographs, blueprints, tables, charts and software, monographs, reference radiographs, and special technical publications.

Section 11 – Water and Environmental Technology

- Volume 11.01/Water (I)
- Volume 11.02/Water (II)
- Volume 11.03/Atmospheric Analysis; Occupational Health and Safety; Protective Clothing
- Volume 11.04/Environmental Assessment; Hazardous Substances and Oil Spill Responses; Waste Management; Environmental Risk Management
- Volume11.05/Biological Effects and Environmental Fate; Biotechnology; Pesticides

American National Standards Institute (ANSI)
7315 Wisconsin Ave. Suite 250-E
Bethesda, MD 20814
Telephone: (301) 469-3363

American Society for Quality Control (ASQC)
611 East Wisconsin Avenue
P.O. Box 3005
Milwaukee, WI 53201
Telephone (800) 248-1946

NSF International
2100 Commonwealth Blvd.
Ann Arbor, MI 48105
Telephone: (313) 332-7333

WATER AND WASTEWATER (TREATMENT AND RESOURCES)

ACR Publications, Inc.
1298 Elm Street SW
Albany, OR 97321
Telephone: (503) 928-5211

Safety and environmental publications and manuals, including water system distribution construction, cave-in protection, simplified math for waterworks operators, pumps and pumping.

American Water Works Association (AWWA)
Member Services
6666 West Quincy Ave.
Denver, CO 80235-9913
Telephone: (303) 795-2449

Large selection of reference books, handbooks, standards, research papers, videotapes, water education, software, and conference proceedings. Topics include water quality and treatment, CD ROM service and training materials also available.

CRC Press/Lewis Publishers
2000 Corporate Blvd. NW
Boca Raton, FL 33431
Telephone: (407) 994-0555

Water/wastewater treatment, environmental science, chemistry, industrial hygiene, air pollution, solid and hazardous waste, groundwater, environmental and occupational health, environmental management and compliance, and general reference. Certified Hazardous Materials Manager (CHMM) preparation database is also available.

National Environmental Training Center for Small Communities (NETCSC)
West Virginia University
P.O. Box 6064
Morgantown, WV 26506-6064
Telephone: (800) 624-8301

NETSCS develops training materials and maintains a database containing information about training organizations, materials, and events. Their library contains training packages and resources for wastewater, drinking water, and solid waste management.

Technomic Publishing Co., Inc.
851 New Holland Ave., Box 3535
Lancaster, PA 17604
Telephone: (717) 291-5609

Water quality management and civil engineering publications and textbooks. Other topics include biological analysis, contamination, groundwater, hydrology, plant operations, reclamation, remediation, and treatment.

Water and Environment Federation
601 Wythe Street
Alexandria, VA 22314-1994
Telephone: (800) 666-0206
Outside the U.S. (703) 684-2452

Water and wastewater treatment, plant design, disinfection, groundwater, hazardous waste, environmental audits, environmental impacts, laboratory, odor control, plant management, safety and health, and regulations. Software and training and certification materials also available.

Water Resources Publications
P.O. Box 260026
Highlands Ranch, CO 80126-0026
Telephone: (800) 736-2405 or (303) 790-1836

Textbooks on hydrology, hydraulics, hydrology papers, reports of studies, U.S. Soil Conservation Service Publications, and other computer programs for the water industry.

PERIODICALS

Air & Waste Management Association (A & WM)
One Gateway Center, Third Floor
Pittsburgh, PA 15222
Telephone: (412) 232-3444
FAX: (412) 232-3450

Canadian Office of A & WM: 155 Quenn Street, Suite 1202
Ottawa, Canada K1P6L1
Telephone: (613) 233-2006
e-mail info@awma.org

> Environmental Manager (ISSN 1079-7343), published monthly and covers international environmental management and compliance topics.

> Journal of the Air & Waste Management Association (ISSN 1047-3289), published monthly and covers a variety of environmental issues and technical papers.

American Water Works Association (AWWA)
Member Services
6666 West Quincy Ave.
Denver, CO 80235-9913
Telephone: (303) 795-2449

> Opflow—monthly water treatment publication for members

> AWWA Journal—a monthly publication since 1914 on safe drinking water treatment and issues

Environmental Resource Center
101 Center Point Drive
Cary, NC 27513-5706

> Beyond Compliance, News for Quality Environmental Compliance

Environmental Protection
5151 Beltline Road, Suite 1010
Dallas, TX 75240
Telephone: (972) 687-6700
FAX: (972) 687-6770

> Environmental Protection (ISSN 1057-4298), a Stevens Publication on management and problem solving for environmental professionals. Published monthly.

Target Group Inc.
Publishing Office
P.O. Box 5244
Glendale, CA 91221-1081
Telephone: (818) 842-4777
target@primenet.com

> Environmental Testing & Analysis (ISSN 1068-7432) – is published bimonthly and covers issues regarding the testing and analysis of environmental samples, including test methods, technology, documentation, news, and products.

Executive Enterprises Publications Company, Inc.
22 West 21st Street
New York, NY 10010-6904
Telephone: (800) 332-8804

- Remediation: The Journal of Environmental Cleanup Costs, Technologies, and Techniques
- Pollution Prevention Review
- Supervisor's Environmental Alert
- Environmental Manager
- Federal Facilities Environmental Journal
- Environmental Claims Journal

IPD P.O. Box 32160
Newark, NJ 07102
Telephone: (800) 545-8398
info@gbhap.com

Some of the available titles:

- Environmental Science Review
- International Journal of Environmental Studies
- International Journal of Environmental Analytical Chemistry
- Toxicology and Environmental Chemistry (international journal)
- Occupational Hygiene
- Radioactive Waste Management and Environmental Restoration

John Wiley & Sons, Inc.
605 Third Avenue
Eleventh Floor
New York, NY 10158-0012
Telephone: (212) 850-6497
subinfo@wiley.com

Environmental Quality Management, published quarterly and is written for environmental managers, engineers, quality directors, and environmental professionals.

VIDEOS

AWT Video Library
Air & Water Technologies Corporation
P.O. Box 1500
Somerville, NJ 08876-1251
Telephone: (800) 998-4298 ext. 98

Training videos for health, safety, and environmental compliance.

BNA Communications, Inc.
9439 Key West Avenue
Rockville, MD 20850
Telephone: (800) 233-6067

Large selection of safety videos, including topics in accident causes and prevention, drug testing, emergency procedures, ergonomics, hazard communication, hazardous materials, laboratory safety, lockout tagout, materials handling, personal protective equipment, and spill response.

Business and Legal Reports, Inc.
39 Academy Street

Madison, CT 06443-1513
Telephone: (800) 727-5257

> Safety videos, including bloodborne pathogens, confined space, hazard communication, and lockout tagout.

Emergency Film Group
225 Water Street
Plymouth, MA 02360
Telephone: (800) 842-0999

> Training videos for hazardous materials emergency response, personal protective equipment, industrial incident management, and air monitoring.

Films for the Humanities and Sciences
Environmental Science
P.O. Box 2053
Princeton, NJ 08543-2053
Telephone: (800) 257-5126, (609) 275-1400

> Videos on cleaning up pollution (earthkeeping), preserving and restoring the environment, geology, weather, energy, urban development, waste disposal, the ozone layer, the greenhouse effect and global climate, radiation, and radioactive waste disposal.

Governments Institute, Inc.
4 Research Place
Rockville, MD 20850
Telephone: (301) 921-2355

> Video series on environmental site assessments, compliance auditing, and RCRA compliance.

Industrial Training Systems Corporation (ITS)
1303 Marsh Lane
Carrollton, TX 75006
Telephone: (800) 568-8788

ITS Canada
3370 South Service Road
Burlington, Ontario L7N 3M6
Canada
Telephone: (905) 333-4310

> Selections in environmental compliance, including environmental law, environmental liability, hazardous waste and hazardous materials transportation, and 40 titles in the safety series.

14 Information Available on the Internet

CONTENTS

ORGANIZATIONS AND ASSOCIATIONS

Air & Waste Management Association
http://www.awma.org/index.html

American National Standards Organization
http://www.ansi.org/home.html

American Society for Testing and Materials
http://www.astm.org/

International Environmental Liability Management Association
http://www.magic.ca/ielma/ielma.home.html

International Standards Organization
http://www.iso.ch/

National Institutes of Health
http://www.nlm.nih.gov/

Environmental Organization Web Directory
http://www.webdirectory.com

GOVERNMENT AGENCIES

AUSTRALIA

Environment Australia
www.environment.gov.au/

Department of the Environment
www.dest.gov.au/dest.html

Department of the Environment, Legislation
www.ea.gov.au/dept_legislation.html

Foreign Affairs and Trade, Environment
www.dfat.gov.au/enviroment/

New South Wales, Environmental Protection Authority
www.epa.nsw.gov.au/index.asp

Queensland, Department of Environment and Heritage
www.env.qld.gov.au/

Tasmania, Department of Environment and Land Management
www.delm.tas.gov.au/

Victoria, Environmental Protection Authority
www.epa.vic.gov.au/

Victoria EPA Information Center
www.eap.vic.gov.au/about/infocentre.htm

CANADA

Alberta Environmental Protection Information Centre
www.gov.ab.ca/env/infolib.html

British Columbia Ministry of Environment, Lands and Parks
www.env.gov.bc.ca/

British Columbia Web Page
www.gov.bc.ca/

Canadian Environmental Assessment Agency
www.ceaa.gc.ca/
www.ceaa.gc.ca/other/index_e.htm
www.ceaa.gc.ca/registry/registry_e.htm

Canadian Wildlife Federation
http://www.toucan.net/cwf-fcf

Canadian Botanical Conservation Network (CBCN)
http://www.science.mcmaster.ca/Biology/CBCN/homepage.html

Federation of Nova Scotia Naturalists
http://ccn.cs.dal.ca/Environment/FNSN/hp-fnsn.html

Government of Canada
http://canada.gc.ca/main_e.html

Manitoba Environment
www.gov.mb.ca/environ/

Manitoba Environment, Departmental Mandates
www.gov.mb.ca/environ/mandate.html

New Brunswick Department of Environment
www.gov.nb.ca/environm/index.htm

Newfoundland, Environment and Labour, Environmental Branch
www.gov.nf.ca/env/env/default.asp

Newfoundland, Environment and Labour, Pollution Prevention
www.gov.nf.env/env/pollution_prevention.asp

Newfoundland, Environment and Labour, Water Resources
www.gov.nf.ca/env/env/water_resources.asp

Newfoundland, Environment and Labour, Occupational Health and Safety
www.gov.nf.ca/env/labour/ohs/ohs.asp

Nova Scotia, Department of Environment
www.gov.ns.ca/envi/

Ontario, Ministry of the Environment
www.ene.gov.on.ca/envision/org/org-moee.htm

Prince Edward Island, Main Web Page
www.gov.pe.ca/government/

Prince Edward Island, Technology and Environment
www.gov.pe.ca/te/index.asp

Quebec, Bureau d' Audiences Publiques sur L' Environnement
www.bape.gouv.qc.ca/

Saskatchewan, Environment and Resource Management
www.gov.sk.ca/govt/environ/progserv.htm

Yukon Department of Renewable Resources
http://rrenres.gov.yk.ca/environ/assess.html

DENMARK

Danish Environmental Protection Agency
www.mst.dk/depa/index.htm

Ministry of Environment and Energy
www.mem.dk/ukindex.htm
www.mst.dk/aakonf/forside.htm

EUROPE

The EU on line, Europa
www.cec.lu/

Central European Environmental Data Request Facility (CEDAR)
http://www.cedar.univie.ac.a

Dolphin Project Europe
http://www.achilles.net/~jamesh/dolphin/

European Centre for Nature Conservation
http://www.ecnc.nl

FRANCE

Minister of Environment
www.environnement.gouv.fr/english/engsom.htm

GEORGIA

Ministry of Environmental Protection
http://gaia.grida.no/prog/cee/enrin/htmls/georgia/soegeor/english/institut/moe.htm

State of the Environment
http://gaia.grida.no/prog/cee/enrin/htmls/georgia/soegeor/hp_soege.htm

GERMANY

National Research Center for Environmental Health
www.gsf.de/gsf/englisch/index.html

Environmental Information Service
www.gsf.de/oa/iu_short.html

DAIN Metadatabase of Internet Resources
http://dino.wiz.uni-kassel.de/daina.html

Federal Environment Ministry
www.bmu.de/englisch/base.htm

Bavarian State Ministry for State Development and Environmental Affairs
www.bayern.de/stmlu/minist/engl.htm

HONG KONG

Planning, Environment, and Lands Bureau
www.pelb.wpelb.gov.hk/content.htm

HUNGARY

Ministry of Environment
www.grid.bp.meh.hu/aszoueg.htm

INDONESIA

Environmental Impact Management Agency
www.bapedal.go.id/index-e.html

ISRAEL

Ministry of the Environment
www.israel-mfa.gov.il/gov/environ/.html

JAPAN

APEC Virtual Center for Environmental Technology Exchange
www.epcc.pref.osaka.jp/apec/eng/index.htm

LUXEMBOURG

Ministry of the Environment
www.meuetat.lu/home.html

New Zealand

Ministry of the Environment
www.mfegovt.nz/

Norway

Center for International Climate and Environmental Research - Oslo
http://www.cicero.uio.no CICERO

Norwegian Pollution Control Authority (SFT)
www.sft.no/e2.html
www.sft.no/e22.html (pollution control)
www.sft.no/e26.html (chemicals and haz waste)

Republic of Mauritius

Ministry of Local Government and Environment
http://ncb.intnet.mu/environ.htm

Department of Environment
http://ncb.intnet.mu/environ/dept_env.htm

South Africa

Department of Environmental Affairs and Tourism
www.gov.za/envweb/

Sweden

Stockholm Environmental Institute
www.tellus.org.seib.html

Swedish Environmental Protection Agency
www.environ.se/www-eng/enghome.htm

Switzerland

National Territory, Landscape and Environment, Swiss Federal Statistical Office
www.admin.ch/bfs/einhalt.htm#the02

Swiss Agency for the Environment, Forests and Landscape
www.admin.ch/buwal/

Taiwan

Environmental Protection Agency (ROC)
www.epa.gov.tw/english/

Thailand

Pollution Control Department
www.pcd.go.th/firstpage.cfm

UNITED KINGDOM

Department of the Environment, Transport and the Regions Environment Agency
    http://www.environment-agency.gov.uk/

UNITED STATES

United States Environmental Protection Agency
    gopher://gopher.epa.gov
    ftp://ftp.epa.gov
    http://www.epa.gov

United States Department of Labor, Occupational Safety and Health Administration
    http://www.osha.gov,

Health and Human Services Department
    http://www.os.dhhs.gov

National Institute for Occupational Safety and Health
    http://www.cdc.gov/niosh/homepage.html

ENVIRONMENTAL ORGANIZATIONS

20/20 Vision
    http://www.2020vision.org/

American Institute of Biological Sciences
    gopher://aibs.org

American Rivers
    gopher://gopher.igc.apc.org/11/orgs/amrivers

American Wind Energy Association
    gopher://gopher.igc.apc.org/11/orgs/awea

Amnesty International
    http://www.traveller.com/~hrweb/ai/ai.html

Aspen Global Change Institute
    http://www.infosphere.com/aspen/agci/index.html

Center for Aquatic Ecology
    http://denr1.igis.uiuc.edu:70/0h/cae/aquatic-intro.html

Center for Biodiversity
    http://denr1.igis.uiuc.edu:70/0h/biod/biodiversity-intro.html

Center for Ecological Health Research
    http://ice.ucdavis.edu/Center_for_Ecological_Health_Research

Center for Energy and Environmental Studies
    http://cees-server.bu.edu

Center for Global Regional Environmental Research
    http://www.cgrer.uiowa.edu

Center for Renewable Energy & Sustainable Technology (CREST)
    http://solstice.crest.org/common/crestinfo.html

Center for Social and Environmental Accounting Research
http://www.dundee.ac.uk/Accountancy/csear.htm

Center for Wildlife Ecology
http://denr1.igis.uiuc.edu:70/0h/cwe/wildlife-intro.html

Centre for Environmental Labelling (CEL)
http://unixg.ubc.ca:780~ecolabel/cel.html

CERES-GKN (CERES-Global Knowledge Network)
http://www.cerc.wvu.edu/ceres/ceres_index.html

Center for Ocean-Land-Atmospheric Studies
http://grads.iges.org/home.html

Consortium for International Earth Science Information Network (CIESIN)
http://www.ciesin.org

Cooperative Institute for Research in Environmental Sciences
http://cires.colorado.edu

CouncilNet, the sustainability network for local government
http://www.peg.apc.org/~councilnet/welcome.html

Earth First! Journal
http://envirolink.org/orgs/ef/index.html

EarthWatch
http://gaia.earthwatch.org

Ecology Action Center
http://www.cfn.cs.dal.ca/Environment/EAC/EAC-Home.html

EcoNet
http://www.igc.apc.org/econet/

EcoNews Africa
http://www.io.org/~ee/ena/

Environmental Bankers Association
http://envirolink.org/orgs/eba/

Environmental Data Research Institute Environmental Grantmaking Foundations, 1995 Directory
http://www.envirolink.org/products/edri

Environmental Hazards Management Institute
http://www.ehmi.org/

Environmental Health Center, National Safety Council
gopher://cais.com:70/11/.nsc

E-Law Environmental Law Alliance (E-LAW)
gopher://envirolink.org:70/00/.EnviroOrgs/.eorgs/

Environmental Law Alliance Worldwide (E-LAW)
gopher://igc.apc.org:70/11/environment/law

Environmental Research Institute of Michigan
http://www.erim.org

Environmental Resource Center
 http://ftp.clearlake.ibm.com/ERC/overview.html

Environmental Working Group
 http://www.ewg.org

Friends of the Earth (International Secretariat)
 http://www.xs4all.nl/~foeint

Global Ecolabelling Network (GEN)
 http://unixg.ubc.ca:780/~ecolabel/gen.html

Global Network for Environmental Technology
 http://www.gnet.org

Global Rivers Environmental Education Network
 http://www.igc.apc.org/green/green.html

GLOBAL 2000
 http://www.t0.or.at/~global2000

Greenpeace International
 http://www.greenpeace.org/

International Arctic Project (IAP)
 gopher://gopher.igc.apc.org:70/11/environment/misc/iap

International Council for Local Environmental Initiatives (ICLEI)
 http://www.iclei.org

International Geosphere Biosphere Programme
 http://igbp1.biogeodis.jussieu.fr

International Institute for Sustainable Development
 http://www.iisd.ca/linkages

International Wildlife Coalition (IWC)
 http://www.webcom.com/~iwcwww

National Audubon Society
 http://www.audubon.org/audubon

National Consortium for Environmental Education and Training (NCEET)
 gopher://nceet.snre.umich.edu:70/1

National Environmental Information Resources Center
 http://www.gwu.edu/~greenu/

National Environmental Policy Institute (NEPI)
 http://nepi.org

National Institute for the Environment
 gopher://envirolink.org:70/11/.EnviroOrgs/.eorgs/National%20Institute%20for
%20the%20Environment

National Parks and Conservation Association
 http://www.npca.com/pub/npca/

Natural Resources Defense Council
 http://www.nrdc.org/nrdc

Nova Scotia Environment and Development Coalition
http://cfn.cs.dal.ca/Environment/nsedc.html

Environmental Institute
http://www.moscow.com/Resources/PCEI>Palouse-Clearwater

Pesticide Action Network
gopher://gopher.igc.apc.org/11/orgs/panna

Rainforest Action Network (RAN)
http://www.ran.org/ran/

Sierra Club
http://www.sierraclub.org

Sustainable Communities Network (SCN)
http://www.cfn.cs.dal.ca/Environment/SCN/SCN_home.html

Institute for Resource and Environmental Strategies
http://www.channel1.com/users/tellus/tellus.html Tellus

UK Centre for Economic and Environmental Development
http://www.pikeperry.co.uk/ppp/cd/ukceed/ukceed.htm

Waste Policy Institute
http://web.wpi.org

Wilderness Society
http://town.hall.org/environment/wild_soc/wilderness.html

World Conservation Monitoring Centre (UK)
http://www.wcmc.org.uk/

World Wildlife Fund
http://www.wcmc.org.uk/aboutWCMC/partners.html#WWF"

UNIVERSITIES AND STUDENT ORGANIZATIONS

Carnegie Mellon University, Dept. of Civil and Environmental Engineering
http://www.ce.cmu.edu:8000/

Center for Environmental Health Sciences at MIT
http://web.mit.edu/afs/athena.mit.edu/org/c/cehs/www/

Georgia Tech, School of Civil and Environmental Engineering
http://howe.ce.gatech.edu/ce.html

Center for Clean Technology at University of California, Los Angeles
http://cct.seas.ucla.edu

University of Florida Civil Engineering
http://www.ce.ufl.edu/brochure/brochure.html

Department of Environmental Science and Engineering, Oregon Graduate Institute of Science and Technology
http://www.ese.ogi.edu/

Environmental Engineering and Science, University of Cincinnati
http://www.cee.uc.edu:80/~eewww/

Hazardous Substance Research Center (Georgia Tech)
http://www.gtri.gatech.edu/hsrc/

Information Center for the Environment - UC Davis
http://ice.ucdavis.edu

Student Environmental Action Coalition (SEAC)
http://www.seac.org/

Students for Environmental and Ecological Diversity (SEED)
www.studorg.nwu.edu/seed/SEEDWeb.html

Sierra Student Coalition
http://ssc.org/SSC

Greenpeace at the University of Glamorgan (UK)
http://www.glam.ac.uk/union/clubs/green/green.exe?green

Green Student Network
http://www.ph.ed.ac.uk/~jonivar/greens

OTHER SITES

Agenda 21
http://www.un.org/esa/agenda21/natlinfo/about.htm

All Environmental Web Resources
http://envirolink.org/envirowebs.html

Biodiversity
http://straylight.tamu.edu/bene/bene.html

Chemical Information, links to hazardous chemical references, MSDSs, etc.
http://www.rpi.edu/dept/chem/cheminfo/chemres.html

Cleaner Technology Information Center
http://www.tei.or.th/bep/ctic/

Cleaner Production Programme
http://www.unepie.org/

Climate Change
http://www.doc.mmu.ac.uk/aric/diction.html
http://www.enn.com/specialreports/climate/countdwn.htm

Hydrogeology Site
http://www.us.net/adept/links.html

The Ecosystem
http://www.gold.net/ecosystem/

EcoWeb
http://ecosys.drdr.virginia.edu/EcoWeb.html

Environment Business Magazine
http://www.ifi.co.uk/ebhtm.htm

The Electronic Green Journal
http://www.lib.uidaho.edu:70/docs/egj.html

Environmental Guide to the Internet
    http://http2.sils.urnich.edu/~/test.html

Environmental Law
    http://www.law.indiana.edu

INFOTERRA
    http://www.cedar.univie.ac.at/unep/infoterra

Environment and Ecology
    gopher://riceinfo.rice.edu/11/Subject/Environment

Environmental Information
    http://boris.qub.ac.uk/cvni/info.html

Environmental Information Services on the Internet
    http://www.foe.co.uk

Environmental Resource Center Home Page
    http://ftp.clearlake.ibm.com

Enviroweb, list of environmental organizations on the web
    http://envirolink.org/start_web.html

ISO 14000
    http://www.iso14000.com/articles.html

National Technology Transfer Center (NTTC) Includes Environmental Technology Gateway
    http://iridium.nttc.edu/nttc.html

Ozone .
    http://unepie-org/ozonaction.html

Waste Audit
    http://vub.mcgill.ca/clubs/qpirg/waste/Guide_to_a_Waste_Audit.html

Index

Milton Keynes UK
Ingram Content Group UK Ltd.
UKHW051943071024
449327UK00026B/2153